Chemistry and Technology of Epoxy Resins

Chemistry and Technology of Epoxy Resins

Edited by

BRYAN ELLIS
Department of Engineering Materials
University of Sheffield

SPRINGER-SCIENCE+BUSINESS MEDIA, B.V.

First edition 1993

Originally published by Chapman & Hall in 1993

Typeset in 10/12pt Times by EJS Chemical Composition, Bath

ISBN 978-0-7514-0095-3 ISBN 978-94-011-2932-9 (eBook)
DOI 10.1007/978-94-011-2932-9

A catalogue record for this book is available from the British Library.

Library of Congress Cataloging-in-Publication Data available.

Preface

Epoxy resins have been commercially available for about 45 years and now have many major industrial applications, especially where technical advantages warrant their somewhat higher costs. The chemistry of these resins is fascinating and has attracted study by many very able scientists. The technological applications of the epoxy resins are very demanding and there are many new developments each year.

The aims of the present book are to present in a compact form both theoretical and practical information that will assist in the study, research and innovations in the field of epoxy resin science and technology. The literature on epoxy resins is so vast that it is not possible to be encyclopaedic and that is not the function of the present text. It is the editor's hope that the selection of topics discussed will provide an up-to-date survey. There is some overlap in the chapters but this is minimal and so each chapter is essentially self contained.

As with all chemicals there are toxicological and other hazards. These are not dealt with in this text since a little knowledge can be dangerous, but material supplied can provide information regarding any safety precautions that may be necessary. However, often these precautions are not onerous and epoxy resins, or more specifically the hardeners, can be handled readily.

It is hoped that this text will provide an up-to-date outline of the science and technology of epoxy resins and stimulate further research into unsolved problems and assist further technological developments.

Bryan Ellis

Acknowledgements

As editor, I would like to thank all my co-authors for their contributions without which there would be no textbook. I am grateful to the secretarial staff of the Department of Engineering Materials, University of Sheffield for their assistance. Finally, but not least, to the editorial staff of Blackie who have transformed the manuscripts into a very well produced book.

B.E.

Note

For ease of reference a general index and a separate index of curving agents and hardeners are provided.

Contents

Contributors

Dr W.R. Ashcroft	Anchor Chemical (UK) Ltd, Clayton Lane, Clayton, Manchester M11 4SR, UK
Dr W.J. Cantwell	Laboratoire de Polymères, Ecole Polytechnique Federale de Lausanne, CH-1014 Lausanne, Switzerland
Dr X.M. Chen	Department of Engineering Materials, University of Sheffield, Sir Robert Hadfield Building, Mappin Street, Sheffield S1 4DU, UK
Mr B. Ellis	Department of Engineering Materials, University of Sheffield, Sir Robert Hadfield Building, Mappin Street, Sheffield S1 4DU, UK
Professor G.P. Johari	Department of Materials Science and Engineering, McMaster University, 1280 Main Street West, Hamilton, Ontario, L8S 4L7, Canada
Dr F.R. Jones	Department of Engineering Materials, University of Sheffield, Sir Robert Hadfield Building, Mappin Street, Sheffield S1 4DU, UK
Professor H.H. Kausch	Laboratoire de Polymères, Ecole Polytechnique Federale de Lausanne, CH-1015 Lausanne, Switzerland
Dr S.J. Shaw	Materials and Structures Department, DRA Aerospace Division, Farnborough, Hants GU14 6TD, UK

1 Introduction to the chemistry, synthesis, manufacture and characterization of epoxy resins

B. ELLIS

1.1 Epoxy resins

The term 'epoxy resin' is applied to both the prepolymers and to the cured resins; the former contain reactive epoxy groups, $\mathrm{R{-}\overset{O}{\overbrace{CH{-}CH_2}}}$, hence their name. In the cured resins all of the reactive groups may have reacted, so that although they no longer contain epoxy groups the cured resins are still called epoxy resins. The relative size of the market for epoxy resins is indicated in Table 1.1 from which it can be seen that they are important industrial polymers. Since they are more highly priced than other resins, they will only find application when they have technical advantages. Many of the applications involve high added value products.

Although the first products that would now be called epoxy resins were synthesized as early as 1891 (see Dearborn *et al.*, 1953; Lee and Neville, 1967a) it was not until the independent work of Pierre Castan in Switzerland and Sylvan Greenlee in the United States that commercial epoxy resins were marketed in the 1940s, although similar resins had been patented in the 1930s. The earliest epoxy resins marketed were the reaction products of bisphenol A and epichlorohydrin and this is still the major route for the manufacture of most of the resins marketed today, although there are many other types of resin available (section 1.3).

Pierre Castan investigated potential resins which could be readily moulded at low pressures for the replacement of vulcanite as a denture base material. The BPA-epoxy resins could be cured by reaction of epoxy groups with phthalic anhydride without the evolution of low molecular species and hence did not require high moulding pressures. An alternative acrylic resin is now used for denture base and the patents were licensed to CIBA in 1942. Epoxy adhesives and casting resins were marketed in the USA in 1946.

Greenlee working for Devoe and Raynolds produced resins which were similar to those of Castan but with a somewhat higher molecular weight with the objective of developing superior surface coatings. The epoxy coatings developed by Greenlee offered improved adhesion, hardness, inertness and thermal resistance compared with alkyd or phenolic resins. Following the

Table 1.1 Epoxy resin sales in the USA, 1990–1991

Resin	1990			1991	
	1000 tonnes† sold	Approximate* relative cost	Approximate relative value	1000 tonnes† sold	Approximate relative value
Epoxy	211	3.8	802	195	741
Phenolic	1286	1.3	1672.8	1163	1512
Unsaturated polyesters	558	2.9	1618	492	1427
Urea and melamine	761	–	–	667	–
Polystyrene	2297	1.15	2.64	2219	2552

Consumption in Japan in 1991 was 166 tonnes of epoxy resin and 386 tonnes of phenolic resin.†
* Relative to low density polyethylene = 1.0, Birley and Scott (1982).
† Modern Plastics International, 1992.

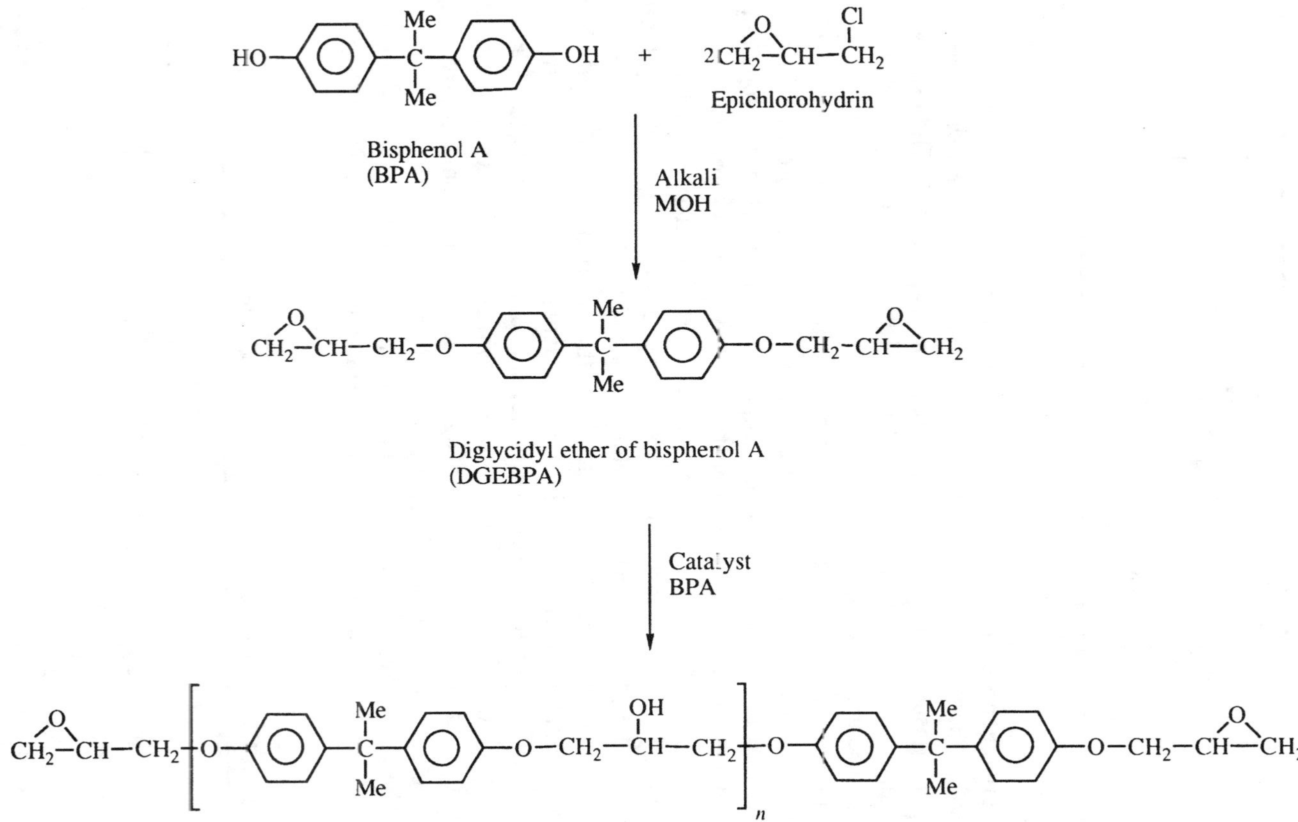
Bisphenol A
(BPA)
Epichlorohydrin
Alkali
MOH
Diglycidyl ether of bisphenol A
(DGEBPA)
Catalyst
BPA

Table 1.2 Number average molecular weight, $\bar{M}_n$, of 'Ideal' epoxy resins

n	$\bar{M}_n$ 340 + 284n	No. of hydroxyl groups	Epoxy equivalent weight
0	340	0	170
1	624	1	312
2	908	2	454
10	3180	10	1590

first patent application (1948) Greenlee obtained about 40 patents for epoxy resins. The major application of epoxy resins is still for surface coatings which consumes about 50% of all epoxy resins produced. The relative use pattern of epoxy resins is indicated in Table 1.3 and applications are discussed in more detail in chapter 9.

Innovation in epoxy resin technology has involved the synthesis of epoxy resins with specific characteristics and some of the more important are discussed in section 1.3. Equally important has been the development of hardeners which depend on their reactivity with epoxy groups. The more important hardener systems are treated in detail in chapter 2. Much of the chemistry of epoxy resins depends on the reactivity of the epoxy groups which will be briefly outlined in the next section (1.2).

The cure of epoxy resins involves the formation of a rigid three-dimensional network by reaction with hardeners which have more than two reactive functional groups, that is, functionality is $f > 2$. Often $f \geqslant 4$ for common hardeners for BPA resins which often have an effective functionality of two, but may be higher when the cure temperature is high enough for the secondary hydroxy groups to react.

The cure of epoxy resins is complicated and it is useful to visualize the process in several stages, which are illustrated in Figure 1.1, although except

Table 1.3 USA applications of epoxy resins (Modern Plastics International, 1992)

	[1990]		[1991]	
	1000 tonnes	%	1000 tonnes	%
Protective coatings	89	49	84	51
Electrical applications	25	14	22	13
Reinforced resins	14	7.5	13	8
Bonding and adhesives	13	7.5	12	7.25
Flooring	12	6.5	11	6.25
Tooling and casting	13	7.5	12	7.25
Other	15	8.3	12	7.25
Total	181	100	166	100

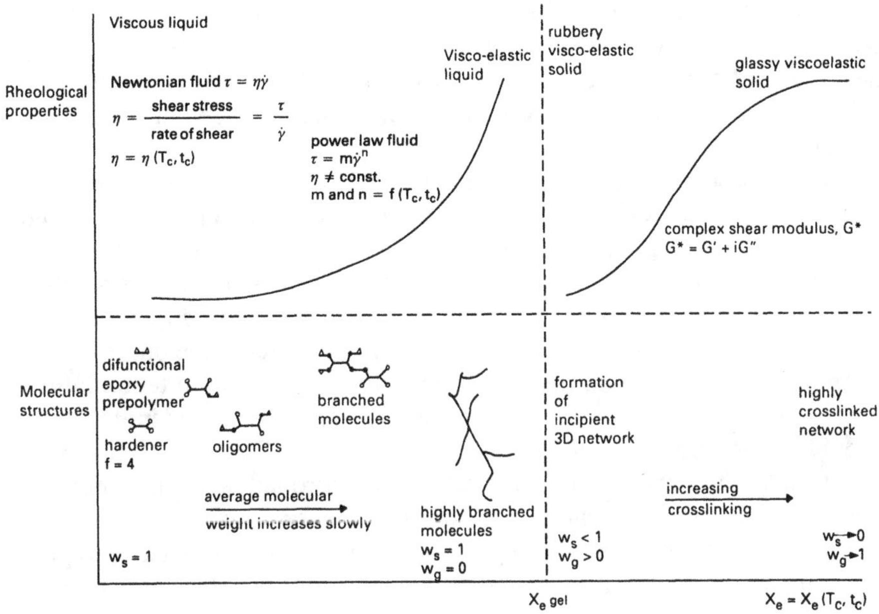

Figure 1.1 The cure of epoxy resins.

$$\text{The extent of reaction } X_e = \frac{E_o - E(t_c)}{E_o}$$

where E_o is the initial concentration of epoxy groups and $E(t_c)$ is their concentration at cure time t_c. T_c is the cure temperature.

$$\text{The sol fraction } w_s = \frac{\text{weight of soluble molecules}}{\text{total weight of the sample}}$$

$$\text{The gel fraction } w_{gel} = \frac{\text{weight of cross-linked network, gel}}{\text{total weight of the sample}}$$

see chapter 3 for further discussion of cure processes.

for gelation, the process is continuous. Initially there is reaction between epoxy and hardener reactive groups so that somewhat larger molecules are formed. As cure proceeds, larger and larger molecules are formed but it should be noted that the average molecular size is still small even when half the reactive groups have reacted. When the molecular size increases as cure progresses, some very highly branched molecules are formed and then more and more highly branched structures develop. The critical point is gelation when the branched structures extend throughout the whole sample. Prior to gelation the sample is soluble in suitable solvents but after the gel point the network will not dissolve but swells as it imbibes solvent. At the gel point small and branched molecules are present which are soluble, hence the curing sample contains sol as well as gel fractions. The gel initially formed is weak and can be easily disrupted. To produce a structural material, cure has to continue until most of the sample is connected into the three-dimensional network so that the sol fraction becomes small and for many cured products it has to be essentially zero. Cure and gelation are discussed in more detail in chapter 3.

As cure proceeds there are major changes in the properties of the epoxy resins. Initially the resin-hardener mixture is fluid and finally an elastic solid is produced. The glass transition temperature of the curing resin increases as cure proceeds and these changes can be represented in a time-temperature transition (TTT) diagram introduced by Gillham (1986). Figure 1.2 is a simplified version which illustrates the dominant effect of the onset of vitrification as T_g increases to the cure temperature T_c. For cure temperatures well above T_g, the rate of reaction between the epoxy and hardener reactive groups is chemically kinetically controlled. When $\Delta T = T_c - T_g$ becomes small the curing reactions become diffusion controlled, and will eventually become very slow and finally stop. For products for which it is necessary to ensure complete reaction of all epoxy groups it is a normal practice to post-cure the resins at an elevated temperature. For successful application of

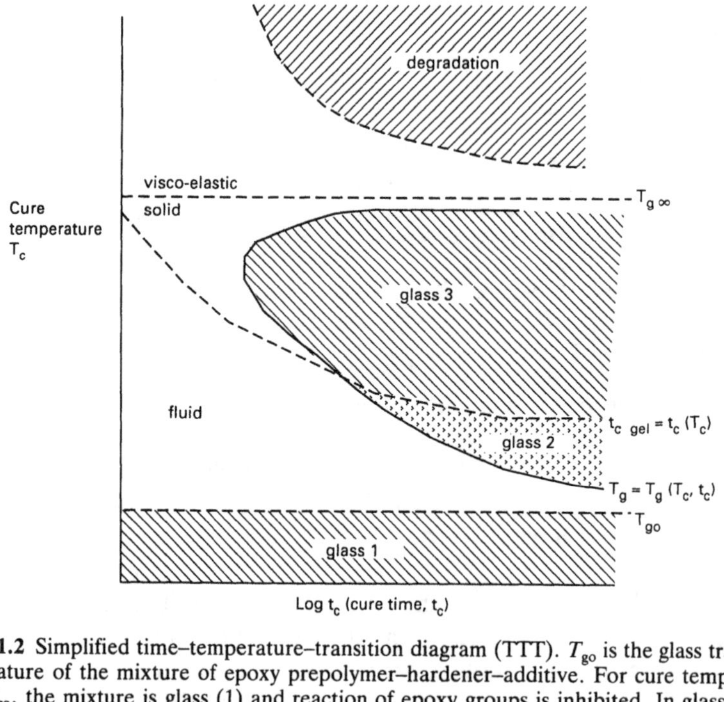

Figure 1.2 Simplified time–temperature–transition diagram (TTT). T_{go} is the glass transition temperature of the mixture of epoxy prepolymer–hardener–additive. For cure temperature $T_c < T_{go}$, the mixture is glass (1) and reaction of epoxy groups is inhibited. In glass (2) the epoxy resin vitrifies *before* gelation. For glass (3) the glass transition temperature increases with increased cross-linking of the network. $T_{g\infty}$ is the limiting glass transition temperature as the concentration of epoxy group $E \rightarrow 0$.

epoxy resins it is necessary to select a suitable hardener (see chapter 2) and then cure the resin to attain a controlled network structure.

1.2 The chemistry of the epoxy group

The original discovery of the parent compound ethylene oxide, or oxirane, is attributed to Wurtz who in 1859 published details of its synthesis from ethylene chlorohydrin by reaction with aqueous alkali.

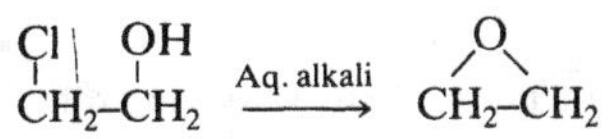

This method is *general* for the synthesis of epoxy compounds but ethylene oxide is now manufactured by direct oxidation of ethylene with air or oxygen and a silver catalyst. Some of the early history of the synthesis and chemistry of epoxy compounds has been discussed by Malinovskii (1965) with especial reference to early Russian work. The synthesis of epoxy rings has been discussed in detail by Gritter (1967) and Lewars (1984) and epoxy resins by Tanaka (1988).

There are many methods for the synthesis of epoxy rings (Rosowsky, 1964), which are classified in Table 1.4. Although not the only ones, the most important routes for the manufacture of epoxy resins are reaction of a halohydrin with hydroxyl compounds, and the oxidation of unsaturated compounds with a peracid. The first method is similar to the original synthesis of ethylene oxide by Wurtz and may be illustrated by the reaction of epichlorohydrin with hydroxyl compounds, such as phenols or aliphatic alcohols.

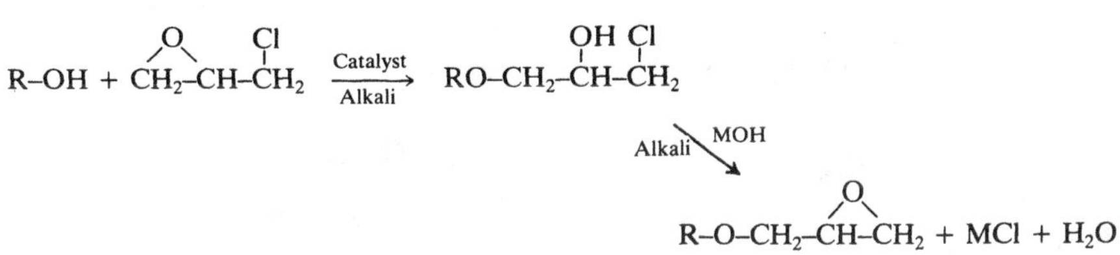

MOH could be sodium or potassium hydroxide and has to be used in stoichiometric concentration to neutralize the halogen acid, HCl in this case, that is produced when the epoxy ring is formed. The application of these reactions for the manufacture of epoxy resins is discussed in more detail in section 1.3.

Unsaturated compounds can be oxidized to yield epoxy groups by the use of peracids such as peracetic acid.

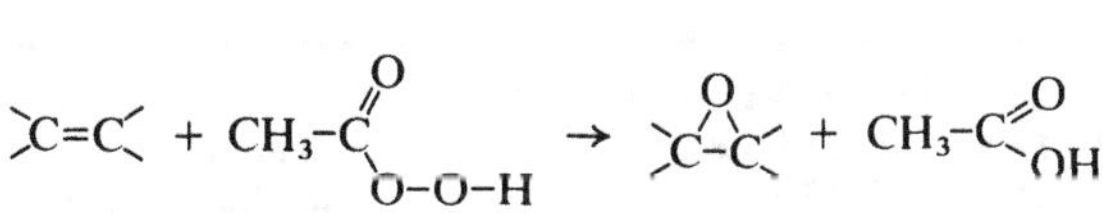

Table 1.4 Outline of the synthetic methods for epoxides

1. Oxidation of alkenes

a. Direct oxidation O_2/catalyst, Ag/100–500°C
b. Inorganic oxidants e.g. chromic acid/permanganate
c. Organic peroxides peroxy acids, $R\text{–}C(=O)\text{–}OOH$
d. Hydrogen peroxide H_2O_2

2. From halohydrin
Hypohalous addition to alkene and then cyclodehydrohalogenation

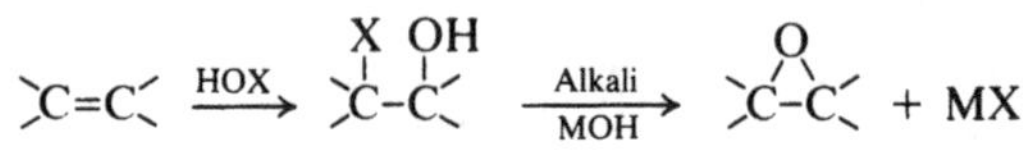

3. From α-halocarbonyl compounds
a. Darzen's condensation

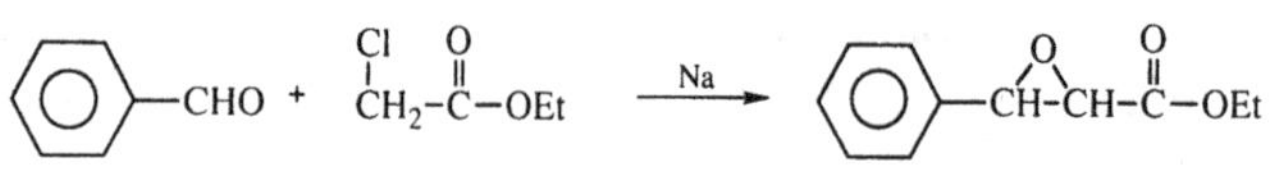

b. Reduction: lithium aluminium hydride

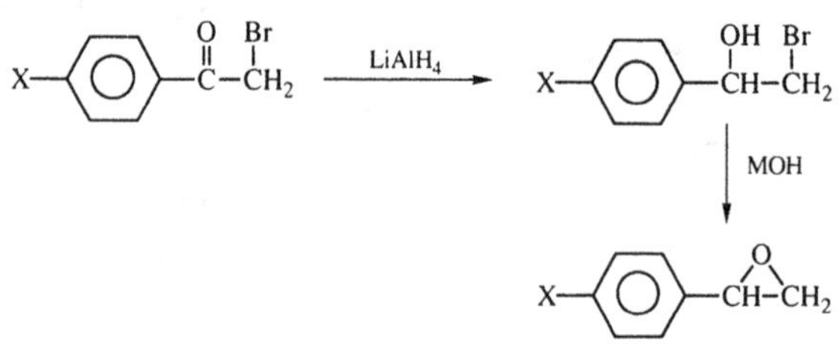

c. Addition of alkoxide ion, followed by ring closure

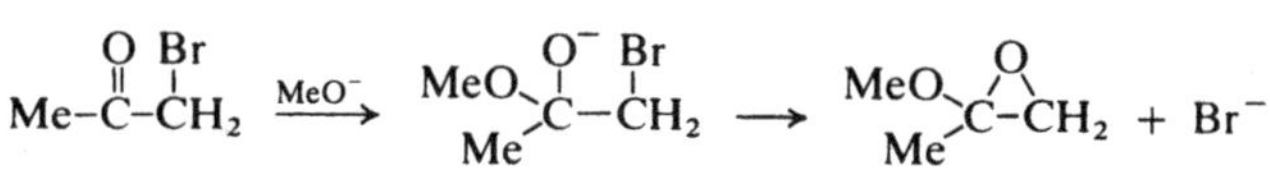

d. Addition of cyanide ion

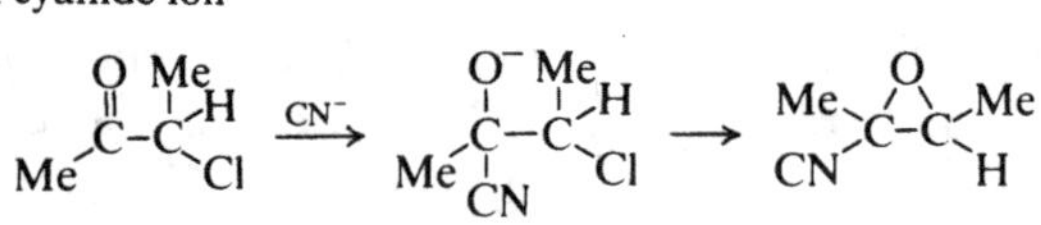

e. Grignard reagent

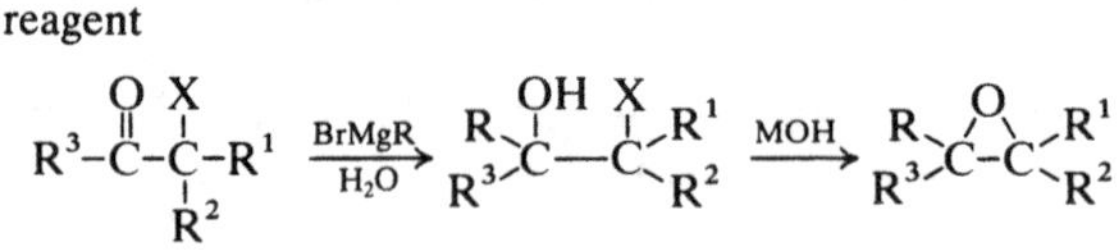

Compiled from: Gritter (1967); Lewars (1984); Malinovskii (1965) and Rosowsky (1964) who give extensive lists of references to the very many routes available for the synthesis of epoxy rings. *Advances in Heterocyclic Chemistry* should be consulted for more recent references.

This is Prileschaiev's reaction which was published in 1912 and has been extensively discussed in several reviews (Rosowsky, 1964; Tanaka, 1988). A possible mechanism for this reaction with epoxidation by perbenzoic acid is

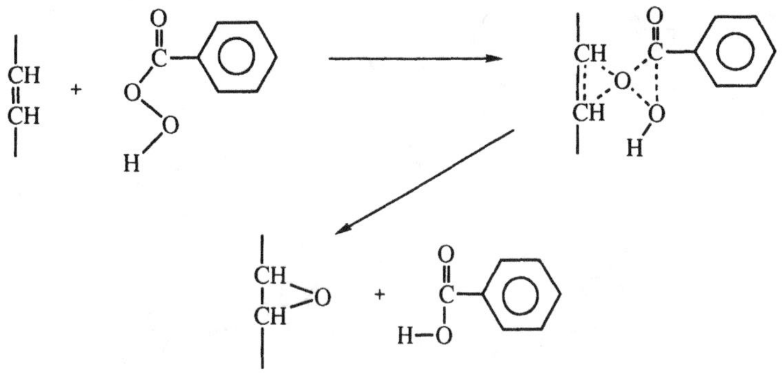

Somewhat alternative mechanisms have been proposed which Gritter (1967) discusses, pointing out that isotopically labelled peroxy oxygen is incorporated into the ring. The mechanism proposed is

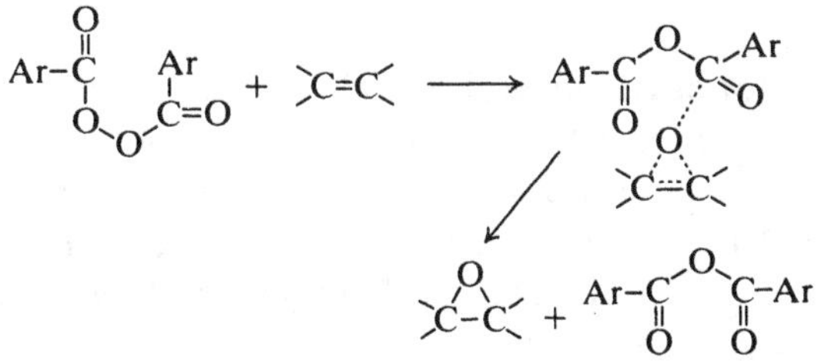

where Ar is aryl group.

Many peracids (Swern, 1970) have been studied as well as reaction conditions, since hydrolysis may occur with the formation of a glycol

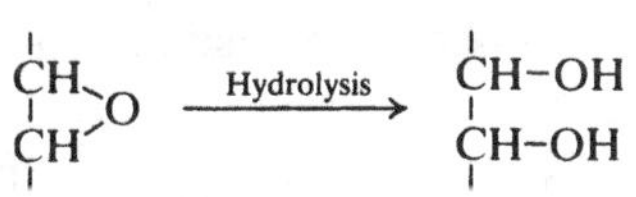

the formation of which has to be avoided to improve the yield of the epoxy compound. The manufacture of epoxy resins from unsaturated compounds by epoxidation with peracids is discussed in section 1.3.3.

The geometrical structure of the epoxy ring is planar, with bond angles and lengths (Figure 1.3) determined from electron diffraction and microwave spectroscopic measurements which have been discussed by Lwowski (1984), Lewars (1984) and Peters (1967). The differences in bond angle from dimethyl ether are considerable as can be seen from Figure 1.3, and there must be considerable ring strain due to angular distortion from the tetrahedral carbon angle of 109°. The dipole moment of simple ethers is 1.1

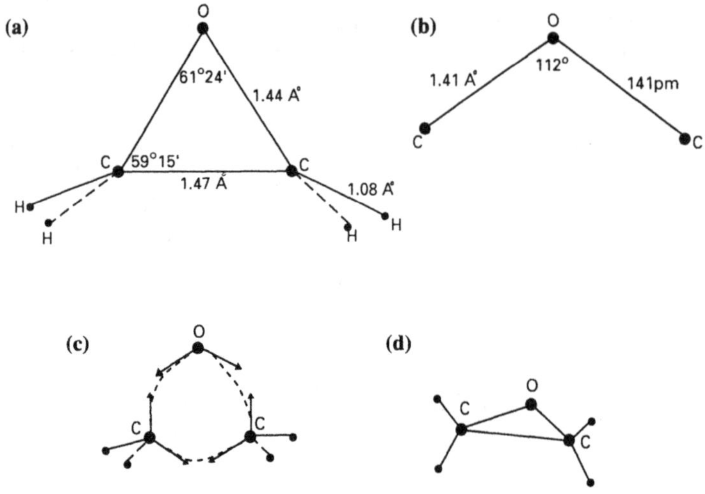

Figure 1.3 The structure of the epoxy ring. (a) Bond lengths and angles for ethylene oxide. (b) C–O bond lengths and angles for dimethyl ether (the positions of the hydrogen atoms not shown). (c) 'Bent' bonds in the strained epoxy ring. (d) Projection view of ethylene oxides.

to 1.3 D and that for ethylene oxide is 1.82 and 1.91 D in benzene solution and the vapour phase respectively. The ionization potential of the oxygen 2pπ lone pair in ethylene oxide is 10.6 to 10.8 eV which is rather higher than that of dimethyl (10.0 eV) and diethyl (9.5 eV) ethers which Peters (1967) has compared with the ionization energies of other simple oxygen compounds.

The electronic structure of three-membered rings poses difficult problems since with a C–O–C or C–C–C bond angle of about 60° the 'normal' sp^3 hybridization with linear bonds between the ring atoms is impossible. The bonding in cyclopropane has been discussed extensively and Halton (1991) in an interesting review of ring strain in cyclic molecules considered the latest evidence. The bonding in cyclopropane is abnormal with the interbond angle compressed to about 60° which is required for ring formation with the nuclei 'moving ahead' of the bonding electron density with the formation of a 'bent' or so-called 'banana' bond as illustrated in Figure 1.3.

The geometry of the epoxy ring is similar to that of cyclopropane but because of the electronegativity of the heteroatom the internal ring bond angles and lengths are not equal (Figure 1.3). Parker and Isaacs (1959) discussed the various structures that have been proposed for ethylene oxide. It has been suggested that the carbon atoms are trigonally hybridized, that is sp^2, and that one such orbital from each carbon atom overlaps with an oxygen atomic orbital to form a molecular orbital which occupies the centre of the ring (Figure 1.3). It is possible that the presence of the 'central' ring orbital accounts for conjugation of the epoxy ring with other delocalized electrons, which is shown by bathochromic shifts in UV (electronic) spectra (Lewars, 1984) and NMR ring currents (Gritter, 1967). Of course such conjugation does not prove that the electrons in the unsubstituted compounds are delocalized, and there has been dispute regarding the possibility of ring currents in these compounds (Gritter, 1967). Although the strain energies of cyclopropane and the epoxy ring are very similar, 27.43 and 27.28 kcal/mole respectively (Gritter, 1967), it may be that the bonding is very different. For instance, from the NMR data compiled by Lwowski (1984) the chemical shifts and coupling constants for the epoxy ring are different from those for cyclopropane.

The many industrial applications of epoxy resins require the formation of three-dimensional networks by reaction with suitable polyfunctional hardeners, which are discussed in detail in chapter 2. Many of these curing reactions depend on the reactivity of the epoxy ring, which is very much more reactive than the 'normal' non-cyclic ethers, R–OR′, where R and R′ are alkyl or aryl groups. In normal ethers the oxygen link is resistant to attack by alkalis, ammonia or amines. Epoxy resins will react with some aliphatic amines at room temperature; these amines may be used as curing agents at ambient temperatures (see chapter 2). This increased reactivity of cyclic ethers is due to the ring strain.

The chemistry of the epoxy ring has been reviewed comprehensively by Parker and Isaacs (1959) and a more recent discussion is that of Lewars (1984). The literature of heterocyclic chemistry (Katritsky and Weeds, 1966; Katritsky and Jones, 1979; Belen'kii, 1988) including that of the epoxy ring has been listed periodically; initially references to epoxy resins were listed but not recently. However, these annotations are a useful source of reference to information on the reactions of the epoxy group. The reactivity of epoxy compounds is summarized in Table 1.5.

The reactions that are most important for the synthesis and cure of epoxy resins involve either electrophilic attack on the oxygen atom or nucleophilic attack on one of the ring carbon atoms. For the unsymmetrically substituted epoxy compound R–CH–CH_2 (epoxide ring, O bridging CH and CH_2), which occurs in most epoxy resins, several factors determine ring opening reactions, such as, the nature of the reagent or catalyst which may be either electrophilic or nucleophilic, the influence of the substituent and the relative steric hindrance at the two carbon atoms. With the general reagent HR′, two possible products of ring opening may be produced:

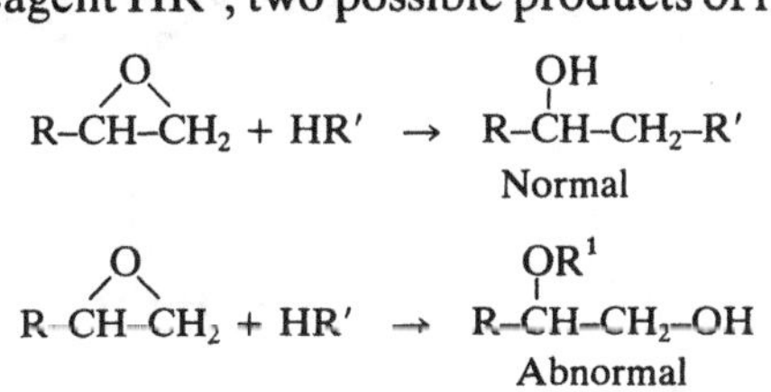

Table 1.5 Some typical reactions of epoxy groups

1. Addition reactions by nucleophilic substitution
   a. Hydroxylic nucleophiles

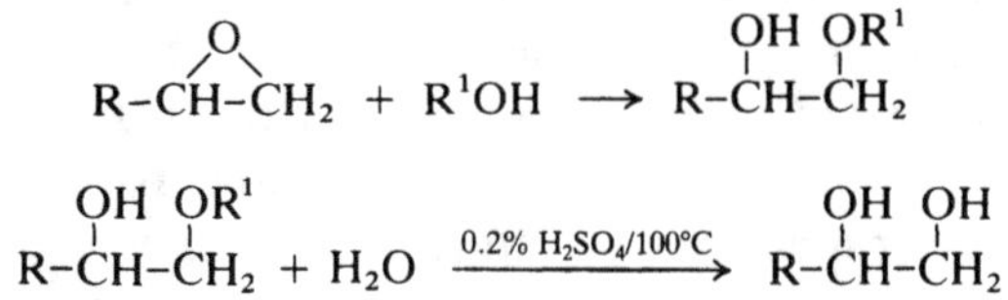

R^1 = alkyl, alcohol; R^1 = aryl, phenol

   b. Acids
      i. Mineral acids

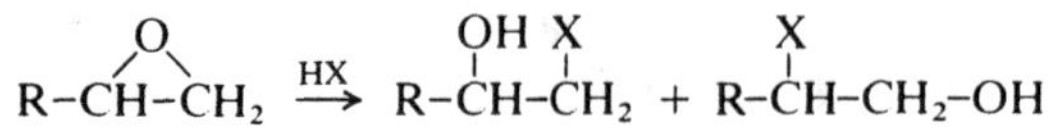

Product ratio depends on reactants and reaction conditions

X = F, Cl, Br, I

      ii. Carboxylic acids

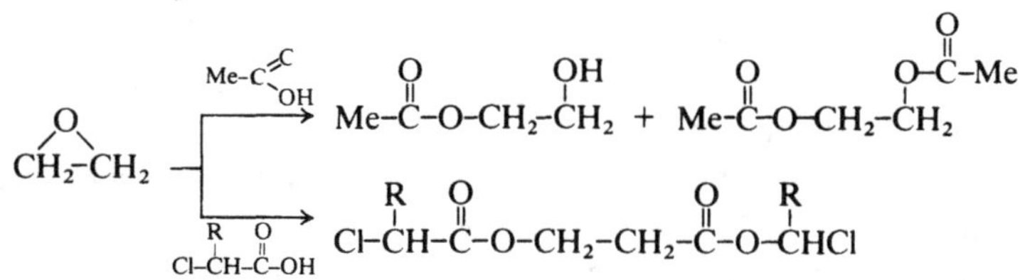

      iii. Ammonia and amines

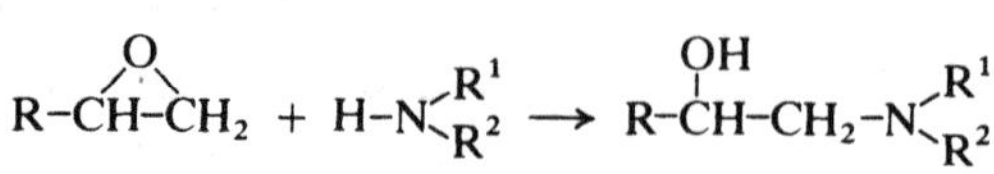

$R^1, R^2 \equiv H$: ammonia
$R^1 \equiv H$; $R^2 \equiv$ alkyl, aryl: primary amine
$R^1, R^2 \equiv$ alkyl, aryl: secondary amine

2. Electrophilic additions
   a. Alkyl halides

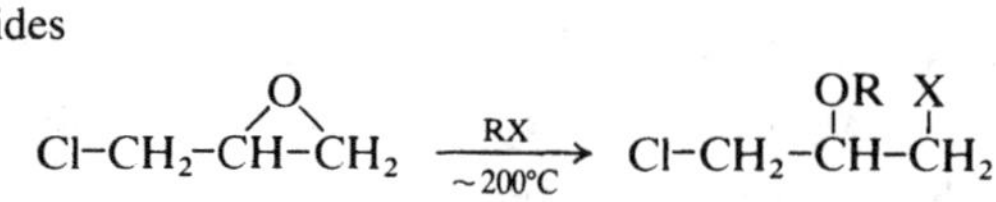

X ≡ Br; R ≡ Et
X ≡ I; R ≡ Me, Et, Pr

   b. Isocyanates

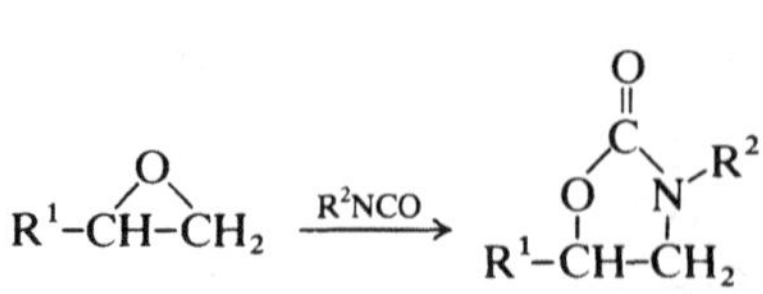

c. Oxides of sulphur

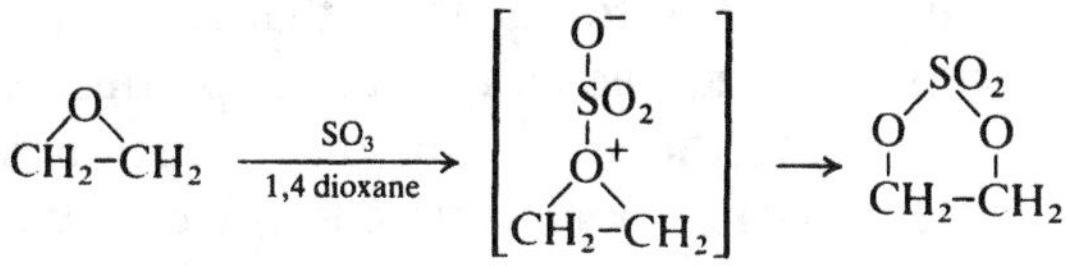

3. Reduction

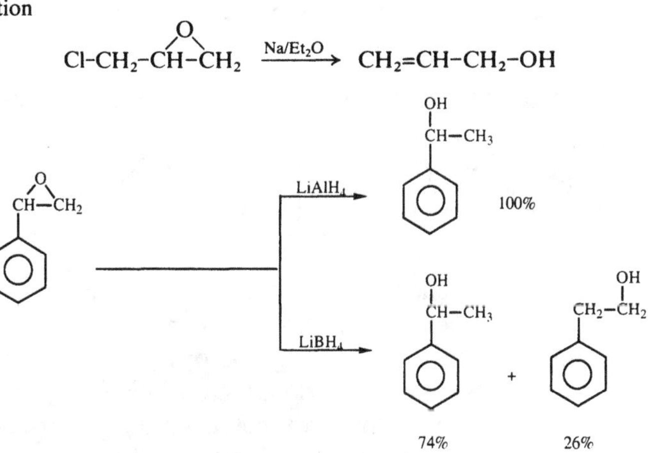

Products depend on the reducing agent

4. Oxidation

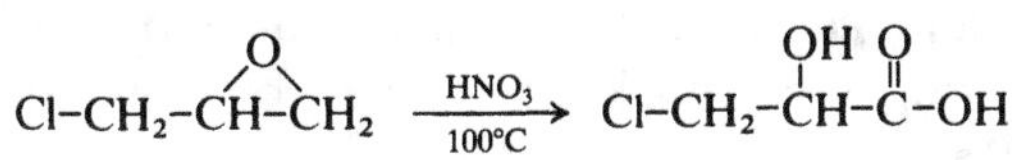

Compiled from: Gritter (1967); Lewars (1984); Malinovskii (1965) and Rosowsky (1964).
This is only a small selection of the very many reactions of epoxy groups that have been reported. Base-catalysed rearrangements of epoxides are discussed by Yandovskii and Ershov (1972), retention of configuration by Akhrem *et al.* (1968) and ring expansion by Grobov *et al.* (1966).

a secondary alcohol or primary alcohol or a mixture. When HR′ is an amine, carboxylic acid or thiol, the 'normal' product, a secondary alcohol, is usually formed.

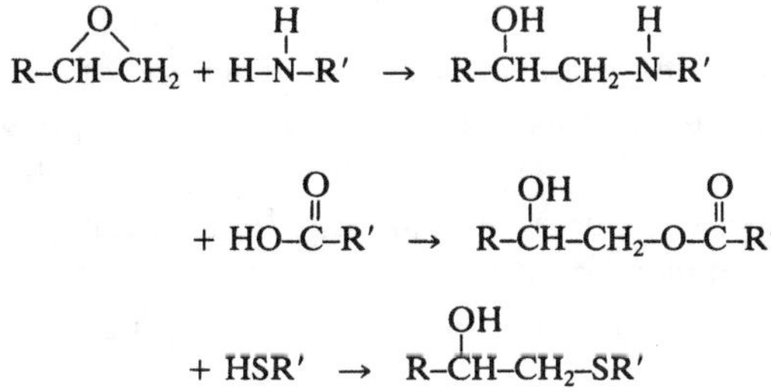

In these reactions the attacking group donates a pair of unshared electrons to the atom with the lowest electron density, that is, the methylene group which is also less sterically hindered and hence the product is the secondary alcohol. A mechanism for the base-catalyzed addition is regarded as 'borderline' S_N2 (Streitweiser, 1956; Parker and Isaacs, 1959).

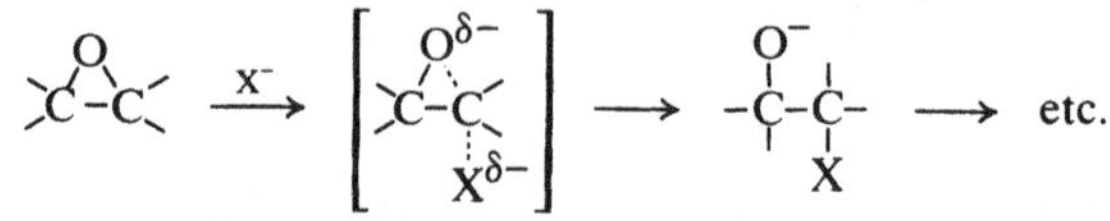

Acid catalysed addition involves proton attack on the ring oxygen atom,

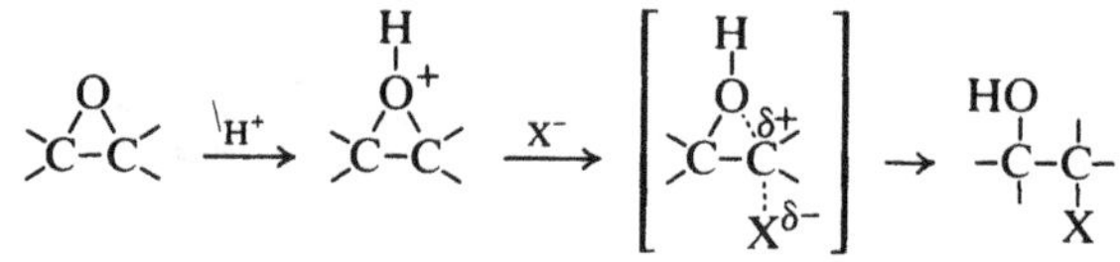

The kinetics of these reactions is discussed in detail by Frost and Pearson (1961) and Parker and Isaacs (1959). The mode of addition may be reversed when R is a strongly electron-attracting group or mixtures may be formed depending on the importance of the various factors involved. Tanaka (1988) discusses the configuration of the protonated epoxy group and Lewars (1984) their basicity and calculations of proton affinity. The latter also discusses other reactions which involve electrophilic attack on the ring oxygen atom. These include Lewis acids, alkyl halides, halides, peroxy acids, aldehydes and ketones. The reaction of epoxides with epoxides initiated by electrophilic attack can lead to dimerization and also polymerization (Price, 1967).

1.3 The synthesis and manufacture of epoxy resins

1.3.1 Synthesis of epoxy compounds

Ethylene oxide can be manufactured by the direct oxidation of ethylene (Kilner and Samuel, 1960)

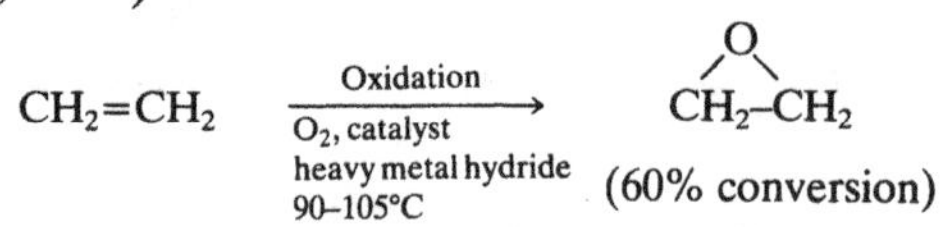

but unfortunately this process is not so efficient for higher olefines. The methods that have been used for the synthesis of epoxy rings have been discussed by Lwowski (1984) and Lewars (1984), the former in a general review of the synthesis of small and large heterocyclic rings (Table 1.4). A comprehensive review of the synthesis of epoxy compounds is that of Tanaka (1988). The most important for the synthesis of epoxy resins are

(i) dehydrohalogenation of halohydrins and (ii) the epoxidation of alkenes with peracids or their esters (section 1.2). Details of many actual syntheses of epoxy resins including reaction conditions and yields are given by Sandler and Karo (1977).

A very important intermediate for the production of epoxy resins is epichlorohydrin, $Cl{-}CH_2{-}\overset{O}{\widehat{CH{-}CH_2}}$, 2,3 epoxypropylchloride. This is because the epoxy ring reacts readily with hydroxyl compounds such as phenols and alcohols

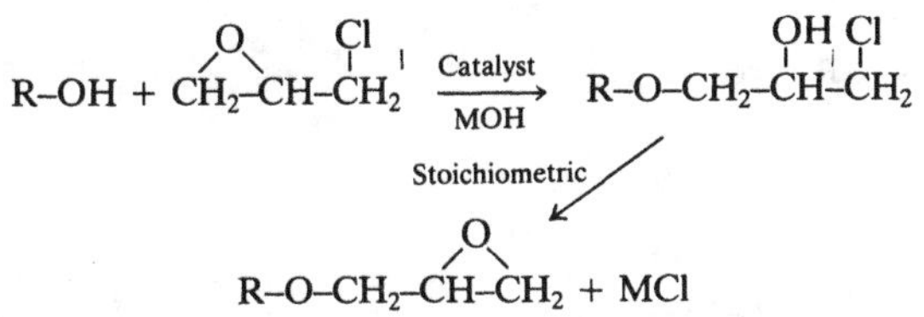

with the formation of a chlorohydrin. The epoxy group is formed by dehydrochlorination with a stoichiometric amount of alkali, such as sodium hydroxide.

The starting compound for the manufacture of epichlorohydrin is propylene which is chlorinated (Kilner and Samuel, 1960).

$$\underset{\displaystyle CH_3}{CH_2{=}CH} \xrightarrow{Cl_2} Cl{-}CH_2{-}\underset{\displaystyle CH_3}{CH}{-}Cl + CH_2{=}CH{-}CH_2{-}Cl$$

1,2 Dichloropropane Allyl chloride

The allyl chloride is converted to dichlorohydrin by reaction with hypochlorous acid and is then dehydrochlorinated with lime to obtain epichlorohydrin (Faith *et al.*, 1965a).

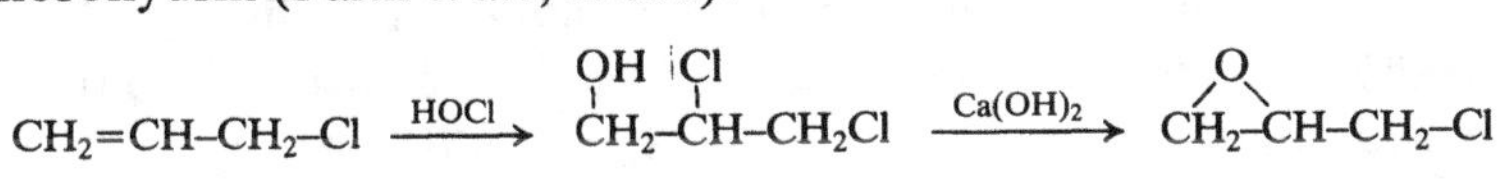

Industrially epichlorohydrin is either converted to glycerol by reaction with sodium hydroxide or isolated by steam stripping and purified by distillation.

An alternative route is from acrolein, produced by oxidation of propylene (Faith *et al.*, 1965a).

$$CH_2{=}CH{-}CH_3 \xrightarrow[\text{Steam catalyst CuO 350°C}]{O_2} CH_2{=}CH{-}CHO + H_2O$$

Acrolein (85% yield)

The acrolein is chlorinated to yield 2,3 dichloropropionaldehyde which is reduced to produce glycerol β,γ-dichlorohydrin, which is then dehydrochlorinated.

$$CH_2{=}CH{-}CHO \xrightarrow{Cl_2} Cl{-}CH_2{-}\overset{Cl}{CH}{-}CHO \rightarrow Cl{-}CH_2{-}\overset{Cl}{CH}{-}\overset{OH}{CH_2}$$

Epichlorohydrin may also be produced from allyl chloride by epoxidation with a peracid.

1.3.2 Epoxy resins manufactured from epichlorohydrin

Epichlorohydrin is used for the production of a range of epoxy resins (Table 1.6) because the epoxy group reacts readily with hydroxylic compounds in the presence of an alkali catalyst, MOH, and then a new epoxy ring can be formed by dehydrochlorination.

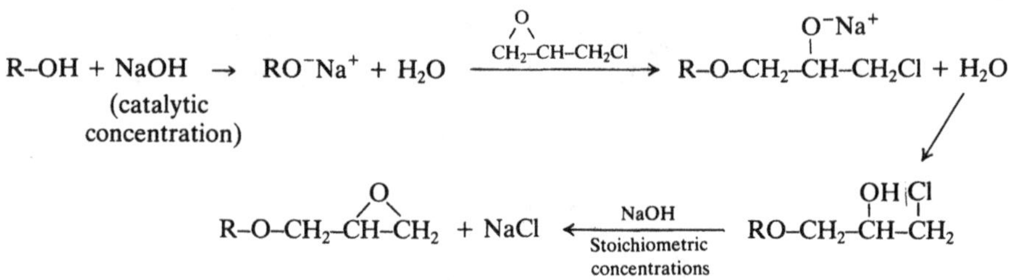

For the production of epoxy resins, the hydroxylic compounds are multifunctional and many such phenols have been studied as possible precursors; the more important are listed in Table 1.7. Also, some monofunctional phenols have been used for the manufacture of resin 'modifiers'. However, the most important phenol used for the manufacture of epoxy resins is the difunctional bisphenol A which was originally studied by Castan (see section 1.1).

1.3.2.1 Resins manufactured from bisphenol A. Bisphenol A or 2,2′bis(*p*-hydroxyphenyl)propane is produced from acetone and phenol with an acid catalyst such as 75% sulphuric acid or dry hydrogen chloride

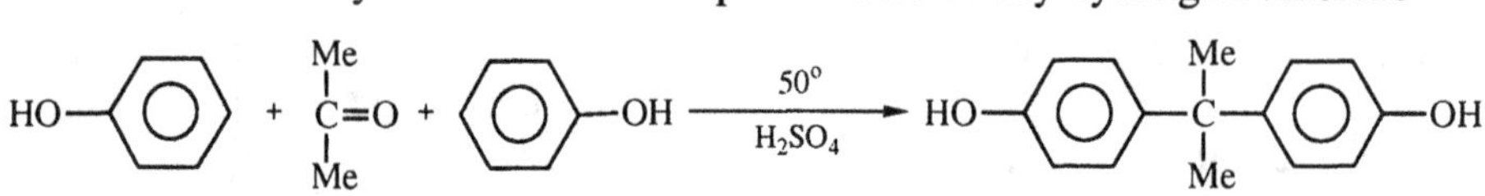

acid gas (Faith *et al.*, 1965b). The reaction conditions will depend on the design of the production unit (Materials and Technology, 1972). The purity of the product is high, >95% *p,p′*-isomer; the other isomers formed are *o,p′* and *o,o′*. For resin manufacture the *p,p′* isomer content should be at least 98%. The light yellow colour of some epoxy resins may be due to trace impurities in the bisphenol A, such as iron, arsenic and highly coloured organic compounds. Other names for bisphenol A are 4,4′-isopropylidene diphenol and diphenylolpropane (DPP).

When a large excess of epichlorohydrin is reacted with bisphenol A with a stoichiometric amount of sodium hydroxide at about 65°C the resin produced contains about 50% diglycidyl ether of bisphenol A, DGEBA

Table 1.6 Epoxy resins derived from epichlorohydrin

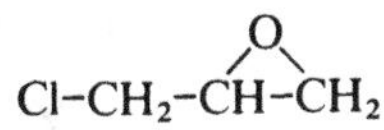

1. Phenols
   a. Many difunctional phenols have been investigated. BPA is the most important, others are listed in Table 1.7
   b. Monofunctional phenols: modifiers for epoxy resins
   c. BPA resins: esterified with fatty acids
   d. Halogen-substituted phenols

2. Alcohols
   a. Multifunctional

1,4-butanediol→ 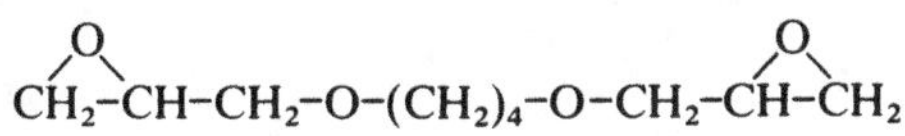

glycerol→ 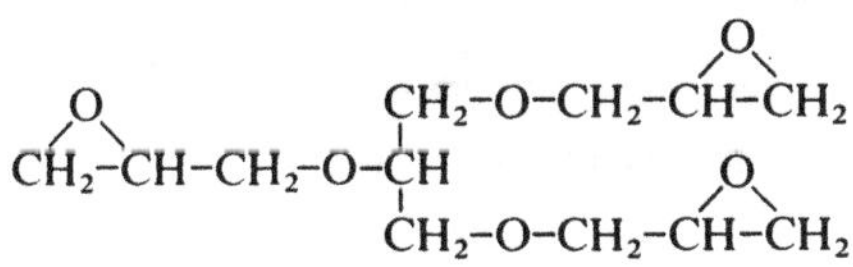

   b. Monofunctional: Resin modifier

Butanol→ 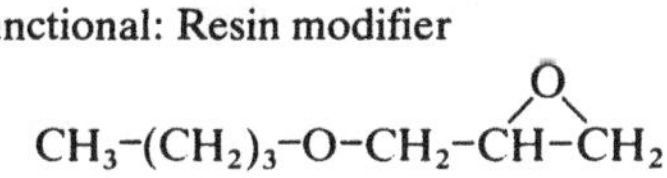

3. Phenolic and related resins
   a. Phenol: formaldehyde novolac
   b. Cresol: formaldehyde novolac

4. Carboxylic and fatty acids
   a. Phthalic acid→glycidyl esters

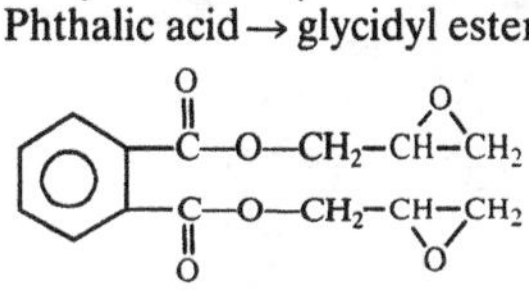

   b. Long-chain acids→epoxy resin esters
   c. Acrylic acid

BPA resins + CH_2=CH(C=O)OH → epoxy acrylates

Acrylic acid

5. Nitrogen compounds
   a. Amines

aniline → 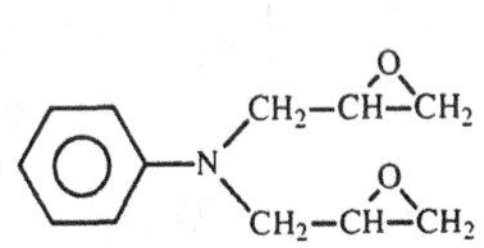

p-amino phenol → 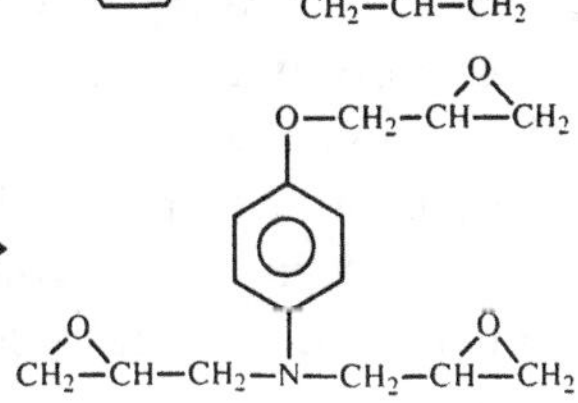

Table 1.6 *Continued*

4,4′ diamino-diphenylmethane →

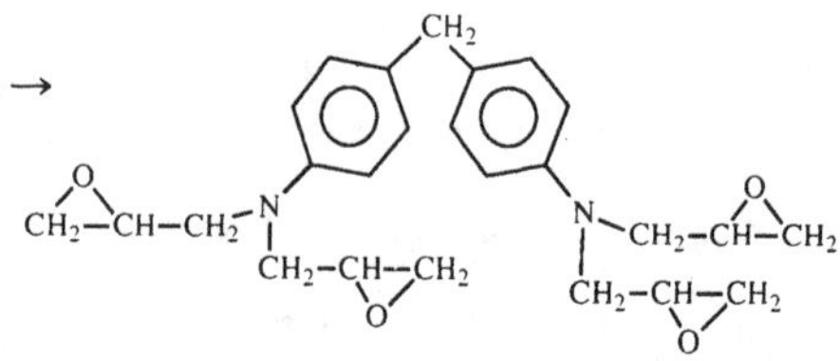

b. Cyanuric acid

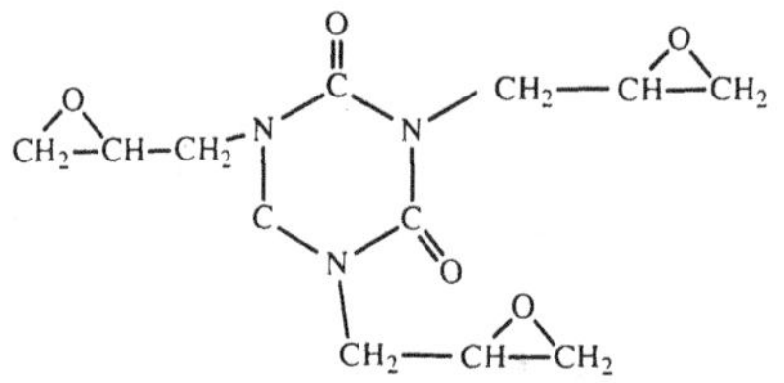

c. Halogenated amines, e.g. tribromo aniline

6. Miscellaneous
   Many other compounds have been reacted with epichlorohydrin. This table is a guide to some of the more important. Other chlorohydrins and halohydrins may also be used, e.g. methylepichlorohydrin $Cl\text{-}CH_2\text{-}\overset{Me}{C}\text{-}CH_2$ (C and CH_2 bridged by O)

(BADGE) and the reaction may be represented formally as

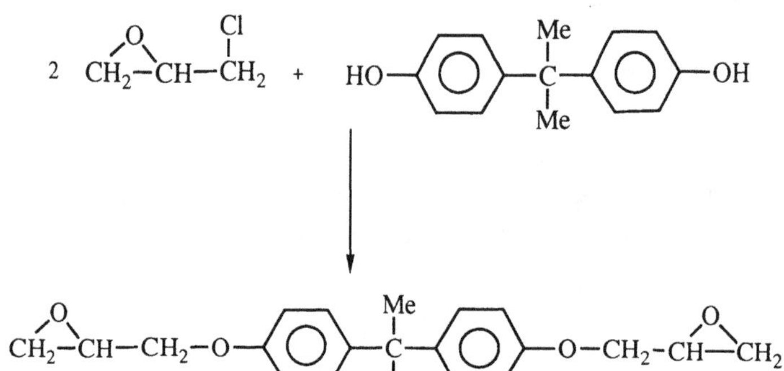

DGEBA

With 'pure' epichlorohydrin it is necessary to add about 5 w/o water to accelerate the reaction. Recovered and recycled epichlorohydrin usually contains sufficient water. The excess of epichlorohydrin is required to limit the production of higher molecular weight products. It is obvious that DGEBA will react with bisphenol A and so on with the formation of higher molecular weight resins, which have the general formula shown on page 20.

Table 1.7 Phenols used for the synthesis of epoxy resins

1. Bisphenol A (BPA)

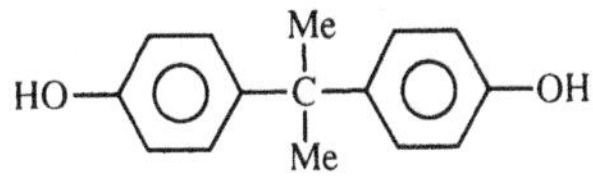

2. Bisphenol F

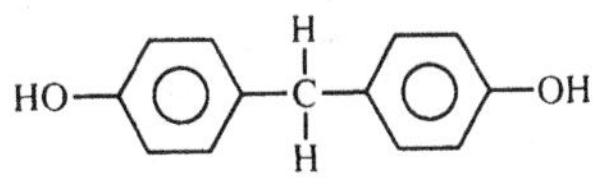

3. Tetrakis phenylolethane

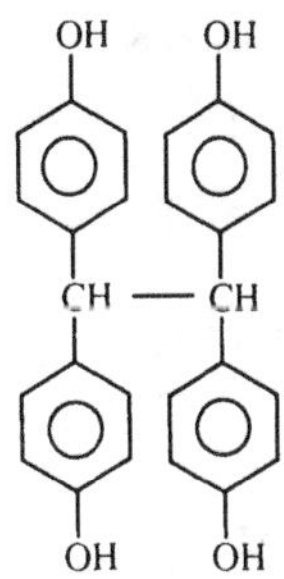

4. Resorcinol

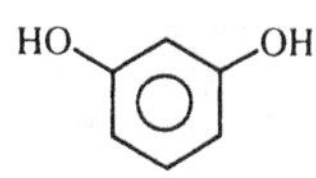

5. Methylolated phenol

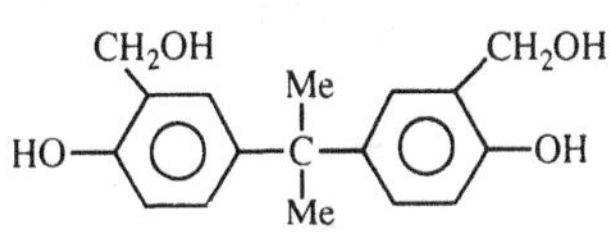

→ diglycidylether the –OH groups accelerate amine cure

6. Brominated and fluorinated phenols

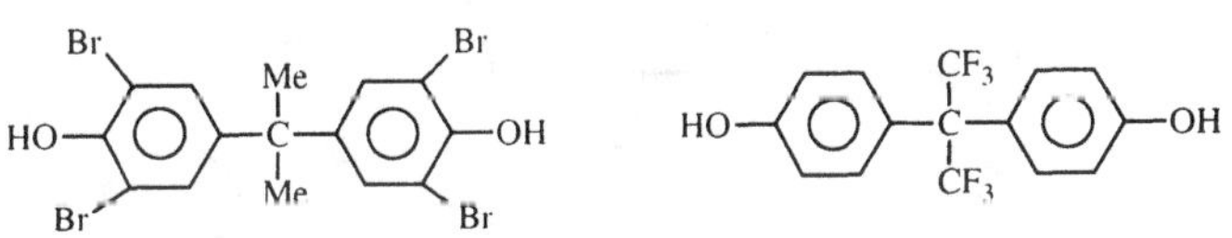

A range of species is present in any specific reaction product and commercial resins are available with $O \leq \bar{n} \leq 14$, and may be higher where $\bar{n}$ is the average number of structure units in the species. A major variable which determines the molecular weight distribution and hence the average molecular weight is the ratio of epichlorohydrin to bisphenol A, but other reaction conditions will affect the product obtained, as will be discussed. For

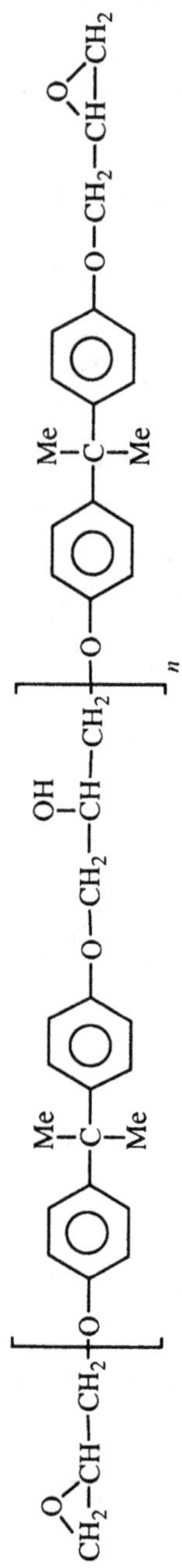

the production of higher molecular weight resins the chain extension process is used (section 1.3.2.4). Batzer and Zahir (1975, 1977) have discussed in detail the molecular weight distributions of epoxy resins prepared by different processes (Ravindrath and Gandhi, 1979).

Although the commercially produced resins have a distribution of chain lengths it is possible to obtain the DGEBA by molecular distillation and it can also be crystallized. Pure DGEBA is a solid which melts at 43°C. Also in liquid commercial resins that have been stored for prolonged periods some crystallization of DGEBA will occur and the resin appears cloudy. Outlines of the production process for the manufacture of a low molecular weight resin are given by McAdams and Gannon (1986) and also Savla and Skeist (1977c). Preparative methods are given by Sandler and Karo (1977).

Not all of the species present in commercial resins are diepoxides; side reactions can occur when epichlorohydrin is reacted with bisphenol A. These can become more important when higher molecular weight resins are produced. The important side reactions are

i. Hydrolysis of epoxy groups

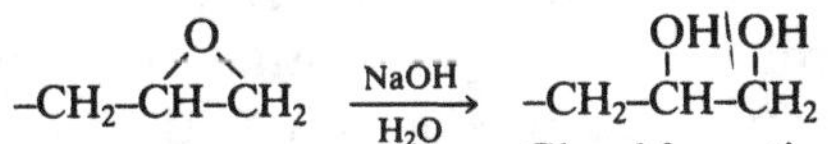

α-Glycol formation

ii. Formation of bound chlorine

a. Reaction of epichlorohydrin with secondary alcohol

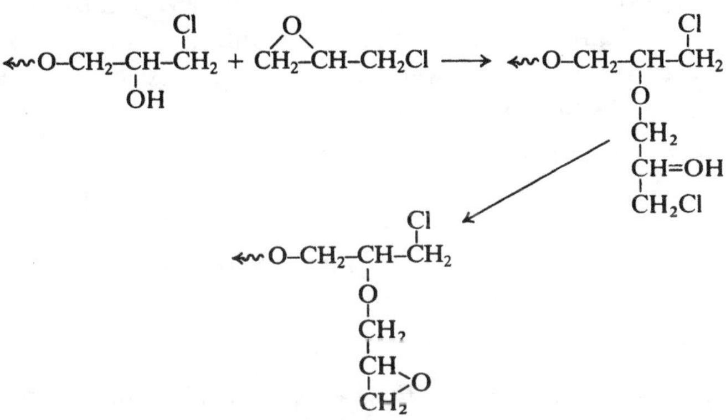

b. Abnormal addition of phenolic hydroxyl

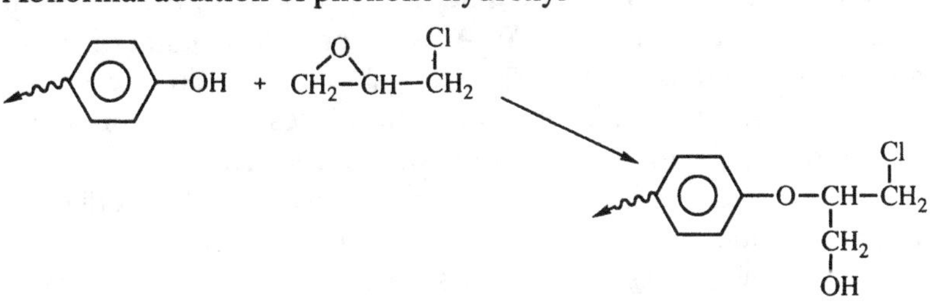

iii. The presence of saponifiable or hydrolysable chloride is due to incomplete dehydrochlorination

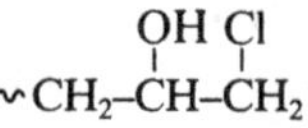

iv. The formation of branched molecules due to reaction of epichlorohydrin with secondary hydroxyl groups.

The last traces of inorganic ions in the liquid resins may be difficult to remove but their concentration can be minimized by suitable washing procedures. The actual concentration of trace amounts of residual material may affect the reactivity of the resins. Also, volatile low molecular weight compounds have to be removed since only small amounts, *ca.* 1 w/o, can significantly reduce the viscosity of the resin. Liquid resins find applications which include coatings, castings, tooling and adhesives, and are also used as starting materials for the manufacture of higher molecular weight and modified resins.

1.3.2.2 Higher molecular weight bisphenol A resins. It is obvious that the bisphenol A/epichlorohydrin ratio is important for control of the average molecular weight of the resins produced with the repeat unit so that the larger the value of *n* the smaller the epichlorohydrin/bisphenol A ratio required. The purity of the reactants is important and monofunctional reactants are chain terminators and hence their concentration has to be controlled. However, it is also necessary to optimize the reaction conditions to achieve the degree of polymerization required. For the production of oligomers with $1 \leq n \leq 4$ the so-called 'Taffy' process may be used but for much higher molecular weight polymers $3 \leq n \leq 20$ the fusion or chain extension process is used.

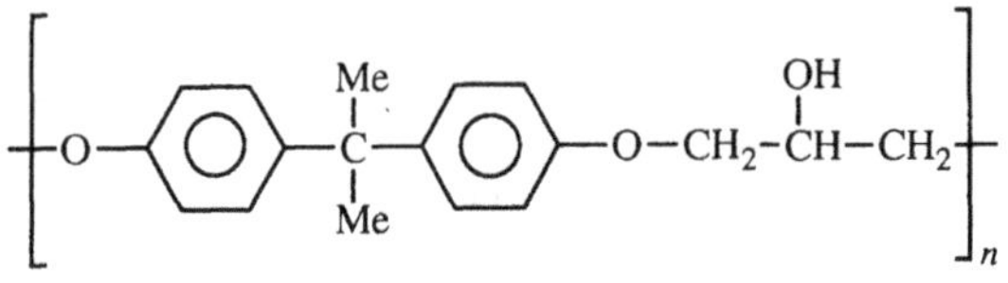

1.3.2.3 The 'Taffy' process. The charge is such that the epichlorohydrin/bisphenol A ratio will yield a resin with the required value of the degree of polymerization $1 \leq n \leq 4$, so that the upper limit for the average molecular weight produced is about 1500. A stoichiometric amount of caustic soda in aqueous solution is added with stirring and the reaction temperature raised to 45–50°C. As the molecular weight increases the reaction temperature is raised to 90–95°C for about 80 min with maybe increased pressure and more vigorous agitation. At the end of the reaction period the product is in a water resin emulsion plus an alkaline brine. The epoxy resin is recovered by separating the phases, washed with water to remove inorganic salts and the water removed by drying at temperatures of up to 130°C and under vacuum. For purification of the resin dissolution in an organic solvent may be advantageous and removal of water may be assisted by the use of methyl isobutyl ketone. However, it is then essential that the level of solvent remaining in the resin is minimized. The recovery of the resin in this process is a major disadvantage especially because of the large amounts of brine that have to be removed. These problems are not encountered in the fusion or advancement process (section 1.3.2.4).

In the 'Taffy' process integral values of n, the degree of polymerization, are usually produced with n values of (0),1,2,3 whereas in the advancement process n is even numbered. A typical product with a weight per epoxide (wpe) *ca.* 500 and $n \simeq 2$ has a softening point of about 70°C and the practical limitation for this process is $n = 3.7$ and a softening point of 95–100°C. These resins will not crystallize as will DGEBA as mentioned previously and the determination of their softening point is discussed in section 1.4.

During the latter stages of this process the temperature is increased up to about 95°C and hydrolysis of the epoxy groups occurs so that there is a relative high α-glycol value of *ca.* 0.5 eq/kg. Although the formation of α-glycol will reduce the molecular weight attained, their presence may be advantageous since they catalyse the reaction between amine hardeners and epoxy groups and hence accelerate cure.

1.3.2.4 Fusion process. This is also known as the chain extension process for reasons which will become clear. The starting materials for the fusion process are bisphenol A plus a liquid epoxy resin which is essentially difunctional in epoxy groups, produced by the processes used for the manufacture of DGEBA. Thus

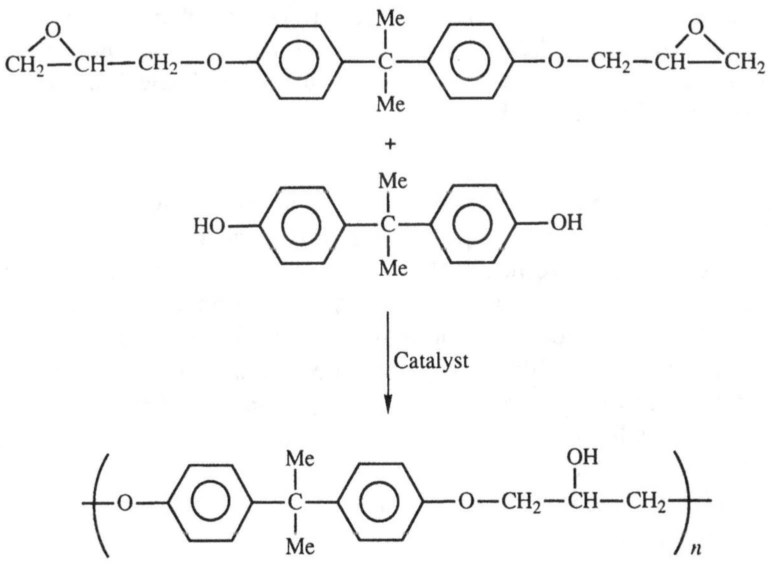

The reaction is carried out at temperatures between 180 and 200°C for about 30 minutes with a nitrogen 'blanket' to minimize oxidative degradative reactions. The reaction is very sensitive to type and concentration of catalyst which must be very strictly controlled to ensure production reproducibility. Thus there has been considerable development of specialized catalysts to minimize chain branching and gelation by promoting reaction between phenolic hydroxyl with epoxy groups and limiting the side reactions, especially between epoxy-epoxy and epoxy-alcoholic hydroxyl groups (McAdams and Gannon, 1986; Savla and Skeist, 1977c). It can be appreciated that the chain extension process is a step-wise polymerization and the relationship between average molecular weight and extent of reaction will depend on the degree of stoichiometric equivalence in the reacting mixture, i.e. the epoxy/–OH ratio and the extent of side reactions which effectively remove either epoxy or –OH groups and hence alter the effective epoxy/–OH ratio. The important parameters were investigated by Batzer and Zahir (1975, 1977).

1.3.2.5 Phenoxy resins. From the previous discussion it would appear that if side reactions are suppressed it should be possible to produce high molecular weight resins from DGEBA and bisphenol A when there is exact stoichiometric equivalence, i.e. epoxy-hydroxyl group ratio is unity, and reaction conditions such that $p \rightarrow 1$, where p is the probability that an epoxy or hydroxyl group has reacted (Reinking *et al.*, 1963). Thermoplastic resins with average values of $\bar{n}$ of about 100 and average molecular weights of maybe 30 000 to 45 000 are available. These resins may not have terminal epoxy groups, but each repeat unit has a secondary hydroxyl group which is reactive, with, for example, isocyanates. Phenoxy resins are available in solution for coating applications which harden when the solvent evaporates. Granular resins may also be used for extrusions and injection mouldings (McAdams and Gannon, 1986).

1.3.2.6 Resins from other phenolic compounds. Any multifunctional, $f \geq 2$, phenolic compound is a potential starting material for the manufacture of epoxy resins. Although many have been studied, only a few have any commercial significance and these are 'formulated' to meet specific requirements. Also, some monofunctional phenols have been reacted with epichlorohydrin to produce monofunctional reactants for use as modifying diluent agents. Some of the phenolic compounds that have been used as starting materials are listed in Table 1.7 and only some of the more important will be briefly mentioned.

The dihydric phenol which is produced by reaction of phenol with formaldehyde is called bisphenol F.

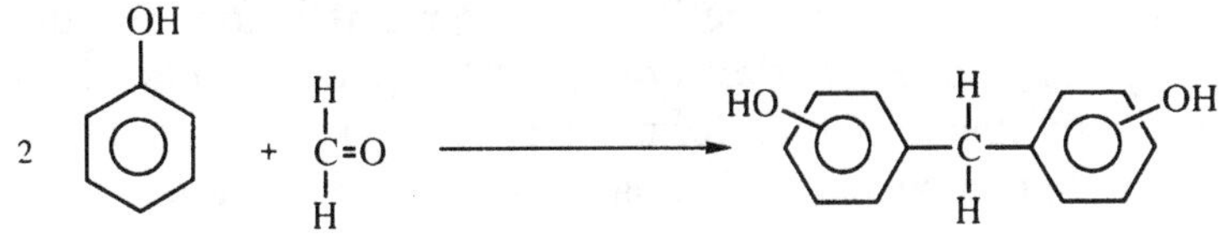

Mixed isomers

Bisphenol F

Resins can be manufactured from bisphenol F by similar methods to those used for bisphenol A and epichlorohydrin with a catalyst such as NaOH. These resins have lower viscosities than the equivalent DGEBA.

Phenolic 'novolac' resins are manufactured by the reaction of phenol with formaldehyde with P/F of ≈ 0.8 with the use of an acid catalyst. These novolac resins may be represented by the 'idealized' structure (Kopf, 1988),

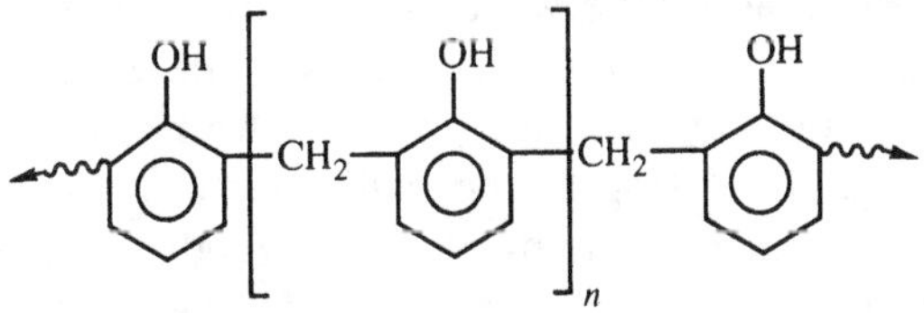

with n depending on reaction conditions and P/F ratio. Not all the substitution is in the *ortho* position but the reactivity of the *ortho* position is higher than that of the *para* because of the 'activating' effect of the phenolic group.

The reaction conditions used to manufacture epoxy-novolacs is similar to those used with BPA resins and the idealized structure of the product is

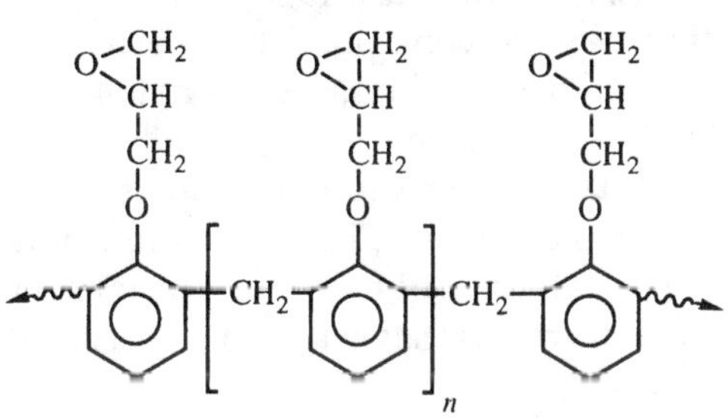

There are considerable 'variants on a theme' with this type of resin; use of excess epichlorohydrin minimizes the reaction of phenolic hydroxyl with the epoxy groups attached to the novolac resin and limits the amount of branching that can occur. Also it is essential that all phenolic hydroxyl groups have reacted because their presence would adversely affect the storage life of the resin and also volatiles would be formed during cure. The epoxy novolacs have improved thermal and chemical resistance compared to the BPA resins.

It is possible to partially esterify novolac resins before reaction with epichlorohydrin or esterify afterwards for ester coating applications. Also the novolacs may be based on chlorinated phenols to obtain improved flame resistance. Other halogenated phenolic compounds have also been used for this purpose (Table 1.7).

1.3.2.7 Other resins derived from epichlorohydrin. Many compounds with reactive hydrogen atoms will react with epichlorohydrin and these have been evaluated as potential resins. Some of the more important are listed in Table 1.6. Formally many react similarly to phenols but there are important differences. For instance, the chlorohydrin formed by reaction of epichlorohydrin with a secondary alcohol is much more sensitive to alkali than aromatic ether-chlorohydrin and hence caustic alkali cannot be used as a catalyst or for dehydrochlorination. A Lewis acid, such as BF_3 or $SnCl_4$, is used to catalyse the reaction between epichlorohydrin and the hydroxylic compounds. Caustic alkali would also catalyse the epoxy-epoxy reaction which leads to polymerization of the product. Aluminates affect dehydrochlorination without catalysing further side-reactions. Some epoxy resins derived from aliphatic alcohols that are commercially available are listed in Table 1.6. Resins have also been prepared from cycloaliphatic alcohols such as hydrogenated bisphenol A, tetracyclohexylethane and hydrogenated novolacs.

Glycidyl esters are manufactured from acids, such as phthalic and hydrogenated phthalic acids. The viscosities of these esters are lower than bisphenol A type resins but have similar reactivities. Also, bisphenol A type epoxy resins can be reacted with fatty acids to produce vehicles for surface coatings (Savla and Skeist, 1977a; Tess, 1988).

Another interesting class of compounds, some of which are available commercially, are derived from amines, such as aniline, or *p*-phenylamine, and these are also listed in Table 1.6.

Other chlorohydrins than epichlorohydrin could be used to produce epoxy resins. Methylepichlorohydrin, derived from isobutylene, has been used to manufacture resins by reaction with bisphenol A and also polybasic acids and novolac phenolic resins (Savla and Skeist, 1977b).

1.3.3 Oxidation of unsaturated compounds

The oxidation of unsaturated compounds has been studied extensively with a variety of reagents (Batzer, 1964) of which the most important for the synthesis of epoxy resins are organic peracids and their esters. The mechanism of such epoxidations has been discussed briefly (section 1.2) and there is an extensive literature (Swern, 1970; Tanaka, 1988) on the methods of preparing the peracids which affect the yields of both desired product and impurities formed. Preparation of the peracid *in situ* often offers advantages.

Numerous peracids, aliphatic, aromatic and cycloaliphatic have been evaluated. Peracetic acid is widely used, either in aqueous, or non-aqueous media in homogeneous or heterogeneous phase systems. It has high epoxidation efficiency and stability at ambient temperatures but there are handling hazards and reaction mixtures can be potentially explosive. A major advantage of the peracid route for the synthesis of epoxy intermediates is that since no species containing chlorine are involved in these syntheses the resins do not contain hydrolysable chlorine, and they are also low in ash and ionic content. Thus these resins have better weathering and ageing properties than 'conventional' epoxy resins and find application where 'good' electrical properties are required.

However, some impurities are formed during oxidation with peracetic acid (Greenspan and Gall, 1953; Lee and Neville, 1967b).

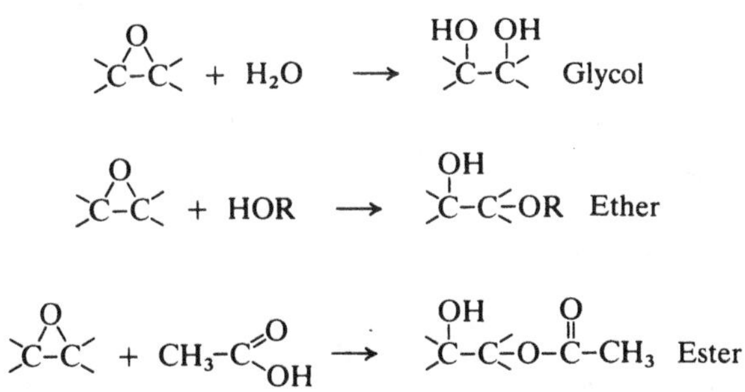

These reactions can be minimized by the use of peracetic acid in which the sulphuric acid used in its synthesis is neutralized. With this reagent natural oils can be epoxidized but it is not satisfactory for the production of compounds with more reactive epoxy groups, even when reaction temperatures are kept low and reaction times short. However, the hydroxyl groups formed are reactive and can be employed in the cure of these resins.

The epoxidation of unsaturated compounds with a peracid is used in the manufacture of cycloaliphatic epoxy resins. Not only are these resins free of hydrolysable chloride and inorganic salts (ash) they do not contain aromatic compounds and hence are more stable to UV exposure than the bisphenol A derived epoxy resins. The presence of aromatic rings in BPA resins increases the UV absorption of the resins and also degradative processes occur by the formation of conjugated structures (Atherton *et al.*, 1982; Bellenger *et al.*, 1986).

The starting materials for the production of cyclo-aliphatic resins may be synthesized by Diels–Alder addition of unsaturated compounds. This can be illustrated by the dimerization of butadiene to yield vinyl cyclohexene which can then be epoxidized with a peracid.

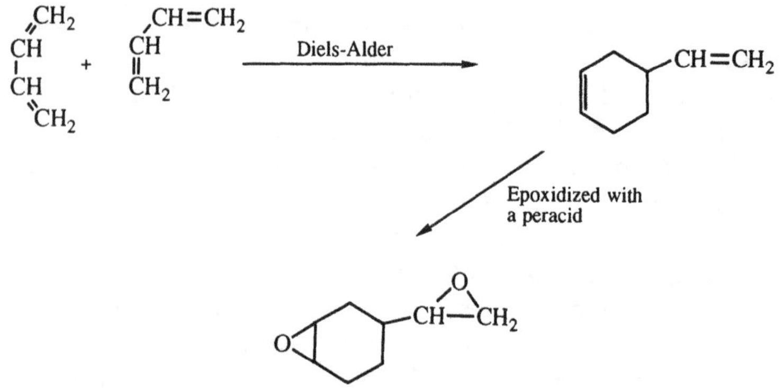

The product, vinyl cyclohexene dioxide is available commercially (Union Carbide, 1989, ERL-4206). The Diels–Alder addition of butadiene with aldhydes or ketones may also be epoxidized. For example, with acrolein two useful products can be synthesized:

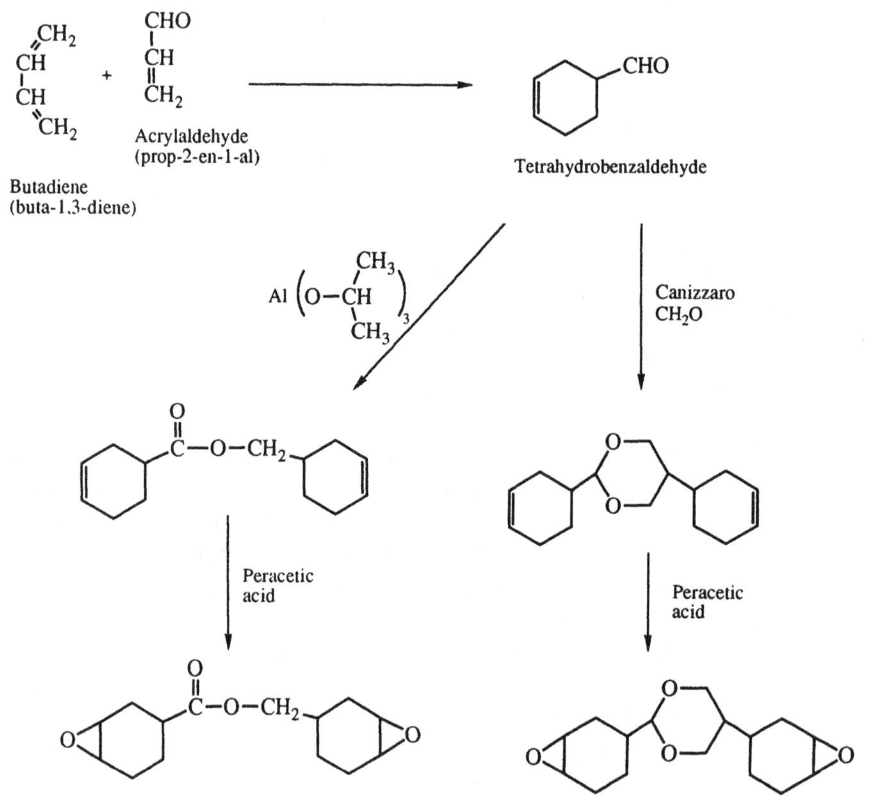

3,4-epoxycyclohexylmethyl - 3,4 epoxy cyclohexane carboxylate

Union Carbide, 1989 ERL-4221
Ciba-Geigy CY-179

2-(3,4 epoxycyclohexyl- 5,5 spiro-3,4 epoxy) cyclohexane metadioxane

Union Carbide, 1989 ERL 4234
Ciba-Geigy CY-175

Other epoxy products can be produced by peracid epoxidation, such as epoxidized polyolefines, oils and fatty acid esters. The last are used as plasticizers and stabilizers for polyvinyl chloride compositions.

1.4 Characterization of uncured epoxy resins

The generalized structure of bisphenol A type epoxy resins can be represented by the structure formula

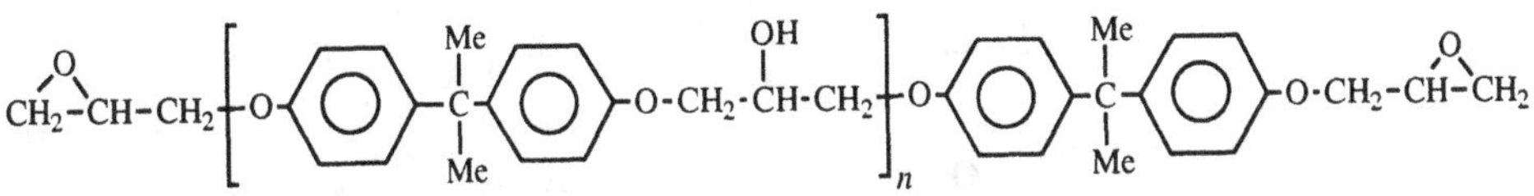

and in general there are different species present with different values of n, the number of structural units. A first approximation in characterizing this resin is to determine the concentration of epoxy groups present since this is needed to calculate the amount of hardener required. A similar consideration will apply to the other types of epoxy resin that are commercially available.

However, the composition of epoxy resins is more complicated than that represented by the general structure and characterization involves much more than only determination of the epoxy content. Thus, the most important parameters that manufacturers of uncured epoxy resins provide as sales specifications include as well as epoxy content, hydrolysable chlorine, specific gravity, colour and for liquid resins their viscosity at ambient temperature. For solid resins the softening point and a solution viscosity are often specified. Some resin manufacturers give detailed instructions for some analytical procedures; for instance Ciba Geigy (1988) give test methods for the determination of epoxy content and readily hydrolysable and total chlorine contents. A detailed treatment of the analysis of uncured resins is given by Urbanski (1977) with experimental details of methods for the qualitative identification of epoxy resins and procedures for their quantitative analysis. The characterization of uncured resins involves determination of their detailed structure. Some of the methods of analysis applied to uncured resins will now be considered.

1.4.1 Chemical analysis

Qualitative test for epoxy resins are available and procedures have been discussed by several authors (Urbanski, 1977; Braun, 1988; Jahn and Geotzky, 1988a). For uncompounded resins infrared methods offer a convenient method of identification. It is useful to establish a set of reference spectra determined with a repeatable set of conditions even though there are compilations of spectra available. Spot tests although simple to execute require cross-checking to ensure reliability. For resin-hardener mixtures and compounded resins, a combination of infrared and NMR spectroscopy (Hadad, 1988a) as well as chromatrographic methods (Mestan and Morris, 1984) may be required.

1.4.2 Quantitative analysis

1.4.2.1 Epoxy content. This is the commonest analytical measurement that is carried out on uncured epoxy resins. Methods for the assay of epoxy content involve the halogen-acid cleavage of the epoxy ring to yield a halohydrin:

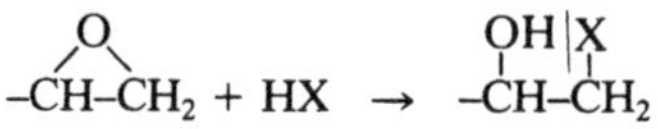

This reaction has been discussed previously and is conceptually obvious as a possible method for the determination of the number of epoxy groups per unit mass present. However, study of the published methods which essentially rely on the reaction being exclusive and quantitative, show that it is by no means easy to ensure the accuracy of these methods. It is essential to follow the detailed procedure implicitly, and to ensure that specified conditions are strictly adhered to (Dobinson *et al.*, 1969; Urbanski, 1977).

1.4.2.2 Determination of hydroxyl groups. For all resin species with $n \geqslant$ there will be a secondary alcohol present unless it has reacted with bisphenol A which of course may lead to the formation of a branched molecule. Hence determination of the concentration of secondary alcoholic hydroxyl groups is important for the characterization of the structure of epoxy prepolymers. Hydroxyl groups may also be potentially reactive with hardeners and their concentration is required to calculate resin/hardener stoichiometry. Also, hydroxyl groups may catalyse the reaction between hardener and epoxy groups. This is very important for the kinetics of cure with amine hardeners which will be discussed in chapter 3.

Other types of hydroxyl group which may be present are α-glycol units formed by hydrolysis of an epoxy group, and also phenolic hydroxyl due to incomplete reaction of the phenol when the resin was manufactured, especially when larger concentrations of bisphenol A are used.

Common methods for the assay of hydroxyl groups are inapplicable because the epoxy groups present interfere (Urbanski, 1977). It is possible to use lithium aluminium hydride which only reacts with active hydrogen atoms.

$$4\ ROH + LiAlH_4 \rightarrow LiAl(OR)_4 + 4\ H_2$$

The hydrogen evolved is determined volumetrically or by gas liquid chromatography. The procedure is difficult to carry out quantitatively and corrections have to be applied to allow for the presence of water or other active hydrogen compounds. A more convenient method is to use an acetylation procedure in which both hydroxyl and epoxy groups react and the total hydroxyl concentration is determined by difference since the epoxy concentration can be determined separately (Jahn and Goetzky, 1988b; Urbanski, 1977).

The α-glycol concentration may be determined by the periodic acid method (Stenmark, 1956).

$$\underset{\alpha\text{-Glycol}}{\mathrm{R{-}\overset{HO\,|\,OH}{C{-}CH_2}}} + IO_3^- \rightarrow IO_4^- + RCHO + HCHO + H_2O$$

Since the α-glycol is usually only present in small amounts it is necessary to take precautions to obtain accurate results. Phenolic hydroxyl is also usually only present in low concentration and although several methods (Jahn and Goetzky, 1988c) have been suggested their accuracy must be assessed to ensure reliable results are obtained (Urbanski, 1977). The esterification equivalent-weight defined as the gram weight of epoxy resin required to esterify one gram-equivalent of carboxylic acid is determined for resins that are converted to epoxy esters. It can be determined by acetylation with acetic anhydride in pyridium chloride solution. The epoxy group will react with two acid groups and the hydroxyl groups will also be converted to esters (Jahn and Goetzky, 1988c).

Infrared spectroscopic methods have also been suggested for the assay of hydroxyl groups but the method requires calibration because of the effects of hydrogen bonding on the spectra (Dannenberg, 1963).

1.4.2.3 Chlorine content. Although 'pure' DGEBA would have zero chlorine content, commercial resins may have up to about 1% chlorine. Self-extinguishing resins may have an organically bound chlorine content of up to 30%, and bromine- and fluorine-containing resins are also available. The resins with high chlorine content may have lower thermal stability particularly when cured with amine hardeners. Flame retardancy is a specialized topic and resin suppliers can formulate compositions which are satisfactory for particular applications.

In bisphenol A type epoxy resins the presence of chlorine will adversely affect the electrical properties as would any polar impurity. The colour and reactivity of these resins may also be adversely affected by the presence of chlorine. The active chlorine may block reaction of lower base catalysts, such as tertiary amines, and hence its concentration may be critical. However, when organically-bound chlorine is present the resin will have a reduced functionality and hence the cured resin will have a less tight network.

The total chlorine present can be determined by the classical method using a Parr bomb where the material is oxidized and then the chloride ion present is determined as silver chloride (Jahn and Goetzky, 1988d). An alternative procedure has been proposed by Urbanski (1977) in which the chlorine is determined hydrolytically by heating for 2 hours with 0.5 N potassium hydroxide in an ethylene glycol-dioxane solution, followed by potentiometric titration with silver nitrate in a dioxane-acetic acid solution. The organically bound chlorine is obtained by correcting for any ionic chloride ions that may be present, which may be determined by a standard procedure. In commercial resins, the inorganic chloride content should be very low, but of course resin manufacturers have to monitor its presence to ensure all of the salt formed by the dehydrochlorination reaction has been effectively washed from the resin.

The organically bound chlorine will be in the forms known as (i) hydrolysable and (ii) inactive. The presence of the former is due to incomplete dehydrochlorination and the latter to either reaction of epichlorohydrin with the secondary alcohol or abnormal addition of the phenolic hydroxyl group as discussed previously (section 1.3). The hydrolysable chlorine, also called saponifiable chlorine, can be determined by reaction with excess caustic soda and then back titration with standard hydrochloric acid (Jahn and Goetzky, 1988d). The inactive chlorine can be calculated by difference from the total organically bound and the hydrolysable chlorine.

1.4.3 Molecular structure

Methods such as infrared and nuclear magnetic resonance (NMR) spectroscopy can be used to identify the type of epoxy resin, that is for determination of the structure of the repeat unit. These methods can also be used to identify the hardeners and other additives present in epoxy resin-hardener mixtures. There are many texts which deal with the theory and practice of IR and NMR spectroscopy and Rabek (1980a) discusses their applications in polymer chemistry.

Hadad (1988b) gives a very good review of the application of various chromatographic methods for the determination of the molecular weight distribution (MWD) of epoxy resins. Batzer and Zahir (1975, 1977) used gel permeation chromatography to analyse the products obtained from the synthesis of epoxy resins using a range of reaction conditions. They compared the MWD actually produced with theoretical predictions and suggested that deviations from the model are due to the higher reactivity of epichlorohydrin with phenolic hydroxyl compared with the arylglycidyl ether group. Chromatographic methods are discussed by Rabek (1980b) and their application to the analysis of epoxy resins by Mestan and Morris (1984).

1.4.4 *Physical properties*

1.4.4.1 Softening point. Although pure DGEBA will crystallize, even when in a mixture with other higher molecular weight species, in general epoxy resins will not crystallize. When cooled they form glassy, amorphous solids with glass transition temperatures (Ives *et al.*, 1971) which have been determined using broad-line NMR spectroscopy (Ellis, 1972). The usual method that is used is measurement of a softening point (Lever and Rhys, 1968) which is defined by the procedure used for its determination. Since the resins soften over a range of temperatures, there will not be exact agreement between methods. There are two standard specifications, the ball and ring and Durran's method which yield somewhat different results. In the ball and ring method the sample is held in a horizontal ring with a steel ball on top of it. The temperature is raised at a prescribed rate and the softening point is that temperature when the ball has moved downwards by a distance of 2.5 cm (1 inch) under its own weight. Durran's 'mercury method' involves warming a known weight of resin in a standard test tube and then cooling it so that it solidifies. A specified quantity of clean mercury is placed on top of the resin, which is then warmed. The Duran's softening point is the temperature at which the mercury sinks to the bottom of the tube, (Stenmark and Weiss, 1956).

1.4.4.2 Viscosity. Control of the processing of epoxy resins requires information on their rheological properties. A good general discussion of rheological measurements is given by Schoff (1988a). For the liquid resins which may have high viscosities it is very difficult to mix with hardeners so that a uniform molecular dispersion is obtained. With some hardeners, reactions occur even at room temperature and hence the viscosity increases before mixing is complete and microgelation may occur and of course locally the epoxy/hardener stoichiometry is out of balance. Even with hardeners which require elevated temperatures it is often necessary to warm the resin to reduce its viscosity so that a molecular mixture can be obtained. Some of the morphological features observed in cured epoxy resins can be attributed to inefficient mixing of the hardener with the epoxy prepolymer. For many applications, such as casting, undiluted epoxy adhesives and laminating resins it is essential that their viscosity during fabrication is low. Also, air bubbles which rise to the surface very slowly, if at all, may be very troublesome when the viscosity increases due to the onset of cure.

The viscosity of epoxy resins depends on their molecular structure, molecular weight distribution, and is often a very sensitive function of temperature. For 'compounded' resins which are liquid at room temperature, their viscosity is measured without dilution but for solid resins the viscosity of a 40 w/o solution in diethylene glycol monobutyl ether is often measured. Solution viscosities (Rabek, 1980c) can be measured by bubble viscometers such as the Gardner–Holdt tubes (Schoff, 1988b), capillary viscometers (Ives *et al.*, 1971) or instrumentally with a spindle viscometer such as the Brookfield (Lee and Neville, 1967c). For these measurements the concentration of the solution and its temperature must be kept to specified values to obtain satisfactory measurements.

For measurement of the viscosity of the liquid resins, the Hoeppler viscometer is often used. The principle of this viscometer is measurement of the time for a ball to descend a tilted tube containing the resin. Both the size of ball, tube diameter and angle have to conform to the specification. Even with these precautions measured viscosities may differ by maybe 15% from the values determined by other methods such as flow through tubes or rotational viscometers. When only one set of conditions is used with a Hoeppler viscometer the viscosity is determined at essentially a single average rate shear. This would be satisfactory for Newtonian fluids but higher molecular weight resins will have a power law relationship between shear stress, τ, and rate of shear, $\dot{\gamma}$

$$\tau = m\,\dot{\gamma}^n$$

where m and n are power law parameters and both are temperature-dependent. When the power law exponent $n = 1$, the fluid is Newtonian with $m = \eta$, the viscosity. For polymer fluids n is less than unity and the fluid is pseudoplastic or shear thinning. For epoxy resins such as Epon 828 the value of n at ambient temperatures is approximately unity (Chen and Ellis, 1992), but with higher molecular weight resins it is necessary to confirm that they are Newtonian. The most suitable instrument is a cone and plate viscometer but use of a Hoeppler viscometer with timing of the rate of descent of the ball for two different angles of tilt may give sufficient data to determine m and n (Barnes, 1980).

The temperature dependence of viscosity of many simple liquids is Arrhenius, $\eta = \eta_e \exp [E\eta/RT]$, so that a plot of $\ln \eta$ vs $1/T$ is linear. However, in some cases it may be satisfactory to represent the temperature dependence on a log-linear plot, so that:

$$\log \eta = A + BT$$

where A and B are empirical constants. Also, the Arrhenius temperature dependence would not be expected to hold when the viscosity is measured at temperatures insufficiently higher than the glass transition temperature of the resin.

The structure of the epoxy resin affects its viscosity, for example the aromatic, bisphenol A type epoxy resins have higher viscosities than the equivalent cycloaliphatic which are themselves higher than linear alophatic.

Other measurements that may be made on uncured resins include, density and refractive index which is very sensitive to the structure and purity of the resin. Colour is also determined by reference to colour standards. The vapour pressure, especially of solutions may be important as is the flash point.

References

Akhrem, A.A., Moiseenkov, A.M. and Dobryin, V.N. (1968) *Russ. Chem. Rev. (Eng. Trans.)* **37**, 448–462.

Atherton, N.M., Banks, L.G. and Ellis, B. (1982) *J. Appl. Polym. Sci.* **27**, 2015–2023.

Barnes, H.A. (1980) In *Rheometry: Industrial Applications*, ed. Walters, K. Research Studies Press, John Wiley and Sons, Chichester.

Batzer, H. (1964) *Chem. and Ind.* **5**, 179.

Batzer, H. and Zahir, S.A. (1975) *J. Appl. Polym. Sci.* **19**, 585–617.

Batzer, H. and Zahir, S.A. (1977) *J. Appl. Polym. Sci.* **21**, 1843–1857.

Belen'kii, L.I. (1988) *Adv. in Heterocyclic Chemistry* **44**, 269–396. Synthesis of epoxy groups p. 304. Reactions of epoxy groups pp. 303–330.

Bellenger, V., Verdu, J., Francillette, J. and Hoarau, P. (1986) *Polymer Commun.* **27**, 279–281.

Birley, A.W. and Scott, M.J. (1982) *Plastics Materials, Properties and Applications*, Leonard Hill, Glasgow.

Braun, D. (1986) *Simple Methods for Identification of Plastics* 2nd edn. C. Hanser Verlag, Müchen, pp. 27, 84 and PIT chart.

Chen, X.M. and Ellis, B. (1992) Unpublished measurements.

Ciba–Geigy (1988) *Araldite for Surface Coatings*, Laboratory Manual, Ciba–Geigy Plastics, Cambridge.

Dannenberg, H. (1963) *Soc. Polym. Eng. Trans.* (Jan. 1963) p. 78.

Dearborn, E.C., Fuoss, R.M., Mackenzie, A.K. and Shepherd, R.G. (1953) *Ind. Eng. Chem.* **45**, 2715.

Dobinson, B., Hofmann, W. and Stark, B.P. (1969) *The Determination of Epoxide Groups*, Pergamon Press, Oxford.

Ellis, B. (1972) *Amorphous Materials* Ch. 38, eds. Douglas, R.W. and Ellis, B. Wiley–Interscience, London. pp. 375–388.

Faith, W.L., Keyes, D.B. and Clark, R.L. (1965a) *Industrial Chemistry* 3rd edn. John Wiley, New York. p. 404. (1965b) p. 152.

Frost, A.A. and Pearson, R.G. (1961) *Kinetics and Mechanism* 2nd edn. Wiley, New York, pp. 288–307.

Greenspan, F.P. and Gall, R.J. (1953) *Ind. Eng. Chem.* **45**, 2722–2726.

Gillham, J.K. (1986) *Encylopedia of Polymer Science and Engineering*, 2nd edn. John Wiley, New York, pp. 519–524.

Gritter, R.J. (1967) *The Chemistry of the Ether Linkage*, Ch. 9, section II, ed. Patai, S. Interscience Publishers, pp. 381–390.

Grobov, L.N., Sineokov, A.P. and Etlis, V.S. (1966) *Russ. Chem. Rev. (Eng. Trans.)* **35**, 671.

Hadad, D.K. (1988a) In *Epoxy Resins, Chemistry and Technology*, 2nd edn, Ed. May, C.A. Marcel Dekker Inc, New York, pp. 1111–1124. (1988b) pp. 1096–1111.

Halton, B. (1991) *Adv. in Ring Strain in Organic Chemistry* **1**, 1–17.

Ives, G.C., Mead, J.A. and Riley, M.M. (1971) *Handbook of Plastics Test Methods*, Iliffe Books, London.

Jahn, H. & Goetzky, P. (1988a) In *Epoxy Resins, Chemistry and Technology*, 2nd edn, ed. May, C.A. Marcel Dekker Inc, New York, pp. 1050–1056. (1988b) p. 1067. (1988c) p. 1069. (1988d) pp. 1070–1072.

Katritsky, A.R. and Jones, P.M. (1979) *Adv. in Heterocyclic Chemistry* **25**, 303–391, see 327–329.

Katritsky, A.R. and Weeds, S.M. (1966) *Adv. in Heterocyclic Chemistry* **7**, 225–299, see 232–233.

Kilner, E. and Samuel, D.M. (1960) *Applied Organic Chemistry*, Macdonald and Evans, p. 54.

Kopf, P.W. (1988) *Enc. Polym. Sci. Eng.* Vol. 11. Wiley–Interscience, New York, pp. 45–95.

Lee, H. and Neville, K. (1967a) *Handbook of Epoxy Resins*, McGraw-Hill, New York, sections 1–2 to 1–4. (1967b) sections 3–6 to 3–8 (1967c) sections 4–13 to 4–15.

Lever, A.E. and Rhys, J.A. (1968) *The Properties and Testing of Plastics Materials*, Temple Press Books, Feltham, U.K. pp. 166–169.

Lewars, E.G. (1984) *Comprehensive Heterocyclic Chemistry*, ed. Katritsky, A.R. and Rees, C.W. Vol. 7, section 5.0.5.4, ed. Lwowski, W. Pergamon Press, Oxford, pp. 114–118.

Lwowski, W. (1984) *Comprehensive Heterocyclic Chemistry*, eds Katritsky, A.R. and Rees, C.W. Vol. 7, Part 5, ed. Lwowski, W. Pergamon Press, Oxford, pp. 1–16.

McAdams, L.V. and Gannon, J.A. (1986) *Ency. Polym. Sci. Eng.* Vol. 6, Wiley–Interscience, New York, pp. 222–382.

Malinovskii, M.F. (1965) *Epoxides and their Derivatives*, Section II. Israel Programme for Scientific Translations, Jerusalem, pp. 38–101.

Materials and Technology (1972) Vol. IV. *Petroleum and Organic Chemicals* Longman, London, pp. 322–324.

Mestan, S.A. and Morris, C.E.M. (1984) *Rev. Macromol. Chem. Phys.* **C24**(1), 117–172.

Modern Plastics International (1982) Jan. p. 48.

Parker, R.E. and Isaacs, N.S. (1959) *Chem. Rev.* **59**, 737–799.

Peters, D. (1967) *The Chemistry of the Ether Linkage*, Ch. 1, ed. Patai, S. Wiley–Interscience, New York (see especially p. 14).

Price, C.C. (1967) *The Chemistry of the Ether Linkage*, Ch. 11, ed. Patai, S. Wiley–Interscience, New York, pp. 501–503.

Rabek, J.F. (1980a) *Experimental Methods in Polymer Chemistry* Wiley, Chichester, UK. Ch. 15, pp. 221–254; Ch. 20, pp. 299–331. (1980b) Ch. 23, 24, 25. pp. 368–442. (1980c) Ch. 9, pp. 123–141.

Ravindrath, K. and Gandhi, K.S. (1974) *J. Appl. Polym. Sci.* **24**, 1115–1123.

Reinking, N.H., Barabeo, A.E. and Hale, W.F. (1963) *J. Appl. Polym. Sci.* **7**, 2135–2144.

Rosowsky, A. (1964) *Heterocyclic Compounds with Three- and Four-membered Rings*, Part I, Ch. 1, ed. Weissberger, A. Wiley–Interscience, New York, Section II, pp. 31–181.

Sandler, S.R. and Karo, W. (1977) *Polymer Synthesis*, Vol. II, Ch. 3. Academic Press, New York, pp. 75–113.

Savla, M. and Skeist, I. (1977a) In *Polymerization Processes*. ed. Schildkrecht, C.E. Wiley–Interscience, New York, Ch. 16, p. 608 and Table 16.8. (1977b) p. 605. (1977c) pp. 589–591.

Schoff, C.K. (1988a) *Ency. Polym. Sci. Eng.* 2nd edn, Vol. 14. Wiley–Interscience, New York, pp. 454–541. (1988b) pp. 64–71.

Stenmark, G.A. and Weiss, F.T. (1956) *Anal. Chem.* **28**, 1784.

Streitweiser, A. (1956) *Chem. Rev.* **56**, 571–752, see pp. 682–690.

Swern, D. (1970) *Organic Peroxides*, Vol. 1, Wiley–Interscience, London, pp. 355–533.

Tanaka, Y. (1988) *Epoxy Resins, Chemistry and Technology*, 2nd edn, Ch. 2. ed. May, C.A. Marcel Dekker, New York, pp. 9–283.

Tess, R.W. (1988) In *Epoxy Resins, Chemistry and Technology*, 2nd edn, ed. May, C.A. Marcel Dekker, New York, pp. 735–742.

Union Carbide (1989) *Cycloaliphatic Epoxide Systems* Union Carbide Chemicals and Plastics Technology Corporation.

Urbanski, J. (1977) *Handbook of Analysis of Synthetic Polymers and Plastics* Ch. 10. ed. Urbanski, J. Ellis Horwood, pp. 295–318.

Yandorskii, V.N. and Ershov, B.A. (1972) *Russian Chem. Rev.* (Eng. Trans.) **41**, 403–410.

2 Curing agents for epoxy resins*

W.R. ASHCROFT

2.1 Introduction

In order to convert epoxy resins to hard, infusible thermoset networks it is necessary to use *crosslinking agents*. These *crosslinkers*, *hardeners* or *curing agents* as they are widely known, promote cross-linking or curing of epoxy resins. Curing can occur by either homopolymerisation initiated by a catalytic curing agent or a polyaddition/copolymerisation reaction with a multifunctional curing agent. Specific curing mechanisms will be detailed within the ensuing sections.

Following publication of the first patents describing cure of epoxy resins and the subsequent commercialisation, the choice of curing agents and catalysts has multiplied. Originally only a relatively few types were known and these included simple amines, acid anhydrides, phenol-formaldehyde and amino-formaldehyde resins. Nowadays there is an extended selection available and there are two reasons for the growth. Firstly, as alternative types of epoxy resin were developed so was the potential for other organic chemicals to function as curing agents recognised. This led to the introduction of polyfunctional amines, polybasic carboxylic acids, mercaptans and even inorganic chemicals: hence the diversity of product types, as previously described by Mika and Bauer (1973), which have been

Table 2.1 Estimated world market for curing agents by type

	Weight (mmlb)	Relative use (%)
Amines – aliphatics	45	10
– cycloaliphatics	30	7
– aromatics	15	3
– dicyandiamide	10	2
Polyamides	75	16
Polyamidoamines	30	7
Phenol- and amino-formaldehyde resins	75	16
Carboxylic acid functional polyesters	100	22
Anhydrides	55	12
Polysulphides and polymercaptans	15	3
Catalysts	10	2
Total	460	100

* Anchor Chemical (UK) Ltd—An Air Products Company.

offered. Secondly, and consecutively with end application development for epoxy systems there has been a need to modify available curing agents to meet specific ultimate requirements – hence the comprehensiveness of the speciality curing agent manufacturers' ranges. Table 2.1 describes the estimated relative world market usage in 1989 for the commercially most significant curing agent types.

2.2 Nitrogen-containing curing agents

2.2.1 Aliphatic amines and derivatives

Ethyleneamines, polyamides, polyamidoamines, imidazolines, adducts with epoxy functional materials, Mannich bases, association salts, ketimines and acrylonitrile adducts, propylenamines, higher alkylenamines, fatty diamines, polyetheramines, araliphatic amines, alicyclic amines, polymeric amines, dicyandiamide and biguanides.

2.2.1.1 Ethyleneamines and derivatives. Ethyleneamine epoxy curing agents are derived from ethylene and ammonia and were amongst the first curing agents reported by Greenlee (1952) for use with epoxy resins. Typical examples are:

$$H_2N(CH_2CH_2NH)_xH$$

x = 1, Ethylene**D**i**A**mine (**EDA***[107-15-3]*)*;
x = 2, **D**i**E**thylene**T**ri**A**mine (**DETA***[111-40-0]*);
x = 3, **T**ri**E**thylene**T**etr**A**mine (**TETA***[112-24-3]*);
x = 4, **T**etra**E**thylene**P**ent**A**mine (**TEPA***[112-57-2]*);
x = 5, **P**enta**E**thylene**H**ex**A**mine (**PEHA***[4067-16-7]*)
* Chemical abstract services number (CAS No.).

The mechanism of cure with ethyleneamines typifies the reaction between epoxy resins and all primary and secondary amines; their speed of reaction is faster, however, than cycloaliphatic and aromatic polyamines particularly at ambient temperatures. Scheme 2.1(a) illustrates the initial step which involves the primary amine active hydrogen adding to the epoxy group. This is followed by the resulting secondary amine adding to another epoxy group. Generally, primary and secondary amines are used at mix ratios that provide one amine active hydrogen for each epoxy group, i.e. the stoichiometric amount. When the epoxy is a diepoxide and when the amine is a diamine the resulting product is an infusible polymer (Scheme 2.1b). The reaction is accelerated by materials that stabilise the alkoxide ion intermediates (Scheme 2.1c). Therefore, hydroxyl groups are accelerators including those generated during cure. They can also be added as accelerators and the more acidic they are, the more effective. Hydroxylic accelerators or hydrogen

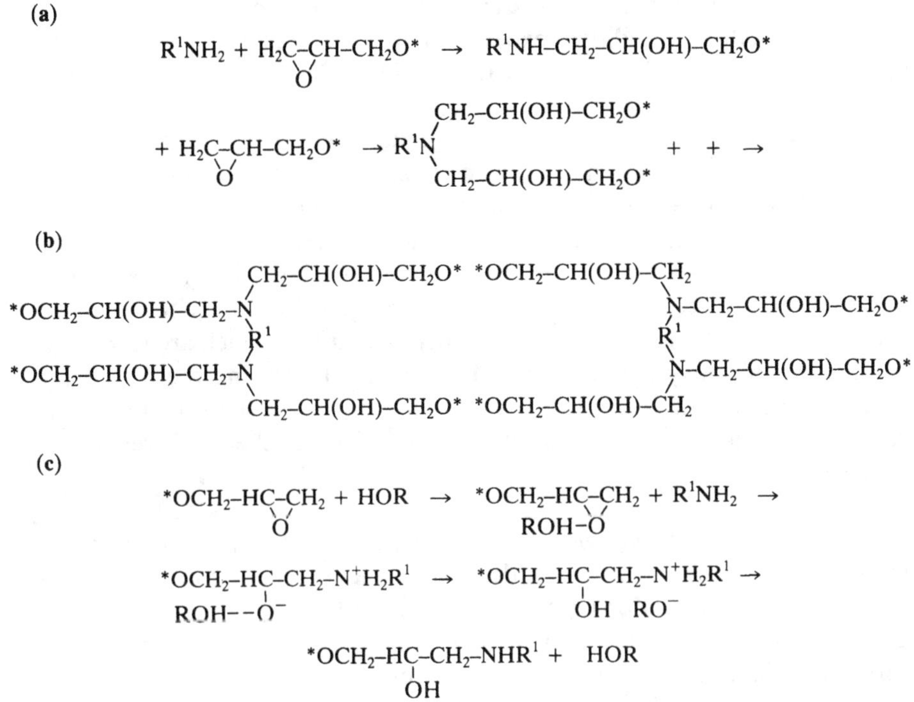

Scheme 2.1 Mechanism of cure of primary amines. $-CH_2O^*$ indicates that only the reactive epoxy group is shown.

donors capable of catalysing the amine epoxy reaction in decreasing order of efficiency are: –OH ($Ar \gg PhCH_2 > RCH_2 > H$); $-CO_2H$; $-SO_3H$; $-CONH_2$; $-SO_2NH_2$.

Ethyleneamines are highly reactive due to their unhindered polyfunctional nature and give tightly cross-linked networks due to the short chain distances between the active sites. This results in cured resins with excellent solvent resistance and mechanical strength, yet limits flexibility. Their low molecular weights mean that they have low viscosities, however it also means high vapour pressures with the outcome that they are corrosive and irritant to the skin. They are also high in odour, are hygroscopic and have poor compatibility with epoxy resins. This last gives them a tendency to exude during cure giving rise to bloom and carbonation. This can result when amines of poor compatibility are used with epoxy resins – they can exude to the surface after mixing and before cure has trapped them into the matrix. *Blooming* or blushing as it is also described, refers to the resulting tacky surfaces. *Carbonation* or *water spotting* refers to the white patches of amine bicarbonate salts which can appear and mar the surface appearance of cured films. The more basic polyamines react with atmospheric CO_2 in preference/competition with epoxy resin to form the unstable carbamic acids and amine carbamates. These can revert to amine and CO_2 which in the presence of water (humidity or excess) generates carbonic acid at the curing surface which can then react irreversibly with amines to form the undesirable amine bicarbonates.

The low molecular weights of the ethyleneamines mean the usage rates are low and therefore they have to be mixed in to the epoxy resin with a high degree of accuracy – apart from the need to avoid excess additions which would exude/bloom, additions under stoichiometric amounts lead to embrittlement. Various modification techniques have been developed, however, to overcome most of the disadvantages and it is the derived products, rather than the unmodified ethyleneamines, which are now used extensively in epoxy applications. Most involve increasing the molecular weight of the amine to reduce the volatility and most lead to significant changes in reactivity and cured mechanical and chemical resistance properties.

2.2.1.2 Polyamides, polyamidoamines and imidazolines. Condensation with carboxylic acids to form amides is a simple technique for overcoming the volatility, irritation potential and critical loading issues of ethyleneamines. Reaction of the higher homologues of the ethyleneamines series with dimerised or polymerised unsaturated fatty acids as recorded by Cowan *et al.* (1948) leads to the popular general curing agent type the *polyamide*.

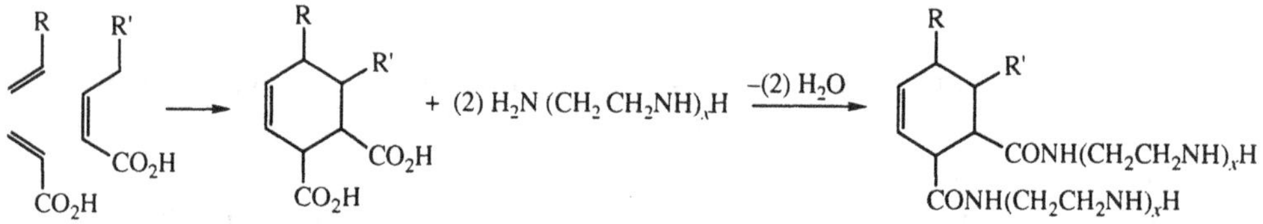

A range of polyamides are available commercially, based upon diverse fatty acid/dimer acid/amine ratios which have been processed under various conditions. They vary in molecular weight, physical form, amine content and reactivity but all are low in volatility. Their mixing ratios with epoxy resins are generally non-critical and cured properties can be deliberately varied from tough and hard at below stoichiometry, to flexible and soft around or above stoichiometry. They give good adhesion to a wide variety of substrates, particularly epoxy-polyamide coatings. This is due to residual secondary amine functionality on the polyamide backbone which is often deliberately not fully converted. This ensures keying-in sites and promotes intercoat adhesion. The fatty acid/dimer acid backbone provides excellent corrosion resistance by its water repelling nature but also leads to a colour stability problem due to its unsaturated nature.

Polyamides are, however, rather slow in curing particularly at lower temperatures and tend to give greasy cures due to exudation, but addition of tertiary amines, phenolic amines or co-curing agents can help to reduce both of the defects. Alternatively, condensation of ethyleneamines with phenolic-containing carboxylic acids to give the so called *phenalkamine* type curing agent, according to Kittridge and Michelli (1968), gives enhanced low temperature reactivity curing agents, though colour stability is sacrificed. An alternative approach to overcoming the exudation problem, which is caused by poor compatibility with epoxy resin of both polyamide and unreacted ethyleneamine which is frequently present from the polyamide manufacturing process, is to pre-react it with an epoxy resin to improve compatibility. In contrast to the polyamides, *polyamide adducts* do not need to be given an induction period prior to use after mixing to ensure that the system is homogeneous. Adducts also cure better at low temperatures under conditions of high humidity and give faster tack-free cure times by virtue of a physical drying phenomenon resulting from their increased molecular weights. Physical drying does not involve a chemical reaction; it is a purely physical effect whereby (pre)polymers, when deposited as a film, upon solvent evaporation appear tack-free or hard. Further improvements in drying rates have also been obtained, according to Wilson (1986), by replacing part of the dimer acid by aromatic polybasic acids.

The lower viscosity, more reactive polyamides are used with liquid epoxy resin in high solids coatings, adhesives, sealants, and cable jointing applications. In contrast the higher molecular weight high viscosity polyamides, adducts and aromatic modified polyamides are used mainly with solid epoxy resin in solvent-based corrosion-resistant coatings. However, the high viscosity of polyamides and adducts means that they cannot easily be used in reduced volatile organic content or high solids coatings. This led Brytus (1986) to the development of new types of low viscosity solvent-free polyamide adducts which give properties in cured coatings comparable to those obtained with conventional polyamides and adducts. They have the added advantage that when used with liquid epoxy resins they give higher crosslink densities and hence improved mechanical properties, chemical resistance and corrosion resistance. Richardson (1973) prepared water dispersible polyamides by modification of standard polyamides and found them to be useful low odour, non-flammable alternatives to solvent-based polyamide solutions with all of their inherent advantages, in concrete primers, new to old concrete adhesives and surface coatings applications.

Reaction of ethyleneamines with aliphatic monocarboxylic acids leads to *polyaminoamides* or *polyamidoamines* or *amidoamines* as they are also frequently called, which can, on further heating, undergo ring closure, forming the cyclic amidines or *polyaminoimidazolines*.

$$\text{R–CH=CH–CH=CH–CO}_2\text{H} + \text{H}_2\text{N(CH}_2\text{CH}_2\text{NH)}_x\text{H} \rightarrow$$

$$\text{R–CH=CH–CH=CH–CONH(CH}_2\text{CH}_2\text{NH)}_x\text{H}$$

polyaminoamide

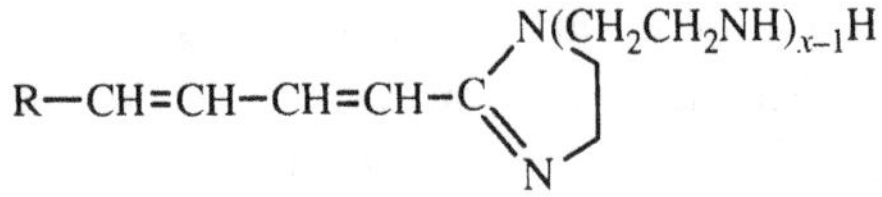

polyaminoimidazoline

Generally tall oil fatty acids are used as the mono-basic acid component and the resulting mixtures of polyaminoamide and polyaminoimidazoline vary in pot-life/reactivity depending upon the imidazoline content. High imidazoline content mixtures are long working life, low reactivity curing agents which in the presence of water, particularly substrate surface humidity, ring open back to the more reactive polyaminoamides. In effect, they function as semi-latent curing agents and are widely used on damp concrete for their excellent wetting characteristics and adhesion in concrete coating, bonding new to old concrete and other adhesives applications. Their low viscosity and excellent corrosion resistance also makes them suitable for electrical potting and filament winding applications.

Mono-basic acids other than fatty acids have also been used by Wilson (1970) in the preparation of pure *imidazolines* for use in fibre-reinforced epoxy lamination applications where low reactivity, high strength and good chemical resistance are important. Additionally, imidazolines, prepared from mono-basic acids, have been further reacted with polycarboxylic acids to form solid salts, and used by Veba Chemie (1977) for their unique ability to provide matt finishes in powder coatings applications.

2.2.1.3 Adducts with epoxy functional materials. One of the earliest methods investigated for the reduction of the irritation potential of ethyleneamines was partial reaction with epoxy functional materials to produce low volatility adducts. Newey (1958) reported that mono-epoxides such as ethylene oxide, propylene oxide and mono-glycidyl ethers react to produce low viscosity mixtures of mono- and di-adducted amines with the expected reduced volatility.

$$H_2N(CH_2CH_2NH)_xH + H_2\underset{\diagdown O \diagup}{C-CH}-R \rightarrow H_2N(CH_2CH_2NH)_xCH_2CH(OH)R$$

$$+$$

$$R = H, CH_3, OC_4H_9 \qquad RCH(OH)CH_2HN(CH_2CH_2NH)_xCH_2CH(OH)R$$

In addition to the reduced irritation potential and increased loading, this type of curing agent also has a faster cure rate due to in-built hydroxy substitution. Although this class of curing agent gives reduced cross-link densities compared to the starting ethyleneamines they are used widely in wet lay-up laminating, patch repair of boats, adhesives and tooling applications. Unfortunately free, or unadducted ethyleneamine is also present as part of the statistical distribution of products and even though it can be removed by distillation the resulting toxicologically safer products in practice fail to give even the same levels of physical properties and chemical resistance as the 'crude' reaction products, let alone the starting ethyleneamines. Adducts with di-glycidyl ethers on the other hand when cured give the same level of physical and chemical properties as the amine adducted as there is no overall loss of active hydrogens/reduction in crosslink density.

$$H_2N(CH_2CH_2NH)_xH + \underset{\diagdown O \diagup}{H_2C{-}CH}{-}CH_2O^* \quad {}^*OCH_2{-}\underset{\diagdown O \diagup}{HC{-}CH_2} + H_2N(CH_2CH_2NH)_xH$$

$$\rightarrow \; H_2N(CH_2CH_2NH)_xCH_2CH(OH)CH_2O^* \quad {}^*OCH_2CH(OH)CH_2HN(CH_2CH_2NH)_xH$$

With liquid epoxy resin adducts prepared using an excess of amine, fairly viscous materials of low volatility and higher mixing ratios result. Higher mix ratios allow small errors to be made in weighing that would be critical in the use of the parent amines. Once again due to the presence of in-built hydroxyl groups, they are faster curing and thus ideal for laminating and adhesives applications. Their ability to give excellent solvent resistance also leads to usage with higher functionality resins for optimum solvent-resistant coatings.

With solid epoxy resin, adducts can be prepared in two ways. *In-situ adducts* are produced simply by addition of amine to a solution of the resin to yield an adduct solution directly. This type is used widely for its ease of manufacture, low cost, good film flexibility and excellent chemical resistance, in solvent based coating applications. *In-situ* adducts do, however, contain low levels of free amine that have a tendency to exude and therefore mixtures with solid epoxy resin solutions often need to be aged prior to application. *Isolated adducts* are prepared by addition of excess amine to the solid epoxy resin solution followed by removal of unreacted amine and reaction solvent by distillation. This yields solid isolated adducts which need to be redissolved and used in the same way as the *in-situ* adduct solutions. They do nevertheless exhibit a number of advantages associated with the lack of any free amine including low odour, and are better suited for application at lower temperatures and higher humidity where bloom, carbonation, loss of gloss and intercoat adhesion criteria are critical.

2.2.1.4 Mannich bases and association salts. An alternative approach developed by Halewood (1962) for overcoming the disadvantages of volatility, irritation and hygroscopicity of the ethyleneamines involves reacting together the amine with phenol and formaldehyde to form low molecular weight condensates or Mannich bases as they are commonly also known.

$$HO{-}Ar + CH_2O + H_2N(CH_2CH_2NH)_xH \;\rightarrow\; HO{-}Ar{-}CH_2HN(CH_2CH_2NH)_xH + H_2O$$

These products cure rapidly at low temperatures and in the presence of moisture, do not blush and are less susceptible to carbonation than other aliphatic amine modifications. They also give excellent solvent and overall chemical resistance and are used for a wide variety of civil engineering, coating, adhesive and laminating applications. Products of this type generally contain 10–20% of unreacted phenols which aid compatibility for coal tar extended epoxy coatings applications. They are causing legislative concern and when phenol rather than an alkyl phenol is used in the manufacture, 'skull and cross-bones' labelling is required for the resulting products, even though the *free* phenol is associated with the amine in salt form.

$$HO\text{–}Ar + H_2N(CH_2CH_2NH)_xH \rightarrow Ar\text{–}O^- + H_3N^+(CH_2CH_2NH)_xH$$

These simple salts or association products provide reduced dermatitic hazard and are also used extensively for optimum reactivity under adverse conditions. Their low viscosities make them ideal for use in the formulation of concrete patch repair mortars and adhesives. When mixed with epoxy resin, however, the salts break down liberating volatile ethyleneamines and their related problems.

2.2.1.5 Ketimines and acrylonitrile adducts. Alternative approaches to increasing the molecular weights other than condensation with carboxylic acids or partial pre-reaction with epoxy functional materials involve condensation reactions with ketones and adduction with acrylonitrile. Ethyleneamines react reversibly with ketones to form *ketimines* as reported by Shell (1964).

$$H_2NCH_2CH_2NHCH_2CH_2NH_2 + 2\ RR'C{=}O \rightarrow RR'C{=}NCH_2CH_2NHCH_2CH_2N{=}CRR'$$

$$+\ PhOCH_2\text{–}\underset{\diagdown O \diagup}{HC\text{–}CH_2} \rightarrow RR'C{=}NCH_2CH_2\underset{\substack{|\\ CH_2\text{–}CH(OH)\text{–}CH_2OPh}}{N}CH_2CH_2N{=}CRR'$$

These polyamines are readily hydrolysed by moisture to regenerate the amine and ketone, the latter of which volatilises out during the normal film forming/cure process. When secondary amine-containing ethyleneamines are used, so called reactive ketimines are formed which can be blocked by adduction with a mono glycidyl ether. Both reactive and blocked ketimines are characterised by their low viscosities, semi-latent nature, and rapid cure when exposed in coatings to atmospheric moisture. Their main application is in high solids coatings. There are, however, a number of limitations. To ensure full through cure, sufficient atmospheric humidity is required and applied film thicknesses need to be restricted to avoid solvent (ketone) entrapment.

Farnham and Caldwell (1956) recorded ethyleneamines react readily to form *acrylonitrile adducts* with variable degrees of substitution depending upon the ratio of reactants and manufacture conditions.

$$H_2NCH_2CH_2NHCH_2CH_2NH_2 + CH_2{=}CH\text{–}CN \rightarrow$$

$$H_2NCH_2CH_2NHCH_2CH_2NHCH_2CH_2CN$$

The cyanoethylated amines are less reactive than the parent amines, low in viscosity and give physical properties generally inferior to those of the parent amines. They are not lower in irritation potential though and care must be taken in operation to avoid the reformation of acrylonitrile from the *retro-Michael* reaction which takes place at around 120°C. Their ability to wet minerals and glass fibres well and low reactivity makes them particularly suitable for laminating and filament winding applications.

2.2.1.6 Propylenamines. Hydrogenation of adducts of acrylonitrile with amines other than ethyleneamines, often secondary amines, leads to another class of hardener – the propyleneamines.

$R^1R^2NCH_2CH_2CH_2NH_2$ R^1,R^2 = Me,Me: **Di**Methyl**A**mino**P**ropyl**A**mine (DMAPA)
= Et,Et: **Di**Ethyl**A**mino**P**ropyl**A**mine (DEAPA)
R^1 = H,R^2 = C_6H_{11}: CyclohexylAminoPropylamine

With the exception of the cyclohexylamino derivative [3312-60-5] all are unsuitable as crosslinking agents in their own right due to insufficient functionality: they give lower cured mechanical and chemical resistance properties than the ethyleneamines. They are low in molecular weights and thus are corrosive and irritant to the skin due to high vapour pressures. In addition their strong basicity leads to a strong tendency to carbonate. However, they do have a number of advantages: low viscosity; high reactivity due to unhindered amine functionality, particularly DMAPA [109-55-7] and DEAPA [104-78-9] which have inherent tertiary amine functionality (see section 2.2.4). In addition they have the ability to act as reactive flexibilisers/toughening agents by virtue of their low functionality which reduces cross-link density and promotes chain extension. These features render them particularly useful as co-curing agents with high functionality (cycloaliphatics) amines in a variety of coatings and flooring applications. They are also used alone for adhesive applications.

2.2.1.7 Higher alkylenediamines. Alternate aliphatic diamines with longer chain lengths between the reactive sites than the ethyleneamines and propyleneamines are:

$H_2N(CH_2)_6NH_2$	**H**exa**M**ethylene**D**iamine (HMD)
$H_2N(CH_2)_3CH(Me)CH_2NH_2$	**M**ethyl**P**enta**M**ethylene**D**iamine (MPMD)
$H_2NCH_2CH_2CH(Me)_2CH_2CH(Me)CH_2NH_2$	**T**ri**M**ethylhexane**D**iamine (TMD)

HMD [124-09-4] which is more widely known for its use as an intermediate for nylon-6,6 manufacture, is derived by tail to tail dimerisation of acrylonitrile and subsequent catalytic reduction of the intermediate adiponitrile. This synthetic route leads to a number of coproducts including MPMD [15520-10-2] which is the hydrogenated head to tail dimerisation product. TMD which was developed for use as a urethane intermediate is actually an equal part mixture [25513-64-8] of the 2,2,4- and 2,4,4-trimethyl hexamethylenediamine isomers.

All three materials have physical characteristics derived from their low molecular weights: low viscosity, corrosivity, irritancy to the eyes and skin. They are high in reactivity and have a strong tendency to carbonate due to unhindered primary aliphatic amine functionality. They also have an ability to give moderate film flexibilities due to the moderate chain distances between the cross-link sites. Additionally, their high hydrocarbon contents make them hydrophobic and, problematically, only moderately compatible with epoxy resin. They thus tend to exude and lead to surface tack in unmodified form. Adduction with epoxy resin, Mannich base formation or salting with acidic accelerators all help to minimise the corrosivity/irritancy hazard, reduce the carbonation tendency and improve their compatibilities with epoxy resins. The principle commercial uses for the various modifications are coatings and concrete repair applications.

2.2.1.8 Fatty amines. Amines derived from C_{16}–C_{18} saturated and unsaturated fatty acids are also used as co-curing agents in epoxy applications. They are used primarily to provide flexibility and enhance water and corrosion resistance but are limited by their low reactivity.

2.2.1.9 Polyetheramines. Of the various polyglycol polyamines types evaluated in epoxy applications only the polyoxypropylene derivatives have achieved prominence. Most other polyglycol polyamines, notably ethylene oxide derivatives, are too hygroscopic to be used extensively. Lee and Winfrey (1966) found that by varying the functionality of the initiating polyol and the molecular weight of the polyoxypropylene derivatives, several different molecular weight amine functional materials could be made for use with epoxy resin.

The 230 and 400 molecular weight diamines and the 403 molecular weight triamine are all low in viscosity and exhibit low reactivity and reduced

Table 2.2 Polyoxypropylene derivatives of varying molecular weight

	x	Molecular weight
$H_2NCH(CH_3)CH_2\text{–}\{\text{–}OCH_2CH(CH_3)\text{–}\}_x\text{–}NH_2$	2.6	230
	5.6	400
	33.1	2000
$CH_3CH_2C(\text{–}CH_2\text{–}\{\text{–}OCH_2CH(CH_3)\text{–}\}_x\text{–}NH_2)(\text{–}CH_2\text{–}\{\text{–}OCH_2CH(CH_3)\text{–}\}_y\text{–}NH_2)(\text{–}CH_2\text{–}\{\text{–}OCH_2CH(CH_3)\text{–}\}_z\text{–}NH_2)$	$x+y+z$ 5.3	403

carbonation tendency compared to other types of aliphatic amines. This is due to the slightly hindered secondary aliphatic amine functionality. When cured they do show good film flexibilities and thermal cycling properties due to the long chain distances between cross-link sites. The flexible aliphatic nature does, however, limit their solvent resistance and heat resistance. This combined with an inconveniently slow cure rate at ambient temperature means that they are commonly used in combination with other amines and/or accelerators, principally for flooring, adhesive, electrical potting and encapsulation applications.

2.2.1.10 Polymeric amines. Many reports have appeared in the patent and academic literature describing amine functionalised polymers but few have achieved commercial prominence. However, McFadden (1972) developed reaction products of ethyleneimine with carboxylic acid terminal prepolymers for water-based coatings applications where they were found to combine the weathering and colour stability properties of conventional acrylic systems with the corrosion resistance of epoxy coatings.

2.2.1.11 Araliphatic amines. The araliphatic amine **MXDA** [1477-55-0] prepared by hydrogenation of isophthalonitrile is a di-primary aliphatic

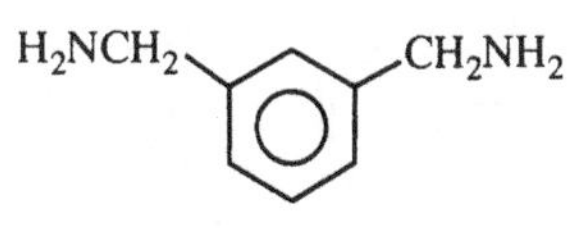

Meta**X**ylene**D**i**A**mine (MXDA)

amine which behaves more like a di-primary cycloaliphatic amine than an aliphatic diamine. This is most likely due to the rigid nature of the non-conjugated aromatic ring backbone.

MXDA has a low viscosity, a light colour, and is insensitive to moisture. The high basicity of the unhindered primary aliphatic amine functionality imparts high reactivity and unfortunately a strong tendency to carbonate. This combined with the corrosive and irritant nature means that it is generally used in preadducted form or as a Mannich base. Resulting products used for high solids and solvent-free coatings and flooring applications, are characterised by their excellent subambient temperature cure rates and high cured chemical resistance.

2.2.1.12 Alicyclic aliphatic amines. One of the very first aliphatic

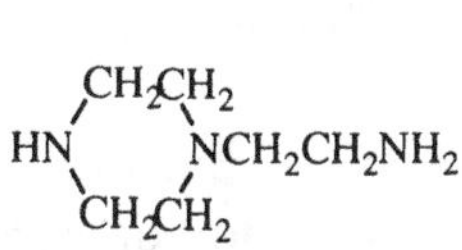

N-**A**mino**E**thyl**P**iperazine (AEP)

polyamines with a (hetero-)cyclic backbone bridging the active amino-hydrogen sites to gain commercial prominence in epoxy applications was *N*-**AEP** [140-31-8]. This higher homologue from the ethyleneamine stream of products contains primary, secondary and tertiary amine groups, although only the three primary and secondary active hydrogens participate in reaction with epoxy resin. At room temperature *N*-AEP behaves like its higher functionality acyclic homologue TETA (see section 2.2.1.1) in terms of corrosivity, handling and in giving rapid gelation and high exotherm. Full cure is not however possible with *N*-AEP due to the formation of the so-called **B-stage**. This term B-staging, adopted from phenolic resin technology, refers to an intermediate stage in the reaction of thermosetting resins in which the partly cured material softens or reflows when heated, allowing it to be moulded into another shape or form. This is desirable in epoxy laminating applications yet undesirable in ambient cure structural coatings where it would be ruinous. Blending with other amines and/or acidic accelerators is required to achieve full cure at ambient temperatures or alternately an elevated temperature post-cure is necessary. Fully cured materials exhibit high rigidity and high impact resistance. Typical applications include: ambient cured patch repair adhesives and flooring applications; heat-cured adhesives and electrical casting applications.

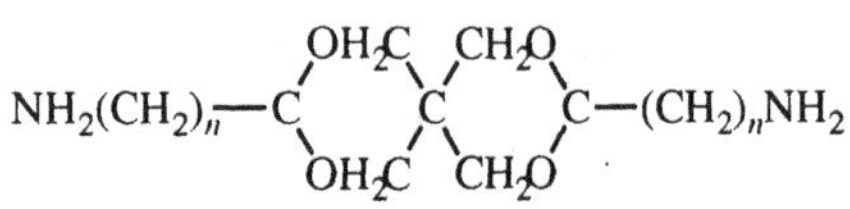

Spiro-acetal diamines
$n = 1–4$

Spiro-acetal diamines developed by Ajinomoto Co. Inc. (1969) from an intermediate in the manufacture of mono-sodium glutamate have gained a market sector presence as curing agents with reduced skin irritancy, reduced amine blush, improved toughness and impact strength and low colour for epoxy applications. The basic spiro-acetal diamines are usually modified by adduction and acceleration and applications include flexible coatings, adhesives and castings.

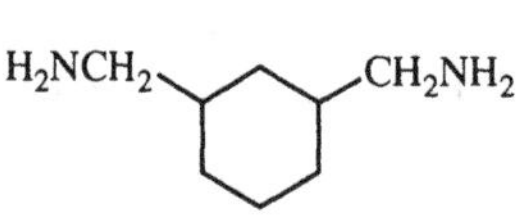

1,3-{**B**is**A**minomethyl}-**C**yclohexane (1,3-BAC)

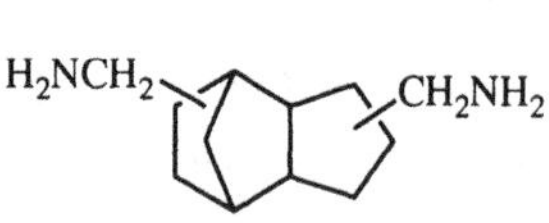

Tri**C**yclo**D**ecane diamine (TCD diamine)

Nor**B**ornane**D**i**A**mine (NBDA)

1,3-**BAC** [2579-20-6] prepared by reduction of MXDA (see section 2.2.1.11) or by exhaustive hydrogenation of isophthalonitrile is one of the few commercially available products clearly recognisable as an alicyclic aliphatic polyamine. This low viscosity material provides superior colour stability compared to its araliphatic precursor but unfortunately inferior low temperature reactivity and sensitivity to atmosphere humidity. The elevated relative cost also limits its utility to high performance coatings applications.

TCD diamine [26655-37-8] which is the hydrogenated hydrocyanation product of dicyclopentadiene is a mixture of *endo* and *exo*, *cis* and *trans* isomeric tricyclic primary diamines. The high viscosity and cost, however, limits its usage to specialist high performance applications where high heat resistance can be attained due to the rigidity of the fused ring backbone. **NBDA** [56602-77-8] which is the related hydrogenated hydrocyanation product of norbornadiene is a lower viscosity alternative with a potential which has yet to be fully explored.

2.2.1.13 Dicyandiamide and biguanide derivatives. **DICY** or **cyanoguanidine** [461-58-5] as it is less commonly known, the agrochemical industry basic intermediate made from cyanamide (H_2NCN), was one of the first curing agents to be used with epoxy resins.

$$H_2N{-}\underset{\underset{NH}{\|}}{C}{-}NH{-}CN$$

Dicyandiamide (**DICY**)

It gives systems with pot lives/shelf-lives of six months with liquid/solid resins by virtue of its insolubility in the resins at ambient temperatures and is considered to be a highly latent epoxy resin curing agent. When heated in the presence or absence of catalysts, rapid polymerisation commences to yield highly cross-linked molecules exhibiting high strength, moderate heat resistance, high solvent resistance and excellent electrical insulating properties. High temperatures, however, are required to initiate reaction, therefore DICY is normally used in conjunction with co-curing accelerators/catalysts as the uncatalysed reaction with epoxy resins begins well above 150°C.

The mechanism of curing epoxy resins with DICY is complex and involves initial reaction between all four active hydrogens with epoxy resin in the presence of tertiary amine catalysts (Scheme 2.2a) and simultaneous epoxy homopolymerisation catalysed by the tertiary amine (Scheme 2.2b). This is followed by a final cure reaction between hydroxyl groups produced in the partially cured epoxy resins and the cyano group in dicyandiamide (Scheme 2.2c). This latter step is also catalysed by tertiary amines.

DICY when mixed with liquid epoxy resin for one component adhesive applications and solventless fibre reinforced laminate applications is used in micronised form. This ensures rapid dissolution at reaction temperature and optimum homogeneity in the cured product. Cure is generally initiated with urea catalysts (see section 2.2.4.3). DICY is also extensively used in both unground and micronised form with solid bis-A epoxy resins in hot melt

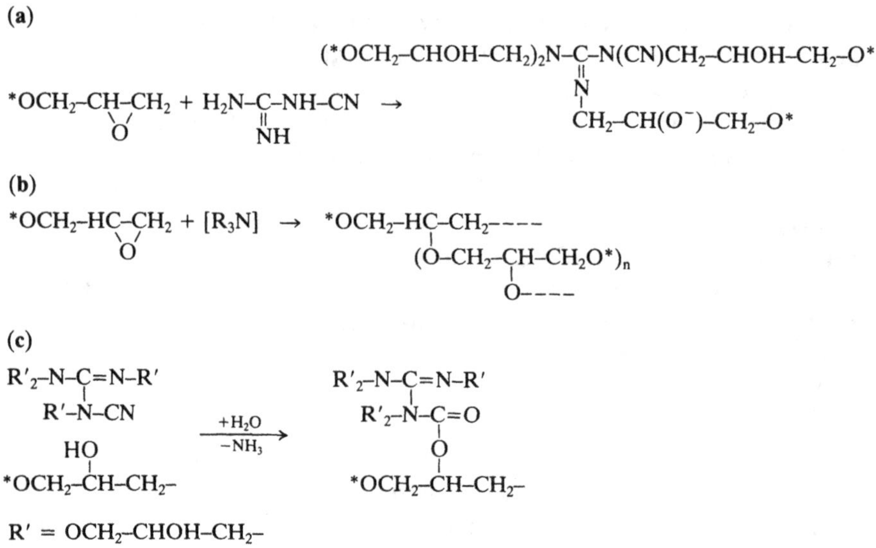

Scheme 2.2 The mechanism of cure of dicyandiamide.

powder coatings applications (unground material being preferred in some instances where the dicyandiamide can be ground down to application particle size at the same time as pigments/resin). For powder coatings applications, **imidazoles** (see section 2.2.4.2) are most frequently used as catalysts. The largest usage of DICY however is in solution with high molecular weight bis-A resins in electrical laminating printed circuit board manufacture. Here the tertiary amines (see section 2.2.4.1) or imidazoles are the preferred accelerators.

Dicyandiamide is also a useful intermediate for the formation of a number of alternative types of curing agents which are likewise used in thermally-cured epoxy resin applications.

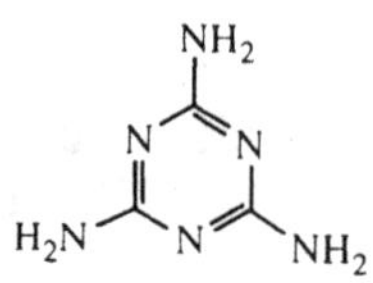

Melamine

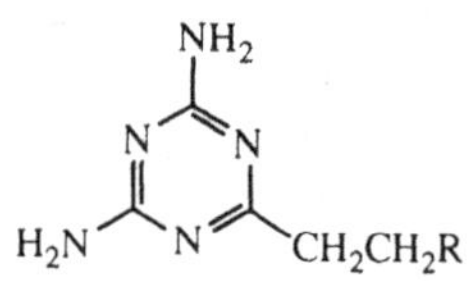

(Fatty) guanamine

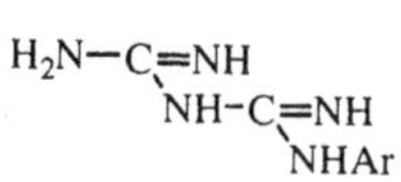

Biguanide

The reactive nitrile group readily trimerises to form the cyclic **guanamine** or **melamine** [108-78-1]. General Mills Inc. (1965) found that reaction also occurs with other nitriles such as those derived from fatty acids as in the production of **fatty guanamines**. Amines add to the nitrile group to form **biguanides** and Gempeler and Zuppinger (1971) discovered that aromatic amine-derived biguanides provided a number of advantages over dicyandiamide for powder coatings, adhesives and moulding power applications. These include improved reactivity due to improved solubility in molten epoxy resin and improved flexibility due to reduced cross-link density.

Joyce (1971) also reported dicyandiamide as a co-curing amine/catalyst for applications at temperatures down to 0°C. Such systems have no latency, and cure rapidly to give excellent adhesion and good chemical resistance; end uses include speciality quick-setting adhesives and wet lay-up laminating.

2.2.2 *Cycloaliphatic polyamines and derivatives*

This group of curing agents includes authentic cycloaliphatic di-primary amines, as well as other di-primary amine types with cycloaliphatic polyamine performance.

2.2.2.1 Authentic cycloaliphatic polyamines. Catalytic hydrogenation of aromatic di-primary amines produces cycloaliphatic di-primary amines.

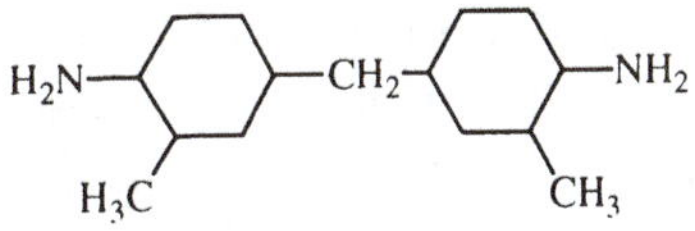

3,3'-Dimethylmethylene-di(cyclohexylamine)

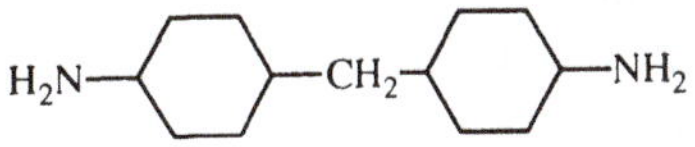

Methylene-di(cyclohexylamine) or
bisParaAminoCyclohexylMethane (PACM)

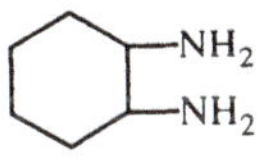

1,2-Cyclohexanediamine or
1,2-DiAminoCycloHexane (1,2-DACH)

Of the many possibilities evaluated with epoxy resin only two have achieved commercial success: **3,3′-dimethylmethylene-di(cyclohexylamine)** [6864-37-5]; and the unhindered analogue **PACM** or methylene-di(cyclohexylamine) [1761-71-3] as it is also known. Two grades of the latter are available: **PACM 48** which is a solid produced under conditions which maximise the thermodynamically stable *trans,trans* [6693-29-3] isomer content; **PACM 20** which is a liquid produced with a low *trans,trans* and high *cis,trans* [6693-30-7] *cis,cis* [6693-31-8] isomer content.

Both the 3,3′-dimethylmethylene-di(cyclohexylamine), which is also a liquid, and PACM 20 are low in viscosity and light in colour. Like aliphatic polyamines they are skin irritants but have lower vapour pressure which reduces the hazard due to inhalation of vapour. When used to cure liquid epoxy resins at temperatures in excess of 100°C they give excellent heat resistance and high mechanical strengths and both are used in filament winding, laminating, tooling and small–medium sized casting applications. The 3,3′-dimethyl substituted material which is relatively hindered and thus lower in reactivity being chosen for longer working life and the PACM 20 due to its higher conformational fluidity chosen for its superior fracture toughness. PACM 20 has been found better suited to epoxy usage than PACM 48 due to the ease of mixing with liquid resins and is also especially suitable for modification as an adduct for ambient cure applications where the flexibility and unhindered nature can be used to advantage.

1,2-DACH [694-83-7] is not produced by catalytic hydrogenation of an aromatic precursor but is available commercially as the principal coproduct from the catalytic reduction of adiponitrile (see section 2.2.1.7). This unique 1,2-diamine has a higher degree of reactivity even compared to the PACM 20 and this coupled with the high volatility and the limited flexibility due to the short distance between cross-link sites makes 1,2-cyclohexanediamine better suited for use in modified form in ambient cure epoxy applications.

2.2.2.2 Other cycloaliphatic polyamines. **1,8-Menthane diamine** [80-52-4] was one of the first cycloaliphatic polyamines to be considered for heat cure epoxy applications. This highly hindered di-primary amine prepared by hydrocyanation of the terpenes pinene or limonene is low in viscosity, provides long working-lives and gives high heat and chemical resistance. 3-Aminomethyl-3,5,5-trimethylcyclohexylamine commonly known as **I**so**P**horone**D**i**A**mine (**IPDA**) [2855-13-2] has however been most widely adopted as the industry standard cycloaliphatic amine.

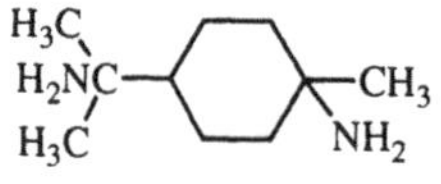

1,8-Menthane diamine

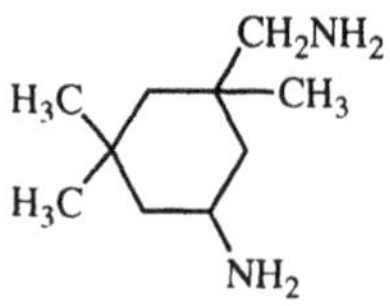

Iso**P**horone**D**i**A**mine
(IPDA)

IPDA which, like its stable-mate TMD (section 2.2.1.7), is also a product of a multi-step synthesis starting from acetone, via isophorone. The low viscosity, light colour, excellent heat resistance and cured mechanical strength mean that in unmodified form IPDA is particularly useful for small–medium sized casting, filament winding, electrical and general laminating applications. The differential reactivity resulting from the presence of primary aliphatic and cycloaliphatic amine functionality content allows selective adduction with epoxy resin and selective formylation/ Mannich base formation. This results in low viscosity, activated derivatives suitable for room temperature curing high performance coatings and civil engineering applications.

2.2.2.3 Accelerated cycloaliphatic polyamines, adducts and Mannich bases. The lower reactivity of the cycloaliphatic primary amine function is caused in part by steric hindrance from the ring(s) and though an undoubted advantage for heat-cure applications it does mean that acceleration is necessary to achieve reasonable room temperature cure rates. As cycloaliphatic polyamines are corrosive, irritant to the skin and have a strong tendency to carbonate/water spot due to their high basicities they are also best modified for ease and safety of handling. Although only a limited number of the modification techniques used with aliphatic polyamines are applicable to cycloaliphatic polyamines a range of curing agents with widely varying properties can none the less be produced. Usage rate, viscosity, reactivity, cured mechanical properties and chemical resistance are all dependent on the formulating technique used.

The simplest modification approach involves blending the cycloaliphatic polyamine with an accelerator and a plasticising diluent. Accelerators which can be used include aliphatic polyamines, simple tertiary amines and organic acids. The most used plasticising diluent particularly in combination with the most used cycloaliphatic polyamine IPDA is benzyl alcohol as this latter reduces brittleness, provides a weak accelerating effect and helps the hindered amine get through the B-stage. Applications for this low viscosity curing agent type include highly filled screed flooring and coatings applications.

Ciba Ltd (1970) found a further useful method of modifying cycloaliphatic amines involving partial pre-reaction with epoxy resin coupled with addition of plasticisers such as benzyl alcohol to reduce the brittleness and lower the viscosity. Accelerators may be added but are not essential as the hydroxyl group formed in the adduction process has an accelerating effect, as does benzyl alcohol. IPDA, 1,2-DACH and PACM 20 are all used to produce this type of curing agent; the IPDA-based products performing as industry standard; the DACH-based materials are used for optimum chemical resistance; the PACM 20 adducts providing the best overall carbonation and water spotting resistance. Applications for these high gloss, blush-resistant, colour-stable curing agents include decorative and chemical spill resistant high solids coatings and flooring applications, as well as laminate/gel coat applications.

Rutgerswerke (1969) reported cycloaliphatic polyamines can also be reacted with formaldehyde and phenols to form Mannich bases and again plasticising diluents such as benzyl alcohol are best added to lower viscosity and reduce brittleness. The Mannich bases are darker in colour than the adducts and blends and are not colour stable, but do however cure better in the presence of humidity and at lower temperatures. IPDA is effectively the only cycloaliphatic polyamine type used, as the other polyamine types give products which are too viscous to be of any value. Applications for the Mannich bases include flooring, old concrete priming, new to old concrete bonding and coal tar modified epoxy coatings.

2.2.3 *Aromatic polyamines and derivatives*

Aromatic amines are less basic than aliphatic or cycloaliphatic polyamines and react only slowly with epoxy resin. Typically they exhibit long pot lives and require long periods at elevated temperature to attain optimum properties.

2.2.3.1 Heat-cured aromatic amines. All heat cured aromatic polyamines provide cured systems with excellent resistance to a variety of chemicals including organic and inorganic acids. They are particularly notable for their high retention of mechanical properties in operation at elevated temperatures. Most of the aromatic amines are solids at room temperature and are often liquified for ease of mixing with liquid epoxy resin by the formation of eutectic mixtures or blending with cycloaliphatic polyamines for heat cure applications.

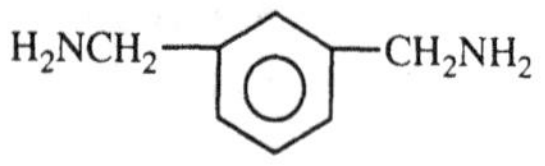

Meta-**P**henylene**D**iamine (MPD)

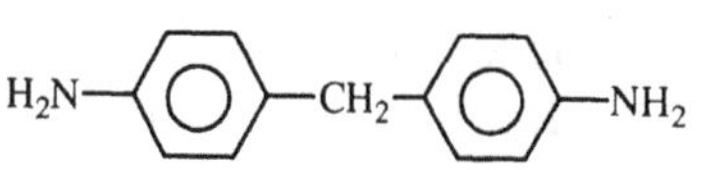

4,4'-**D**iamino-**D**iphenyl**M**ethane (DDM)
or **M**ethylene**D**i**A**niline (MDA)

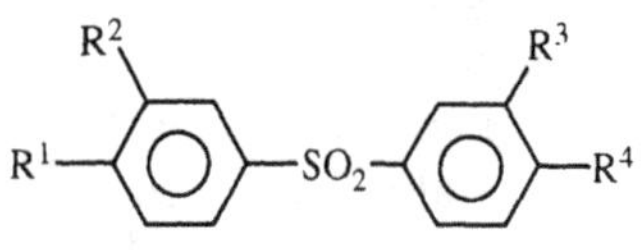

$R^1,R^4 = NH_2$; $R^2,R^3 = H$ 4,4′-**D**iamino-**D**iphenyl**S**ulphone
$R^1,R^4 = H$; $R^2,R^3 = NH_2$ 3,3′-**D**iamino-**D**iphenyl**S**ulphone
(DDS)

Of all the aromatic amines, **MPD** *[1477-55-0]* provides the tightest cross-link density and exhibits the best solvent resistance. Philipson (1959) found MPD, when blended with other solid aromatic amines, formed low viscosity eutectics useful in the manufacture of chemically-resistant filament wound pipe. **DDM** or **MDA** as it is also known, prepared by condensation of aniline with formaldehyde exists in a number of forms ranging from the pure 4,4′-diamino-diphenylmethane solid *[101-77-9]* with a melting point of around 90°C to low softening-point mixtures of methylene bridged polyphenyl polyamines which are easier to melt/mix with liquid epoxy resin, as reported by Seeger and Fauser (1954). The low polarity of the pure DDM curing agent which imparts excellent electrical insulation properties combined with high mechanical property retention even under conditions of high humidity, make this the most suitable candidate for electrical casting, laminating and adhesives applications. Although the most widely used of the aromatic amines, DDM is now classified as a potential human carcinogen and for many applications is being replaced by other types of curing agents.

4,4′-DDS [80-08-0] has the principle advantage over other aromatic amines in providing the highest heat resistance. It has become the standard curing agent, for use in combination with the BF_3-MEA catalyst (see section 2.5.1), with specialised epoxy resins such as the tetra-glycidyl ether of DDM, in high temperature tooling and high performance military and aerospace laminating applications. **3,3′-DDS** [599-61-1], although giving reduced heat resistance compared to the 4,4′- analogue, has been adopted for its enhanced honeycomb peel strength in aerospace laminating applications.

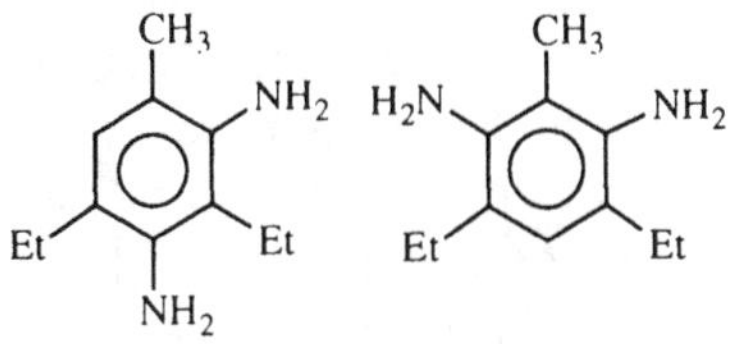

DiEthylTolueneDiAmines (DETDA)

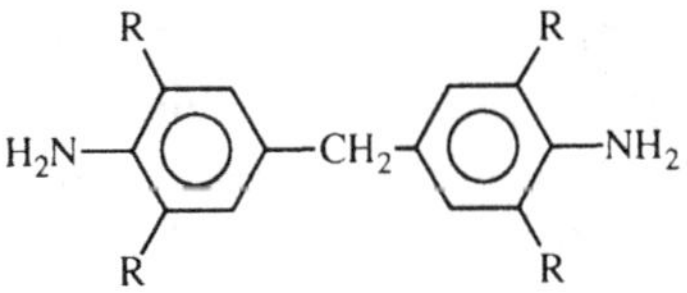

Tetra-alkyl DDM

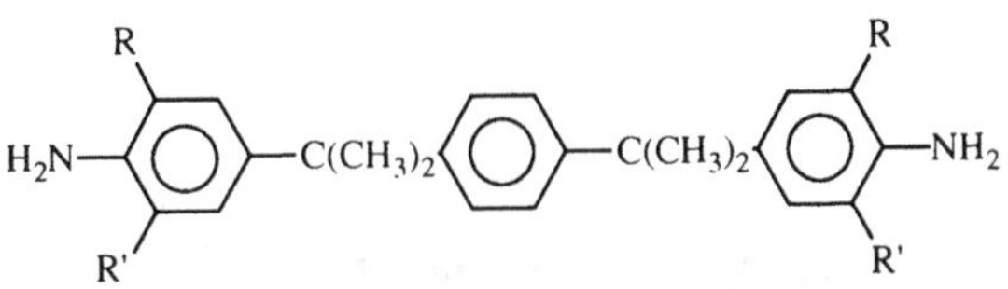

Tetra alkylated di(4-aminophenyl) di-isopropyl benzene

DETDA which is the 2,4-diamino-3,5-diethyltoluene/2,6-diamino-3,5-diethyltoluene isomer blend [68479-98-1] is a low viscosity non-crystallising liquid. Urech *et al.* (1982) reports that it is much less reactive than (unhindered) DDM and needs to be used in conjunction with accelerators or co-curing (cycloaliphatic) polyamines to provide acceptable cure rates even at relatively high cure temperatures. Applications for the accelerated variant include filament winding, casting and tooling, and for the cycloaliphatic polyamine co-curing blends, syntactic foam and casting. Toxicity however is now also a cause for concern with the DETDA. Alkylated DDMs such as **tetraethyl DDM** [13680-35-8], and the tetra-alkylated di(4-amino phenyl)-di-isopropyl benzenes such as the tetramethyl derivative [5432-99-5] developed by Bauer (1989) are ‘safer’ than DDM from a toxicological point of view being Ame’s test negative (a test providing an idea of the potential for a mutagenic effect which is some indication of carcinogenicity). They are used in a variety of laminating applications for their long working life/ready compatibility with liquid epoxy resin and long out-times/hot–wet resistance respectively.

2.2.3.2 Room temperature cured aromatic amines. Sinclair and Tilley (1967) found aromatic polyamines, such as DDM, can be dissolved in phenols to produce curing agents which will cure liquid epoxy resins at temperatures below 0°C. Dunn and Mebes (1963) revealed that addition of organic acid accelerators to solutions of partially adducted DDM in plasticisers such as di-butyl phthalate also produced effective room temperature curing agents. By changing the plasticiser types/concentrations, and the accelerator amounts, a range of curing agents with pot-lives from a few minutes to several hours, and with differing mechanical and chemical resistance properties are obtained. All exhibit lower heat resistance compared to unmodified aromatic amines yet are still tolerant to moisture, are resistant to water spotting and provide excellent resistance to chemicals upon spillage and immersion. The high abrasion resistance, organic acid resistance and non-tainting behaviour of certain plasticised types has led to their use in the internal coating of beer and wine storage tanks. These, and others, are also used for the internal coating of gas and oil pipe lines and chemically-resistant flooring, marine and maintenance coatings where dark colours are acceptable. Toxicity and colour stability are current issues with room temperature curing aromatic amines however, and for most applications they are being replaced by formulated cycloaliphatic amines.

2.2.4 Catalysts and co-curing agents

Tertiary amines, imidazoles, ureas.
Catalytic epoxy resin curing agents can function in a variety of ways. They can function as: *catalytic curing agents* which achieve cross-linking by homopolymerisation; as *accelerators* and *co-curing agents* for polyamines, polyamides or anhydrides; or as *activators*.

2.2.4.1 Tertiary amines. The reaction rates of most tertiary amines with epoxy resins are low at room temperature. Tertiary amines are therefore generally used catalytically as accelerators in elevated temperature cured systems. They are rarely used as sole curing agents. The mechanism of cure can follow two mechanisms. In the absence of hydrogen donors, epoxy resin polymerisation proceeds directly via an anionic mechanism initiated by the t-amine/epoxy zwitterionic adduction product (Scheme 2.3a). In the presence of molar quantities of hydrogen donors, e.g. hydroxyl groups, polymerisation also proceeds anionically, but through a two step initiation process (Scheme 2.3b).

The most commonly used tertiary amines **Dimethylaminomethyl phenol** [25338-55-0] and **Tris(dimethylaminomethyl) phenol** [90-72-2] contain phenolic hydroxyl groups as an integral part of the structure of the amine. As such they are particularly effective room temperature accelerators/co-curing

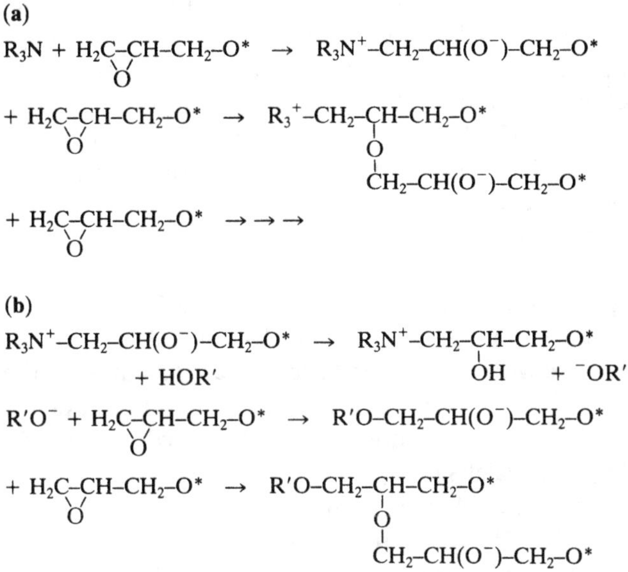

Scheme 2.3 Mechanism of cure of tertiary amines.

agents for polyamine and polyamide-based coatings and adhesives applications.

Along with **BDMA** [103-83-3] and **DBU** [6674-22-2], the phenolic amines are also efficient activators for: polysulphides and polymercaptans (see sections 2.4.1 and 2.4.2 respectively) in room temperature cure adhesives and sealants applications; for dicyandiamide in elevated temperature cure electrical, sports and industrial laminate manufacture; and for anhydrides in filament winding and electrical casting applications. In all cases the amounts of the different t-amines added to a system are determined empirically to achieve an optimum balance of cure rate, working life and cured mechanical properties. Too much catalyst in the formulations may help achieve faster cure rates but usually at the expense of working-life, embrittlement and/or shrinkage.

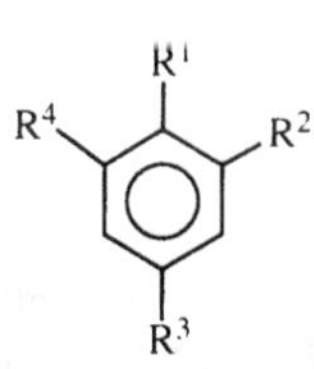

$R^1 = OH$; $R^2 = CH_2NMe_2$; $R^3, R^4 = H$ Dimethylaminomethyl phenol

$R^1 = OH$; $R^2, R^3, R^4 = CH_2NMe_2$ Tris(dimethylaminomethyl)phenol

$R^1 = H$; $R^2 = CH_2NMe_2$; $R^3, R^4 = H$ **B**enzyl**D**i**M**ethyl**A**mine (BDMA)

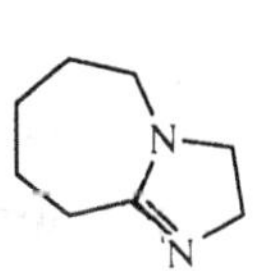

Diaza**B**icyclo**U**ndecene (DBU)

Salting of tertiary amines is a commonly used technique to prolong the working lives particularly for electrical casting and lamination applications where vacuum degassing needs to be carried out without any danger of premature gelation. The tri-2-ethyl hexanoate salt of tris (dimethylamino-methyl) phenol [51365-70-9] which is a liquid and readily miscible with liquid epoxy resin is actually used as a sole curing agent for electrical casting applications. DBU, which is an extremely strong base (pKa 11.5), when reacted with organic and organophosphorus acids forms salts with varying activation temperatures and pot-lives such as the long latency 4-methylbenzenesulphonate salt [51376-18-2]. The mechanism of cure of the tertiary ammonium salts involves initial esterification of the blocking acid group with epoxy resin. The liberated amines are then able to react with the epoxy group in the case of the tri-2-ethylhexanoate salt of tris(dimethylaminomethyl) phenol, or activate anhydride co-curing agents in the case of the DBU salts.

Quaternary ammonium salts such as **B**enzyl **T**rimethyl **A**mmonium **C**hloride (**BTAC**) [56-93-9] also find use as latent accelerators for anhydride-cured epoxy casting applications. They are unable to dissociate on heating to the tertiary amine and little is known about their mechanism of action.

2.2.4.2 Imidazoles. Imidazoles are obtained by dehydrogenation of imidazolines which are themselves generally prepared from reaction of 1,2-aliphatic diamines and nitriles. Many examples of 1-H-imidazoles or 1-(N)-unsubstituted imidazoles as they are also known, have been commercialised and include: 2-**M**ethyl-**I**mida**Z**ole (**2MI/2MZ**) [693-98-1]; **2**-**E**thyl-**4**-**M**ethyl-**I**mida**Z**ole (**2E4MI/2E4MZ**) [931-36-2] and **2**-**P**henyl-**I**mida**Z**ole (**2PI/2PZ**) [670-96-2]. They are efficient accelerators for anhydrides and dicyandiamide and also act as catalytic epoxy curing agents at moderate to high temperatures.

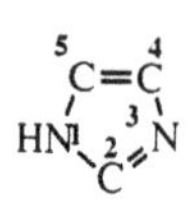

1-H-Imidazole

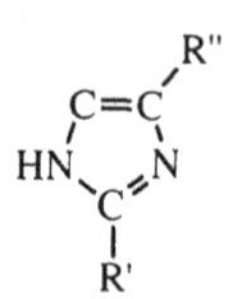

R'= Me; R''= H (2MI/2MZ)
R'= Et; R''= Me (2E4MI/2E4MZ)
R'= Ph; R''= H (2PI/2PZ)

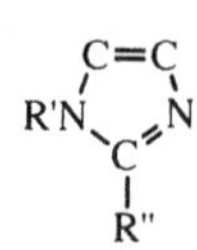

R' = $PhCH_2$; R'' = Me
1-Benzyl-2-methylimidazole

When used as a unique curing agent they react with epoxy resin at the 3-N position to form a 1 : 1 molar adduction product (Scheme 2.4a). They then react at the 1-N position with a second molecule of epoxy resin (Scheme 2.4b) to form a 2 : 1 adduct which contains a highly reactive alkoxide ion which initiates rapid anionic polymerisation of epoxy resin. For 1-(*N*)-substituted imidazoles, 1 : 1 molar adducts with highly reactive alkoxide ions

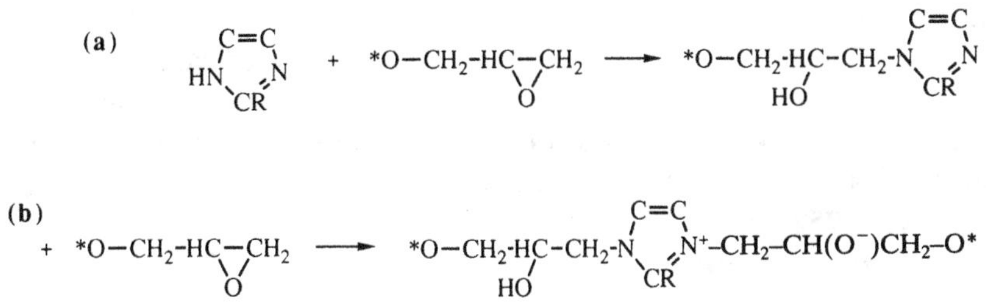

Scheme 2.4 Mechanism of cure of imidazoles.

are formed directly therefore these tend to be somewhat more reactive catalysts and accelerators than the 1-(*N*)-unsubstituted analogues. 1-Benzyl-2-methyl-imidazole [13750-62-4] which is one of the few examples of a 1-(*N*)-substituted imidazole to be commercially exploited is a highly reactive catalyst for anhydrides and epoxy resins at elevated temperatures.

The 2-methyl-imidazole, 2-ethyl-4-methyl-imidazole and 2-phenyl-imidazoles are used exclusively as accelerators for dicyandiamide-cured high molecular weight epoxy resin in powder coating and electrical laminating applications. These 1-H-imidazoles, unlike the *N*-substituted variants which are liquid, vary in physical form (low to high melting points) depending upon the molecular weight. Although they are poorly soluble in liquid epoxy resin at room temperature and have pot-lives of several hours, for many applications including one-pack adhesives, solvent-free laminating and filament winding, various analogues with considerably extended latency/working lives have been developed. Essentially, improved latency has been achieved by: Sawa and Gohda (1978) by general insolubilisation using c-methylolation; Winslow and LaBelle (1979) by incorporation of electron-withdrawing nitrile groups; and again Sawa *et al.* (1980) by salting with sterically-demanding isocyanuric acid.

2.2.4.3 Ureas. Trisubstituted ureas, or **urons** as they are also known, derived by blocking isocyanates with dimethylamine have been used widely as accelerators for the dicyandiamide cure of epoxy resin. The most common types are:

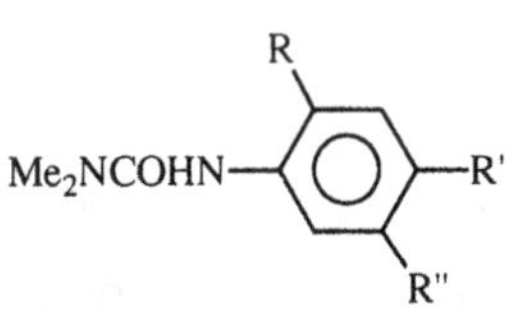

R = H; R' = Cl, R'' = H 3-(4-chlorophenyl)-1,1-dimethylurea [150-68-5]

R, R' = Cl, R'' = H 3-(3,4-dichlorophenyl)-1,1-dimethylurea [330-54-1]

R, R', R'' = H 3-phenyl-1,1-dimethylurea [101-42-8]

R = Me, R' = H; R'' = $NHCONMe_2$: 2,4-toluene di-isocyanate derivative

All exhibit outstanding latency at normal ambient temperatures and are widely used in one-pack adhesive applications. Although the mechanism of cure is not fully explained there is clearly a synergistic effect between dicyandiamide and certain trisubstituted ureas. It is possible that dicyandiamide activates deblocking *in situ*, causing release of dimethylamine, which then functions as a conventional tertiary amine accelerator. Toxicity issues with the traditional chlorinated aromatic substituent components means that for most applications the 2,4-toluene di-isocyanate derivative [17526-94-2], which is a more efficient accelerator, is presently the preferred replacement.

2.2.5 *Hydrazine and hydrazides*

$$H_2N\text{–}NHCO\text{–}R\text{–}CONH\text{–}NH_2$$

$$\begin{aligned} R = {} & \text{–}(CH_2)_4\text{–: Adipyl} \\ & \text{–}(CH_2)_7\text{–: Azelayl} \\ & \text{–}(CH_2)_8\text{–: Sebacyl} \end{aligned}$$

Dihydrazides derived from flexible di-carboxylic acid esters and hydrazine hydrate provide a class of curing agent which exhibits an unusual combination of attributes according to 3M Co. (1960). These include long shelf-lives, moderate pot-lives at advanced temperatures and the ability to be heated and subsequently cooled without substantial curing. Furthermore rapid cure at application temperatures, yielding heat-resistant, high mechanical and adhesive strength products makes this class of curing agent particularly suitable for coatings and laminating applications.

2.3 Oxygen-containing curing agents

2.3.1 *Carboxylic acids and anhydrides*

Carboxylic acid functional polyesters and anhydrides are the second most important family of curing agents for epoxy resins. Although the acids are only practical in heat-cured surface coatings, or stoving enamels, they are nonetheless the most widely used epoxy curing agent type (Table 2.1). Anhydrides are suited to most other heat cure applications.

2.3.1.1 Carboxylic acid functional polyester resins. The basic structure of the carboxylic acid functional polyester resin is:

$$n\text{HOOCR}^1\text{COOH} + (n\text{-}1)\text{HOR}^2\text{OH} \rightarrow \text{HOOCR}^1\text{CO-(OR}^2\text{O-OCR}^1\text{CO)}_{n-1}\text{OH} + 2(n-1)\text{H}_2\text{O}$$

According to polyester resin technology, the epoxy resin component of a hybrid epoxy/carboxyl functional polyester system is regarded as the cross-linking or curing agent; di-glycidyl ethers of bisphenol A are furthermore considered as modifiers. The reason for the remarkable success of carboxyl functional polyesters according to Husbands (1987) was and continues to be

(a)

$$*{-}OCR^1COOH + \underset{O}{H_2C{-}CH}{-}CH_2O* \rightarrow *{-}OCR^1COO{-}CH_2{-}CH(OH){-}CH_2O*$$

(b)

$$*{-}OCR^1COOH + HO{-}CH(CH_2O*)_2 \rightarrow *{-}OCR^1CO{-}OCH(CH_2O*)_2 + H_2O$$

Scheme 2.5 Mechanism of cure of hybrid epoxy/polyesters.

the availability of a wide range of relatively cheap raw materials. These can be combined together easily to give the necessary properties of correct softening point and reactivity for powder coatings applications.

The mechanism of cure involves two stages: addition of the carboxyl group to the epoxy functionality (Scheme 2.5a) and esterification with the secondary hydroxyls on the epoxy backbone (Scheme 2.5b). The water released in this latter condensation reaction then volatilises from the film during the cure process.

The poor resistance to weathering of the bisphenol A, di-glycidyl ether epoxies has precluded widespread use in external topcoat systems, but their excellent corrosion resistance and adhesion make them ideal reaction partners for carboxyl functional polyesters in a number of appliance and metal-finishing powder coatings applications. Epoxy resins with high aliphatic contents such as **Tri-GlycidylIsoCyanurate (TGIC)** on the other hand provide improved weatherability and are used with carboxyl functional polyesters for architectural powder coatings.

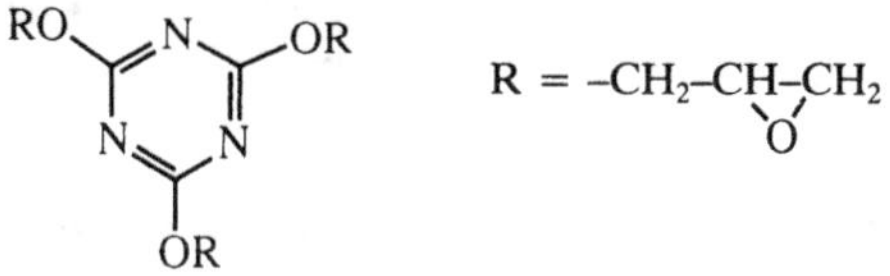

TGIC

2.3.1.2 Carboxylic acid anhydrides. Dicarboxylic acid anhydrides were amongst the first curing agents used with epoxy resins. They have achieved widespread commercial importance due to a combination of their cure and cured properties. Reaction with epoxy resin is characterised by a low exotherm and although long periods at elevated temperatures are required to achieve full cure the resulting low shrinkage, stress-free systems provide excellent electrical insulation properties. These properties are retained even under moderate to high continuous operating temperatures due to the good thermal stability and have therefore helped in their adoption in the heavy electrical industry.

The mechanism of cure of anhydrides is more complex than that of amines due to several competing reactions which can occur, especially when accelerators are added to enhance cure rates. In the absence of added accelerators or catalysts (Scheme 2.6a) the anhydride ring is opened by a hydroxyl group from the backbone of the epoxy resin forming a half-ester. This is followed by the half-ester carboxylic acid group initiating reaction with epoxy resin to form a di-ester-alcohol which can continue the polymerisation process either by esterification with another anhydride molecule, or etherification with an epoxy group. The latter etherification reaction is the accepted route and in practice only 0.85 equivalents of anhydride are required to provide optimum cross-link densities and cured properties. Lewis bases such as tertiary amines and imidazoles are widely used as anhydride accelerators. These are able to open anhydride rings to form internal salts (betaines) which then act as initiators of cure (Scheme 2.6b). The resulting carboxylate ions react with an epoxide group to yield alkoxide esters which undergo reaction with further anhydride molecules to form carboxylate anion functional esters. These can then react with further epoxide groups and continuation of this alternating sequence leads to the formation of a polyester. In usage, optimum properties are produced when stoichiometric equivalents of epoxy and anhydride are employed, in line with this mechanism which does not involve etherification reactions.

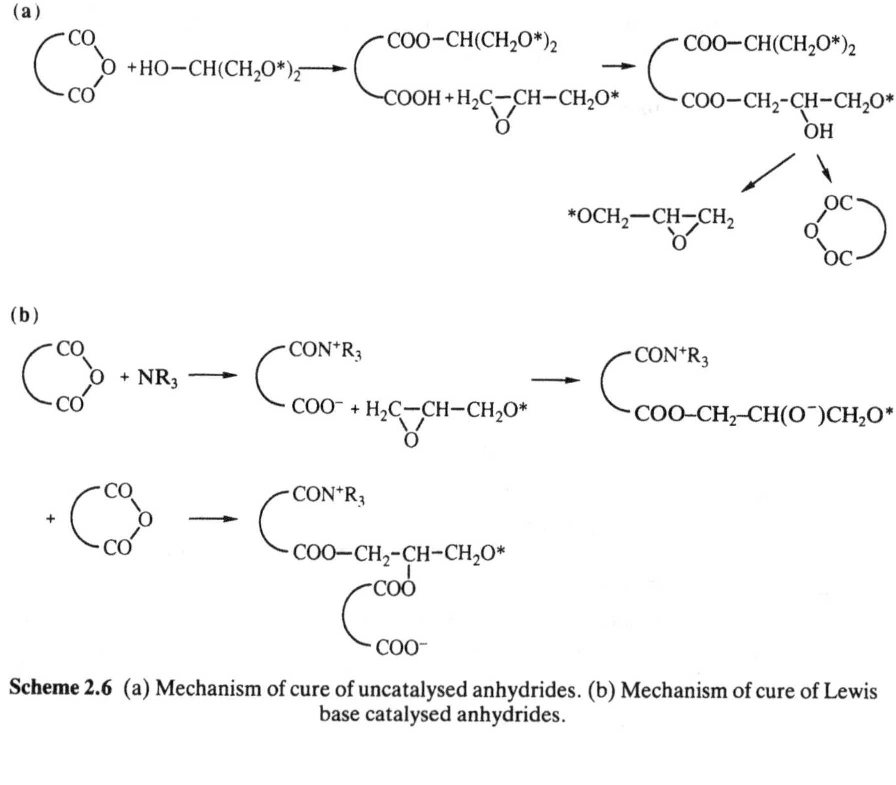

Scheme 2.6 (a) Mechanism of cure of uncatalysed anhydrides. (b) Mechanism of cure of Lewis base catalysed anhydrides.

Lewis acids such as BF_3-amine complexes and tetra-alkylammonium salts are also catalysts for the epoxy-anhydride reaction although no fully satisfactory mechanisms have been proposed. This form of catalysis does however strongly favour the etherification reaction and as little as 0.55 equivalents of anhydride per epoxy equivalent are found to provide optimum cured properties.

Apart from the effects of stoichiometry and accelerator or catalyst choice, widely differing cured mechanical properties can obviously be obtained by varying the anhydride curing agent type itself. The anhydride types vary from dicarboxylic acid cyclic anhydrides to tetracarboxylic acid cyclic di-anhydrides (and beyond) providing increasing cross-link density and thus heat, chemical and solvent resistance respectively. Linear aliphatic polyanhydrides are used as flexibilising modifiers.

2.3.1.3 Dicarboxylic acid cyclic anhydrides. With the exception of **P**hthalic **A**nhydride (**PA**) [85-44-9] the most important dicarboxylic acid anhydrides are cycloaliphatic. PA which is the least expensive and the most difficult to handle due to a strong tendency to sublime during use, and the hydrogenated derivative **H**exa-**H**ydro**P**hthalic **A**nhydride (**HHPA**) [85-42-7] are used chiefly for casting and laminates for electrical applications. **T**etra-**H**ydro**P**hthalic **A**nhydride (**THPA**) [26266-63-7] gives properties very similar to HHPA though darker coloured products result. HHPA which is a low-melting point solid does not sublime and is used with other anhydrides to make low viscosity liquid eutectic mixtures.

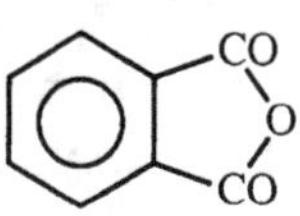

PA

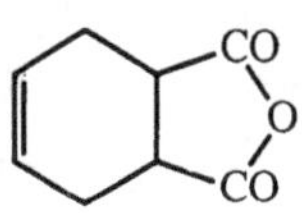

THPA

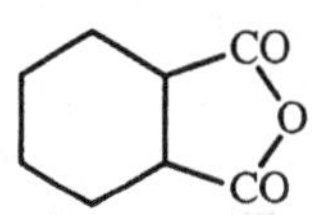

HHPA

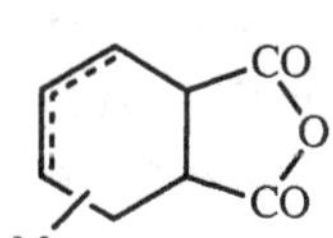

MTHPA

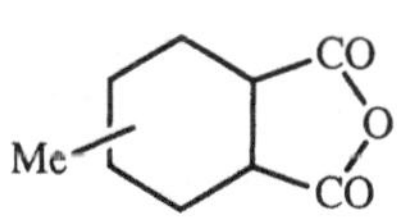

MHHPA

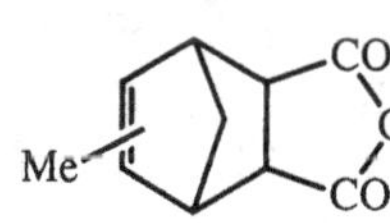

MTHPA
or NMA

Methyl**T**etra**H**ydro**P**hthalic **A**nhydride (**MTHPA**) [11070-44-3] prepared by Diels–Alder reaction between isoprene and maleic anhydride is a liquid eutectic mixture of regio- and stereo-isomers and is now one of the more commonly used anhydride curing agents for filament wound pipe manufacture. Other applications include electrical casting, potting, encapsulation and impregnation – here however the more exclusive hydrogenated derivative **M**ethyl **H**exa-**H**ydro**P**hthalic **A**nhydride (**MHHPA**) [25550-51-0] is preferred, and used with cycloaliphatic epoxy resins for optimum light stability.

Methyl Endomethylene Tetra-HydroPhthalic Anhydride (**METHPA** or Nadic Methyl Anhydride (**NMA**) [25134-21-8] as it is also known), prepared by Diels–Alder reaction between methylcyclopentadiene and maleic anhydride is used in components where electrical property retention at high temperatures is required. NMA is the anhydride of choice here due to the high rigidity of the fused cycloaliphatic ring backbone.

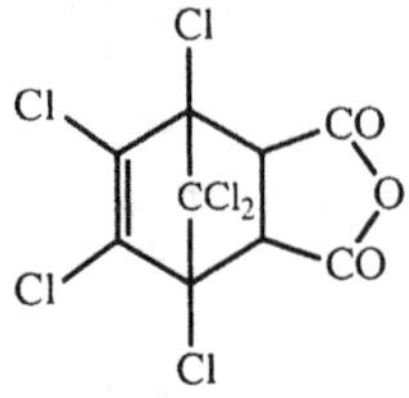

HET

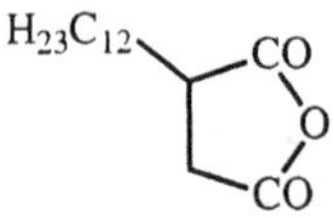

DDSA

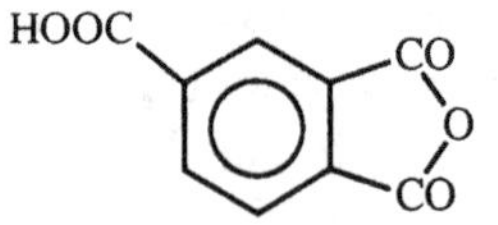

TMA

Further examples of dicarboxylic acid anhydrides less frequently encountered but none the less offering unique properties include; **H**exachloro **E**ndomethylene **T**etra-hydrophthalic anhydride (**HET**) [115-27-5], or chlorendic anhydride as it is also known, like METHPA/NMA has a fused ring structure and in addition substituent chlorine atoms which confer flame-retardency for electrical laminate and casting applications; **D**o**D**ecyl **S**uccinic **A**nhydride (**DDSA**) [51546-46-7] imparts flexibility to castings and is generally used in blends with other anhydrides for improving thermal shock resistance; **T**ri-**M**ellitic **A**nhydride (**TMA**) [552-30-7] is a very reactive anhydride due to the free carboxyl group which tends to accelerate cures with epoxy resins and is normally used with solid epoxy resins for high performance laminate applications. Toxicity concerns have limited wider usage of TMA.

2.3.1.4 Tetracarboxylic acid di-anhydrides. **P**yro**M**ellitic **D**i-**A**nhydride (**PMDA**) [89-32-7] and **B**enzophenone **T**etra-carboxylic acid **D**i-**A**nhydride (**BTDA**) [2421-28-5] are aromatic di-anhydrides which are used to achieve high cross-link densities and high heat resistance.

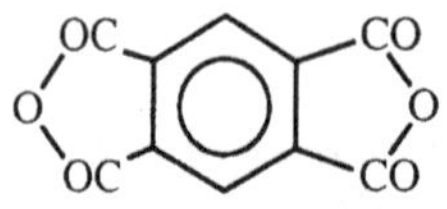

PMDA

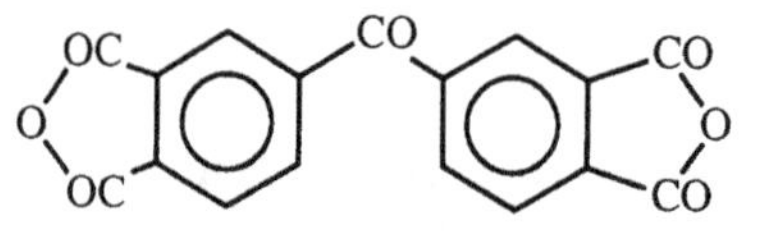

BTDA

PMDA which is the more compact cross-linking agent of the two gives the better chemical resistance and the best resistance of all anhydride curing agents. It is however a high melting solid and very reactive towards epoxy resin at high temperatures. It is used therefore in combination with other anhydrides such as maleic anhydride to make it easier to dissolve and more convenient to handle for a variety of electrical casting applications. Maleic anhydride by itself is not a particularly useful curing agent as it produces brittle cured epoxies. BTDA, also a high melting point solid, is somewhat easier to handle due to its lower reactivity and is mostly used, alone or in combination with other anhydrides, for high temperature-stability electronic moulding powder and adhesive applications.

2.3.1.5 Linear poly-anhydrides. Black *et al.* (1964) have reported the use of linear aliphatic poly-anhydrides derived by dehydration polymerisation of adipic, azelaic and sebacic acids which contain terminal carboxylic acid groups at either end of long chain aliphatic polymers. They are used essentially as flexibilising modifiers for other anhydrides where improvements in thermal shock resistance are required.

2.3.2 Phenol formaldehyde resins

Phenols can be reacted with formaldehyde to give resinous products, **P**henol-**F**ormaldehyde (**P-F**) resins, with a wide variety of commercial applications; Oldring (1987) records certain types being suitable as cross-linking agents for epoxy resins.

2.3.2.1 Phenol novolac resins.

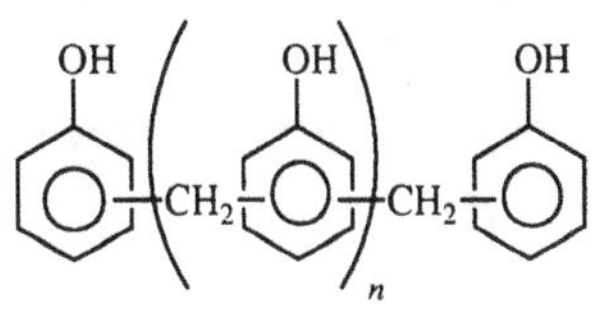

The novolac resins, which are the reaction products from formaldehyde and excess phenol under acidic catalysis, when co-cured with high molecular weight solid bis-A epoxy resins result in coatings with excellent adhesion, film strength, flexibility and chemical resistance.

They are especially useful in powder coatings applications for corrosion resistant pipe/reinforcing bars (rebars) and with brominated epoxy resins for FR3 electrical laminate production. The cure mechanism (Scheme 2.7) involves poly-addition to epoxy resin and is activated by acids such as *p*-toluenesulphonic acid.

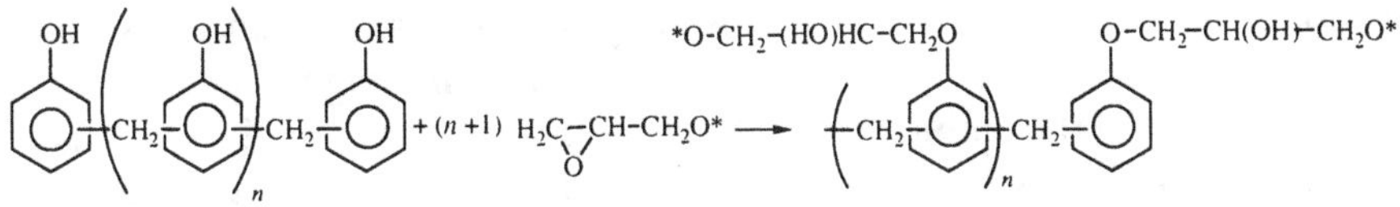

Scheme 2.7 Mechanism of cure of novolac resins.

2.3.2.2 Resole resins.

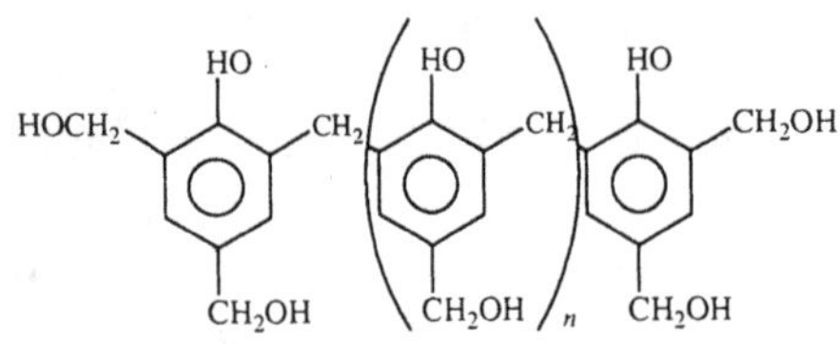

The resole resins are the reaction products from excess formaldehyde and phenol under basic catalysis. When co-cured with high molecular weight solid bis-A epoxy resins in a poly-etherification reaction (Scheme 2.8) because of the secondary hydroxyls on the epoxy backbone they yield even higher cross-link densities and higher chemical resistance than novolac resins. As such they are well-suited to drum and pail coating applications where the high stoving temperatures easily drive off the water produced in the condensation cure reaction.

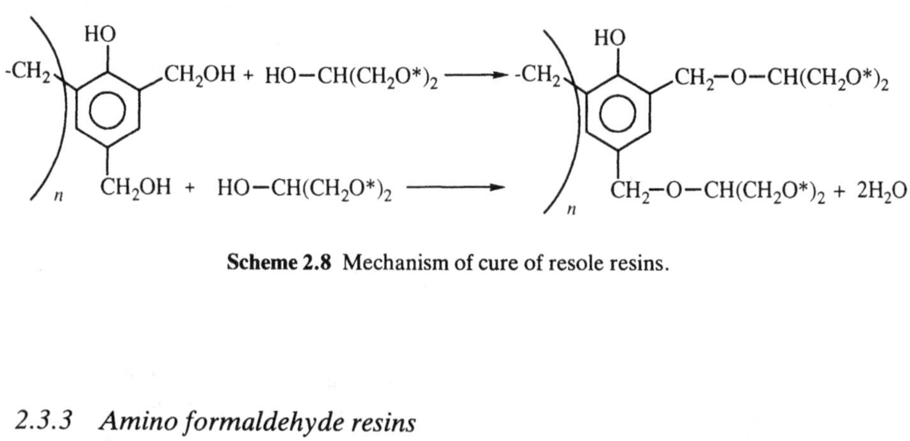

Scheme 2.8 Mechanism of cure of resole resins.

2.3.3 *Amino formaldehyde resins*

Even more tightly cross-linked structures than attainable with phenol-formaldehyde cured epoxy resins can be derived by heating **Amino-Formaldehyde (A/F)** resins and epoxy resins at temperatures in excess of 150°C. Oldring (1987b) describes the most commonly used amino-resins produced from: urea, the **U**rea-**F**ormaldehyde (**U-F**) resins; and melamine, the **M**elamine-**F**ormaldehyde (**M-F**) resins. Amino-resins are invariably also alkylated-etherified with alcohols under acidic conditions to restrict possible methylol-methylol condensation/polymerisation reactions. The degree of alkylation is an important factor in determining the performance characteristics of the amino resins. As the degree of alkylation increases, the viscosity and reactivity of the resin decreases, while compatibility with epoxy resin increases.

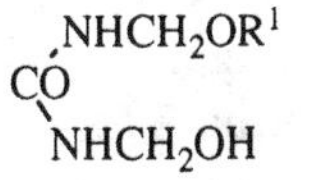

U-F resin

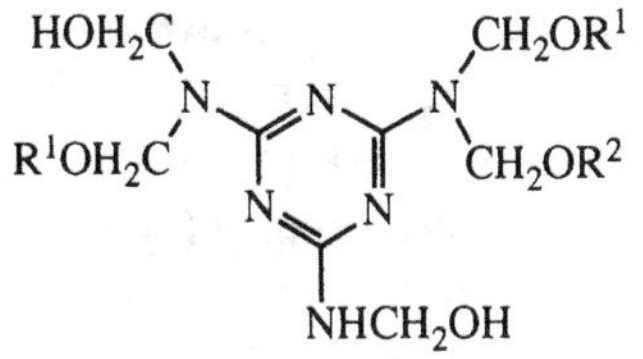

M-F resin

$R^1 = n\text{-}C_4H_9$, $iso\text{-}C_4H_9$, C_2H_5, CH_3; R^2 = another M-F resin

The mechanism of cure with epoxy resin involves: etherification with loss of water or alcohol through the many secondary hydroxyls on the backbone of the high molecular weight solid bisphenol A epoxy resins (Scheme 2.9a) and, addition of *N*-methylol groups to the epoxy functionality (Scheme 2.9b). M-F amino-resins will also catalyse homopolymerisation of the epoxy resin.

(a)

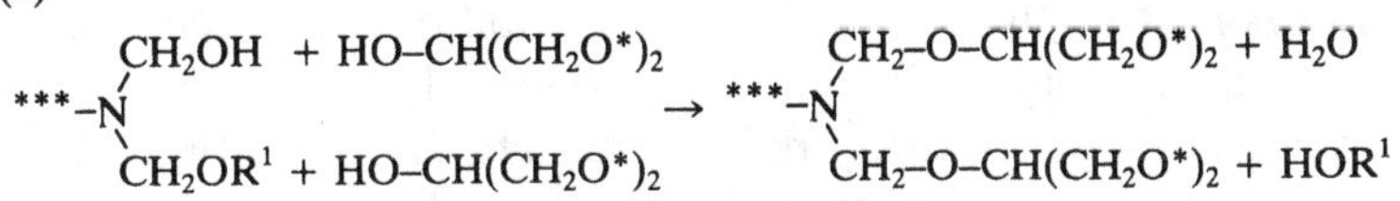

(b)

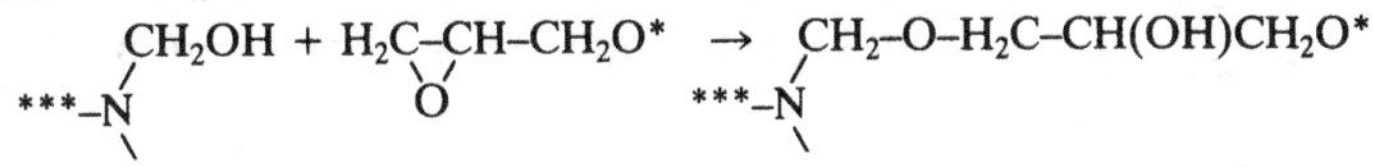

Scheme 2.9 Mechanism of cure of amino-resins.

The resulting cured films offer a combination of the best features of epoxy and amino resins namely high adhesion, chemical resistance, gloss retention, colour and colour retention and, with correct choice of amino-resin, very good flexibility for metal decorating applications. M-F resins are used for their superior film hardness and gloss properties where high performance demands outweigh cost considerations, notably varnish and moulding powder applications. U-F resins offer cost performance benefits for fast bake enamel, stoving primer and can and drum top coating applications.

2.4 Sulphur-containing curing agents

The thiol or mercaptan group (-SH) is able to react with epoxy resin in an addition reaction. This requires catalysis at room temperature by amines which promote production of reactive mercaptide ions (Scheme 2.10). With

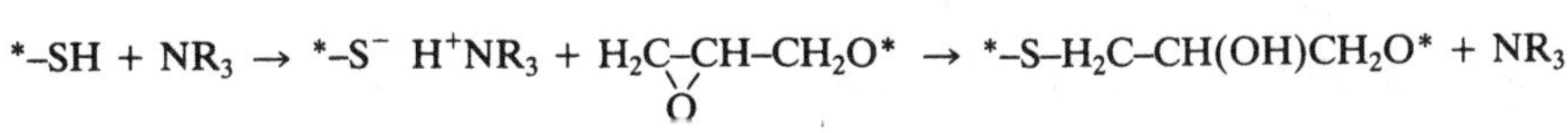

Scheme 2.10 Mechanism of cure of thiols/mercaptans.

mercaptan terminal **polysulphides**, whose functionality (typically 2) is too low for efficient cross-linking of di-functional bisphenol A epoxy resin, primary amines are used as initiators/co-curing agents. With **polymercaptans** where the functionality (typically 3) is adequate for efficient cross-linking, strongly basic tertiary-amines are used as initiators and rapid cure times result.

2.4.1 Polysulphides

$$H(S{-}CH_2CH_2{-}OCH_2O{-}CH_2CH_2{-}S)_x(SCH_2CH(SH)CH_2S)_yH$$

Fettes and Gannon (1957) found low mercaptan content polysulphides useful curing agents for epoxy resin. As the stoichiometric loading with liquid epoxy resins is high, this leads to highly flexible products with very low cross-link densities and poor mechanical properties. In practice however not more than half the stoichiometric quantity is used with the proportion varying according to the degree of flexibility required, and co-curing polyamides, polyamidoamines, cycloaliphatic amines and tertiary-amine catalysts are used to upgrade mechanical properties. Main applications for epoxy-polysulphide systems are in building adhesives, primarily because they give excellent adhesion to both concrete and glass and also because they cure considerably faster than amine curing agents.

2.4.2 Polymercaptans

$$R{-}[O{-}\{CH_2CH_2CH_2O{-}\}_nCH_2{-}CH(OH){-}CH_2{-}SH]_3$$
$$n = 1 - 2;\ R = \text{aliphatic hydrocarbon}$$

The attractive reactivity characteristics of the mercaptan group with epoxy resin have led to several attempts to commercialise other mercaptan-curing agents. Diamond Chemicals (1968) designed certain tri-functional polymercaptans to meet special application situations. Here the inherent *beta*-hydroxy groups activate the mercaptan groups to give high reactivity at low temperature and in thin films. When used at high loadings relative to tertiary-amine co-catalyst, these polymercaptans also give the high lap-shear adhesion required for *Do It Yourself* patch repair adhesive and structural concrete bonding applications. The moisture resistances of the cured systems are low however, due to the *beta*-hydroxy functionality, and this with the disagreeable odour and skin sensitisation tendency has limited their widespread adoption.

2.5 Miscellaneous curing agent types

2.5.1 Amine-boron trihalide complexes

$$R^1R^2R^3N^+{-}B^-X_3 \qquad R^1,R^2,R^3 = \text{alkyl or aryl};\ X = \text{F or Cl}$$

Greenlee (1955) reported that amines can react with boron trihalides to form stable donor-acceptor complexes, which act as latent catalysts for epoxy homopolymerisation. The reactivity of the complexes is controlled by either the boron trihalide, the trichloride being a stronger acceptor than the trifluoride, or the amine selection. The choice of amine also determines physical properties such as solubility and miscibility with the epoxy resin, as well as hygroscopicity: mono-ethylamine-boron trifluoride (**BF_3-MEA**) is strongly sensitive to water; benzylamine-boron trifluoride is not. Amine-boron trihalide complexes cure epoxies to highly cross-linked, high heat and chemical resistant materials which are somewhat brittle in nature; plasticising carrier diluents are often used to improve compatibility with liquid epoxy resin, and to reduce brittleness.

When activated by heating in the presence of epoxy resin, rapid reactions occur accompanied by considerable evolution of heat therefore mass sizes are limited. The complexes are used principally as catalysts with other curing agents such as anhydrides for thc curing of large masses for lamination, casting, encapsulation and electrical insulating varnishes. Applications as sole curing agents include adhesives, vacuum pressure impregnation, encapsulation and electrical varnishes.

2.5.2 *Quaternary phosphonium salts*

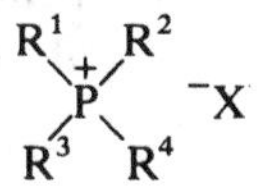

R^1,R^2,R^3,R^4 = alkyl or aryl; X = Cl or I

Quaternary phosphonium salts such as the alkyl-triphenylphosphonium halogenides are recorded by Smith (1978) as effective latent accelerators for anhydride and phenolic resin cured epoxy resins. They have indefinite shelf-lives, confer rapid cure rates at only moderately elevated temperatures, and have no adverse effect on cured resin mechanical and electrical properties. Although it is not clear exactly how they function their mechanism of cure does not involve decomposition to phosphine. Typical applications with anhydride/liquid epoxy resin include electrical casting, potting, encapsulation and 'vacuum pressure' impregnation. Quaternary phosphonium salts according to Martin (1976) are also effective catalysts for the advancement of liquid epoxy resins with phenolic materials, in, for example, the preparation of powder coating resin systems.

2.5.3 *Cationic salts*

$$Ar_3S^+ \ {}^-X$$

$$Ar = \text{Phenyl};\ {}^-X = {}^-BF_4,\ {}^-PF_6,\ {}^-AsF_6,\ {}^-SbF_6$$

Cationic salts such as the **tri-arylsulphonium salts** have been found by Crivello (1977) to be exceptionally efficient radiation cure photoinitiators for cationic polymerisation of various resin types including epoxy. Only cycloaliphatic epoxy resins however are highly sensitive to acid-catalysed cationic polymerisation at ambient temperatures. Glycidyl ethers and epoxidised oils require thermal treatment. The mechanism of cure with these catalysts involves breakdown to a source of strong acid, HX where X is BF_4^-, PF_6^-, AsF_6^- SbF_6^- and various radical fragments. Flexibility, hardness, solvent and water-resistance properties depend upon the exact nature of the resins and diluents selected. Cure rates depend upon the catalyst concentration, coating thickness and irradiating light intensity.

Applications thus far developed for this type of curing system include can coatings and a variety of specialist plastic coatings where adhesion is difficult to obtain with other resin systems.

2.6 Summary

It can be readily appreciated that the precise choice of curing agent type can markedly affect the properties of an epoxy resin when cured. As it is unusual for a single catalyst or curing agent to provide optimum characteristics in any given application, the identification and selection of a satisfactory curing agent is often, therefore, in itself a complex procedure. Dunn (1984) has however successfully rationalised the technique for surface coatings applications and Ashcroft (1991) has attempted civil engineering/construction applications.

References

Ajinomoto Co. Inc. (1969) Brit. Patent 1,155,973.
Ashcroft, W.R. (1991) *Eur. Coatings J.* **4**, 229–241.
Bauer, R.S. (1989) US Patent 4,855,386.
Black, R.G. *et al.* (1964) *SPE Antec*, Jan.
Brytus, V. (1986) *J. Coatings Technology* **58**(740), 44–48.
CIBA Ltd (1970) British Patent 1,204,224.
Cowan, J.C. *et al.* (1948) US Patent 2,450,940.
Crivello, J.V. (1977) US Patent 4,058,401.
Diamond Chemicals (1968) *Diamond Chemicals Technical Bulletin* 3-800LC-610.
Dunn, P.A. (1984) *Polymers Paint Colour Journal* Aug. 8/22.
Dunn, P.A. and Mebes, K.H. (1963) *Farbe Und Lacke*, **69**, 346–353.
Farnham, A.G. and Caldwell, N.J. (1956) US Patent 2,753,323.
Fettes, E.M. and Gannon, J.A. (1957) US Patent 2,789,958.
Gempeler, H. and Zuppinger, P. (1971) US Patent 3,631,149.
General Mills Inc. (1965) British Patent 988,410.
Greenlee, S.O. (1952) US Patent 2,585,115.
Greenlee, S.O. (1955) US Patent 2,717,885.
Halewood, P. (1962) British Patent 886,767.
Husbands, M.J. (1987) In *Resins for Surface Coatings*, ed. Oldring, P. and Haywood, G. SITA Technology **III**, 63–167.
Joyce, J.D. (1971) British Patent 1,219,069.

Kittridge, J.B. and Michelli, A.L. (1968) US Patent 3,390,124.
Lee, J.M. and Winfrey, J.C. (1966) US Patent 3,236,895.
McFadden, R.T. (1972) US Patent 3,634,372.
Martin, P.H. (1976) British Patent 1,429,076.
Mika, T.F. and Bauer, R.S. (1973) In *Epoxy Resins: Chemistry & Technology*, ed. May, C.A. Marcel Dekker, New York, pp. 465–550 and reference cited therein.
Newey, H.A. (1958) US Patent 2,864,775.
Oldring, P. (1987) In *Resins for Surface Coatings* ed. Oldring, P. and Haywood, G. SITA Technology **II**, 29–46. (1987b) *idem*, 46–71.
Philipson, J. (1959) US Patent 2,891,927.
Richardson, F. (1973) *Pigment And Resin Technology* **May** 41–43.
Rutgerswerke (1969) French Patent 1,568,011.
Sawa, N. and Gohda, K. (1978) US Patent 4,104,473.
Sawa, N. *et al.* (1980) US Patent 4,189,577.
Seeger, N.V. and Fauser, E.E. (1954) US Patent 2,683,730.
Shell International Research (1964) British Patent 960,236.
Sinclair, J.H. and Tilley, C.G. (1967) British Patent 1,083,657.
Smith, J.D.B. (1978) *Org. Coatings & Plastics Chem.* **39**, 42–46.
Urech, K. *et al.* (1982) US Patent 4,365,108.
Veba Chemie (1977) British Patent 1,460,505.
Wilson, A.F. (1970) British Patent 1,212,276.
Wilson, A.F. (1986) UK Patent Application GB 2,170,813A.
Winslow, J.M. and LaBelle, G.E. (1979) Deutsches Patentamt 28 34 371.
3M Co. (1960) British Patent 842,867.

3 The kinetics of cure and network formation

B. ELLIS

3.1 Cure of epoxy resins

Although the cure of an epoxy resin involves reaction between epoxy and hardener reactive groups, as discussed in chapter 2, the full characterization of the cure process involves many other factors. During cure, a liquid or fluid resin–hardener mixture is converted to a solid (chapter 1, section 1.1 and Figure 1.1). Critical features are gelation and the onset of vitrification. The latter effect occurs as the glass transition temperature, T_g, approaches the cure temperature, T_c. The determination and alternative definitions of T_g will be considered later (section 3.2). However, it should be noted that the glass transition temperature is a function of the extent of reaction, X_e,

$$X_e = \frac{E_o - E_{(t_c)}}{E_o}$$

with E_o and $E_{(t_c)}$ being the concentrations of epoxy groups present initially and at cure time t_c and has been used as a measure of the extent of reaction (Wisanrakkit and Gillham, 1990). When the difference between the cure temperature and the glass transition temperature becomes 'small', that is, $\Delta T = T_c - T_g(X_e)$ is small, curing reactions become diffusion controlled because molecular mobility is rapidly reduced as $\Delta T \rightarrow 0$. Cure may continue very slowly even when ΔT becomes negative. Then the glass transition temperature exceeds the cure temperature.

Initially when $T_c > T_g$ (0) the rates of the curing reactions are chemically kinetically controlled, and the rate expressions which have been proposed for the early stages of cure will be discussed in section 3.4. With the formation of larger more highly branched molecules, a critical point is reached, the gel point, and then there is the formation of an infinite three-dimensional network. If T_c is too low, vitrification may occur before gelation and then further reaction may be inhibited. An elevated temperature may be used to reduce the viscosity so that the hardener can be mixed into the resin and then the curing reactions can be quenched by rapidly cooling the mixture to a storage temperature below $T_g(0)$, the glass transition temperature of the 'compounded' resin. However, it is necessary to ensure that mixing is complete before gelation, since the viscosity rises rapidly at the gel point and the resin will not flow after formation of an incipient network. Thus, information on gelation and vitrification is required to characterize an epoxy resin cure system and to specify an efficient cure programme to ensure that the properties of the cured resin are optimized for a specific application.

For the technological application of epoxy resins it is essential that mixing of resin with the hardener and any other components should be complete before the onset of gelation. The origin of some of the morphological features that have been observed in cured resins may well be due to incomplete mixing, so that there are regions rich in hardener and hence *vice versa*. When the stoichiometry of the resin and hardener is considered, such heterogeneity will limit the extent of chemical conversion and hence the attainment of optimum physical and mechanical properties.

The 'pot-life' of the mixed resin and hardener is the time after mixing that the resin can be processed for such applications as impregnation, casting or other operations which involve 'flow'. Such operations have to be completed before the viscosity of the resin starts to increase as the gel point is approached. The viscosity could obviously be reduced by increasing the temperature, but then curing reactions can occur and hence the pot life-time will be reduced. For some applications diluents are added to reduce the viscosity so that processing can be completed expeditiously. The pot-life may be determined by use of so-called 'gel-timers'; commercial instruments with a probe which 'prods' the resin periodically. When penetration is prevented by gelation the time is recorded. Such measurements can be sensitive to the sample size since the heat evolved in an exothermic reaction causes the temperature to rise and hence the reaction proceeds more rapidly and the curing resin may exhibit an 'exotherm'. For similar reasons the pot-life will depend on the batch size, since the heat evolved will depend on the volume which increases more rapidly than the surface area from which heat can be lost.

In the cure schedule it is usual to specify a cure time t_c and a cure temperature T_c. However, the resin temperature may not ever be constant at T_c due to its low thermal conductivity and subsequently to an exothermic heat of reaction (Amdouni *et al.*, 1990). With a resin sample of finite thickness there will be a time-lag before the whole sample attains this impressed temperature, T_c. The time required for the centre of the resin to attain T_c will depend on the thermal diffusivity of the resin/hardener, the product thickness and the heat transfer coefficient between for example a metal mould and the initially liquid resin. This heat transfer coefficient may change during cure because of shrinkage of the resin away from the metal mould surfaces. For the cure of thicker sections the temperature of the centre of the moulding may well exceed that of the mould surfaces, because of the heat of reaction. To avoid excessive exotherms it is often necessary to cure the resin at a relatively low cure temperature, e.g. 100°C for BADGE + DDM, with a subsequent increase in the temperature for a post-cure of up to maybe 180°C.

When the temperature of the curing system is near to $T_{g(t_c)}$, i.e. $\Delta T = (T_c - T_{g(t_c)})$ becomes small, the rate of cure can become very slow and hence post-cures at elevated temperatures are normally used to ensure that all of the epoxy groups are consumed. Also, for many applications it is advantageous to maximize the glass transition temperature; this is also achieved by a suitable post-cure treatment.

Information regarding the chemical reactions occurring during cure are difficult to monitor for the whole of the curing process. It is only possible to carry out a 'wet' or similar chemical analysis on soluble resin hardener mixtures. So alternative methods are required to monitor cure beyond the gel point. The many methods that have been used will be discussed in section 3.3. Even with the extensive choice available it is not easy to monitor the cure of epoxy resins *in situ*. Maybe measurement of the change in electrical properties, either conductivity or dielectric properties, should be explored more extensively.

3.2 Gelation, network structure and glass transition temperature

In this section the critical conditions necessary for gelation will be outlined. After gelation the sol fraction, w_s decreases to essentially zero and the gel fraction increases so that $w_g \rightarrow 1$, where w_s and w_g are the weight fractions of sol and gel respectively. With increasing extent of reaction the structure of the network changes with more and more chains being 'tied' into the network. Also the length of the network chains decreases as the degree of cross-linking increases. Many of the physical and mechanical properties of the cured resin depend on the details of the network structure. Because of its crucial importance for the final stages of cure the glass transition temperature will be considered, after discussion of gelation. There are still many theoretical papers each year on the nature of the glass transition but here only the essential features will be outlined which are relevant to the cure of epoxy resins.

For an initial analysis of the change in macroscopic properties of epoxy resins during cure it is usual to assume that the curing resin is homogeneous. That is, the property changes are uniform throughout the sample. However, there is some evidence for uneven curing (Ghaemy *et al.*, 1982), and there have been a number of studies of the morphology of cured resins. However, with BADGE cured with DDM it was not possible to identify any structure larger than a few bisphenol A repeat units but with other hardeners which were less soluble in BADGE some features could be identified (Rashid, 1978). It would appear that at least some of the morphological features observed in the cure of resins are due to poor mixing and dissolution of the hardener in the resin. In other cases there may be either phase separation or the formation of microgel during the curing reaction. Ideally, before interpreting macroscopic property changes the morphology of the cured resin should be examined. Some aspects of the inhomogeneity in cured epoxy resins have been reviewed by Dušek (1986). However with these cautionary remarks, further discussion in this chapter will be based on the premise that the inhomogeneities that may be formed during cure are unimportant.

3.2.1 Branching theory

From study of condensation reactions with monomers which have functionalities greater than two it became clear that there was a critical point, the gel point. With extents of reaction less than that required for gelation the sample was soluble in suitable solvents and had measurable viscosities. Beyond the gel point the sample was a visco-elastic rubbery solid and would only swell with imbibed solvent. Carothers (1936) suggested that the critical extent of reaction for gelation is $p_g = 2/f_{av}$ where p_g is the extent of reaction at the gel point and f_{av} is the average functionality with

$$f_{av} = \frac{\Sigma N_i f_i}{\Sigma N_i}$$

where N_i is the number of molecules with a functionality f_i and the summation is over all monomers in the system. For kinetic studies the extent of reaction is defined as $X_e = [E_o - E_{(t_c)}]/E_o$ but for branching theory it is given defined by p_e = the probability that a reactive group has reacted. Also, α is the branching coefficient. In this chapter usual terminology has been used.

Unfortunately the Carother's equation is only approximate and predicts higher extents of reaction at the gel point than are observed experimentally (Pinner, 1956). An outline of gel point prediction is given in standard textbooks on polymer chemistry (see for example Elias, 1976; Billmeyer, 1984; Labana, 1986).

The theory of gelation was developed systematically by Flory who gives an account of these developments in his standard text (Flory, 1953). The problem is illustrated in Figure 3.1 where the spatial configuration of highly branched molecules is represented. It is necessary to predict the extent of reaction required for the formation of an infinite molecule. A diepoxide molecule $f = 2$ can be cured with an f-functional hardener and for simplicity the reactivities of these hardener functional groups are regarded as equal. Since the formation of an infinite molecule occurs at the gel point any repeat unit on such a molecule must be connected to another unit via a branched unit, in this case the hardener. The probability that this occurs can be represented by a branching coefficient α which is defined by

α = probability that a functional group on a branch unit is connected to another branch unit

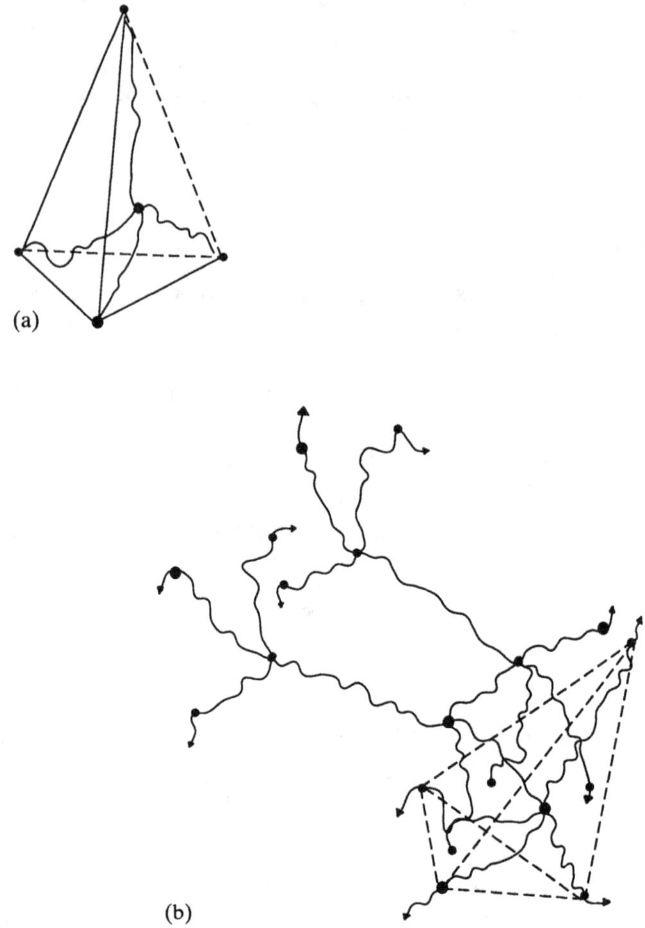

Figure 3.1 The spatial configuration of highly-branched moledules. (a) Tetra functional junction point. (b) Small portion of a three-dimensional network with tetra functional junction points. → indicates that the junction point is connected to other network chains – for a tetra functional junction point there will be three such connections.

The formation of branched molecules can be represented as a branching tree (Figure 3.2). A randomly chosen hardener branched unit with functionality f is placed at the base. With a branching coefficient α there will be αf reacted hardener groups in Rank 1, that is the number of reacted branches in Rank 1 is $B_1 = \alpha f$. In Rank 2 one of the functional groups has been consumed connecting it to Rank 1 so that the number of new reacted branches is $B_2 = \alpha(f - 1) . \alpha f = \alpha^2 f(f - 1)$. Generally:

$$B_n = \alpha^n . f . (f - 1)^{n - 1}$$
$$= \alpha f[\alpha(f -)]^{n - 1} \quad (3.1)$$

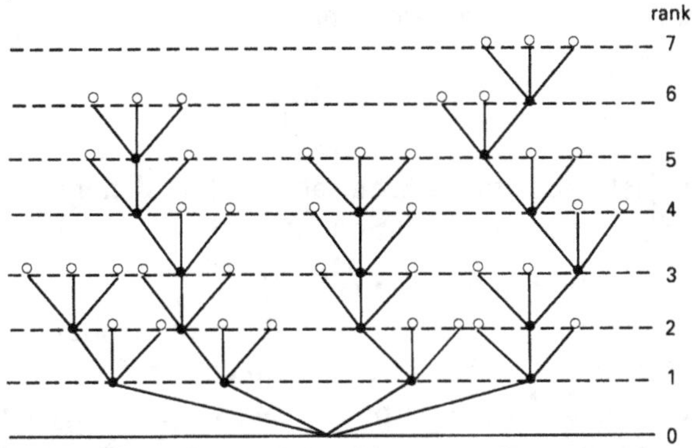

Figure 3.2 Formation of branched molecules. (•), Reacted functional group; (○), reactive functional group. For alternative representations and discussion of molecular trees and networks see Gordon and Ross-Murphy (1975).

when $[\alpha(f-1)] < 1$ the number of branches in the nth rank decreases as n increases, $B_n \to 0$ as $n \to \infty$. However, when $[\alpha(f-1)] > 1$ then $B_n > 0$ for all n. Thus, there is finite probability that an infinitely large molecule will be formed, i.e.

$$\alpha_c = \frac{1}{f-1} \tag{3.2}$$

where α_c is the critical branching coefficient required for incipient network formation.

It should be noted that this prediction of the critical condition for gelation excludes the possibility of ring formation and hence the observed gel point may occur at a higher extent of reaction.

The relationship between α and the extent of reaction for a simple case may be obtained by the following argument. The calculation of α is equivalent to the probability that a branched unit is part of the structure

$$\leftarrow \text{HE} \longrightarrow \text{EH} \rightarrow$$

where HE and EH are reacted epoxy and hardener groups. This structure must be present when a chosen reacted hardener group is part of a link joining two branched units. This chosen hardener group can only be part of such a structure if it has reacted and is also attached to an epoxy group which has also reacted. Thus

$$\alpha = p_H \cdot p_E \tag{3.3}$$

where p_H = probability that a hardener group has reacted

$$p_H = \frac{[\text{H}]_o - [\text{H}]_t}{[\text{H}]_o} \tag{3.4}$$

p_E = probability that an epoxy group has reacted

$$p_E = \frac{[E]_o - [E]_t}{[E]_o} \tag{3.5}$$

where $[H]_o$ and $[E]_o$ are initial concentrations of hardener and epoxy groups and $[H]_t$ and $[E]_t$ are the concentrations at time t.

It should be noted that equal reactivity of hardener functional groups must apply for this definition of p_H to be exact. This is often not the case as will be discussed in section 3.4 (Durand and Bruneau, 1983). When the initial concentrations of hardener and epoxy groups are exactly equal and only reactions between such groups occur $p_H = p_E = p$ and $\alpha = p^2$. The use of equation 3.1 to calculate B_n, the number of branches in Rank n is only valid when all reactions are intermolecular. Thus, it becomes more and more inaccurate after gelation. However, it does illustrate the small degree of branching for low extents of reaction. Also, the number-average molecular weight only increases slowly with extent of reaction and is finite at the gel point (Figure 3.3). The number of average molecular weight can be readily calculated since it is the total mass divided by the number of molecules present.

$$\bar{M}_n = \frac{\text{Mass}}{N_t} = \frac{\text{Mass}}{N_o - N_b} \tag{3.6}$$

where N_o is the number of molecules present initially and N_b is the number of bonds formed, since for each bond formed the number of molecules present decreases by one. The general case is treated by Macosko and Miller (1976) who give relations for calculation of number, weight and Z average molecular weights as a function of the extent of reaction. They have given a general consistent treatment of non-linear polymerizations which is now frequently cited in papers dealing with the cure of epoxy resins. Similarities with the classical treatments of Flory and Stockmayer are pointed out so that the relationship of their approach to that of others can be appreciated. In Figure 3.3 the weight average molecular weight is also plotted and it can be seen that it becomes infinite for $p > p_c$ at the gel point. For $p < p_c$ the weight fraction of sol is unity, $w_s = 1$. For extents of reaction greater than p_c the weight fraction of sol, w_s, decreases monotomically and the weight fraction of gel, w_g, increases correspondingly ($w_s + w_g = 1$), as shown in Figure 3.3. Network formation, in terms of $P(X_3)$ and $P(X_4)$ is also shown in Figure 3.3, the former has maximum for $p \sim 0.85$, and $P(X_4)$ approaches unity only when $p \to 1$. Thus, for the network to be fully 'cross-linked' cure has to be essentially complete.

A general treatment of gelation was given by Case (1957) and re-examined by Tanaka and Kakiuchi (1965) who assessed its applicability to the cure of epoxy resins with acid anhydrides. Other work on the application

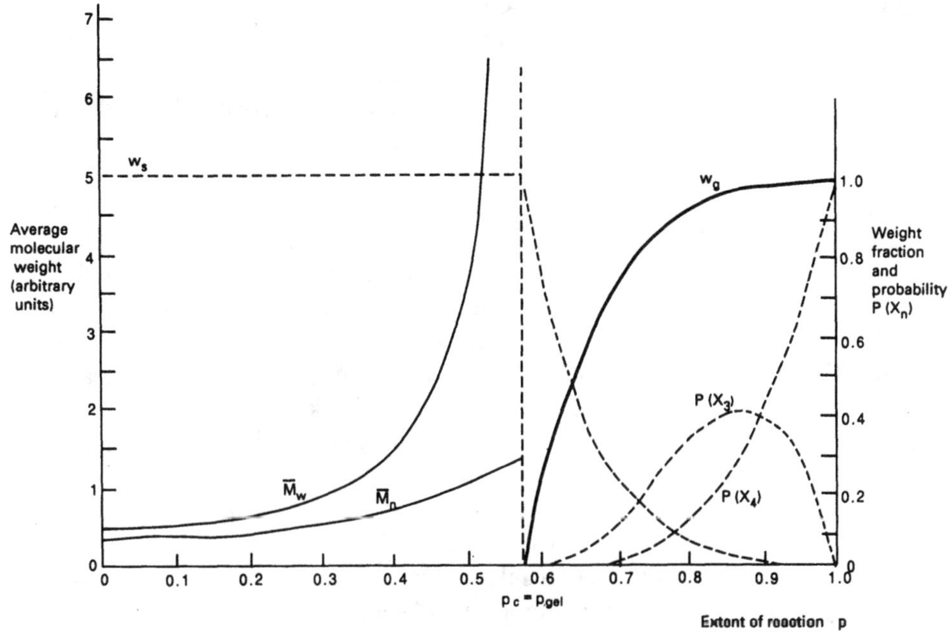

Figure 3.3 Effect of the extent of reaction on molecular properties for a stoichiometric mixture of a diepoxide with a tetra function hardener, e.g. 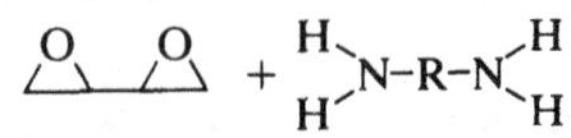 When the amine hydrogens have equal reactivity (but see section 3.4) and epoxy groups only react with amine hydrogens. (---), weight fraction of sol (w_s); (——) weight fraction of gel (w_g); $w_s + w_g = 1$. $\overline{M}_n$ = number-average molecular weight = $(M_A + 2M_E)/(1 + 2 - 4p)$, $\overline{M}_w$ = weight-average molecular weight. For BADGE + DDM initially $\overline{M}_n$ (p = 0) = 261 and at the gel point $p_c = p_{gel} = 0.58$ (ideally) $\overline{M}_n(p_c) = \overline{M}_n(0)/(1 - \tfrac{4}{3}(p_c)) = 4.4\ \overline{M}_n(0) \simeq 1150$ (see Macosko and Miller, 1976).

of branching theory to the cure of epoxy resins has been discussed by Tanaka and Mika (1973) who annotate research published by Japanese workers. A more general review is that of Dušek (1986) who shows that it is necessary to consider the kinetics of the epoxy reactions to model the curing process when there are substitution effects. The statistical theory yields only approximate results due to neglect of stochastic correlations when a substitution effect is important. The importance of the gelation and network formation has led to the development of several different mathematical models. Gordon has published a large number of papers using graph theory to model the structure of networks, and a good starting point is the paper by Gordon and Ross-Murphy (1975). The sol-gel transition is a critical phenomenon and its relationship to other critical processes is outlined by Stanley (1986). Ziman (1979) considers macromolecular disorder and deals with gelation as part of a general discussion of models of homogeneously disordered systems. Scaling concepts have been used to predict the sol-gel transition by de Gennes (1979) who introduces the percolation model. The origin of percolation theory (Grimmett, 1989) is attributed to Flory (1941). His theory of gelation which was referred to earlier is well worth reading for an analysis of the basic concepts that any model must satisfy, although it is essential that a refined theory should allow for the formation of loops due to intramolecular reactions, especially for accurate prediction of the gel point. Staufer (1985) has written an introduction to percolation theory and also reviewed its application to gelation (Staufer *et al.*, 1982). The bond percolation model is applied to sites or monomers which are linked together on a d-dimensional lattice by random bonding with a probability p. At a critical bond probability p_c an infinite cluster appears. Thus $p = p_c$ is the critical condition for gelation. For $p > p_c$ the largest cluster became attached to the infinite molecules to form a network, whilst the sol fraction w_s and the average size of the sol molecules decreases as $p \rightarrow 1$. Near the critical condition as $p \rightarrow p_c$, the properties of the system diverge as powers of $\varepsilon = |p - p_c|$. For instance the viscosity is related to p_c by $\eta \sim \varepsilon^{-k}$ as $p \rightarrow p_c$ and for $p > p_c$ the elastic shear modulus increases, $G \sim \varepsilon^z$. The experimental determination of a gel point is not as facile as this discussion of a mathematical discontinuity might imply. Although for $p < p_c$ (or $\alpha < \alpha_c$), $w_g = 0$, the extent of reaction is continuous through the gel point.

Measurement of a sol/gel ratio is inaccurate for p only slightly greater than p_c and the increase in viscosity is not satisfactory, even when it is due to an increase in molecular complexity and not due to $\Delta T = (T_c - T_g)$ becoming small. A method that does yield a good estimate of the gel point is the change in slope of a curve of log G'' vs t_c where t_c is the cure time and G'' is the imaginary part of the complex shear modulus ($G^* = G' + iG''$). This method was proposed by Harran and Laudourd (1986) and has been confirmed by Chen and Ellis (1992a) for log E'' vs t_c. Recently, percolation theory has been applied to account for the changes in the visco-elastic properties of an epoxy resin during cure. Harran and co-workers (Serrano *et al.*, 1989, 1990) determined a relationship between the viscosity of the curing resin and the extent of reaction X,

$$\eta = A(1 - X/X_c)^{-\nu} \tag{3.7}$$

where X_c is extent of reaction at the percolation threshold, that is the gel point, A is a constant and ν is the critical percolation exponent. X_c is treated as an empirical parameter which is determined by a curve fitting procedure. For cure of tetraglycidyldiaminodiphenylmethane with diaminodiphenyl sulphone (TGDDM-DDS) they found that X_c and ν were temperature-dependent and concluded that this was due to a change in reaction mechanism with cure temperature. This could be due to a temperature dependence of the relative rates of reaction of primary and secondary amines with epoxy groups (see section 3.4). They exclude the possibility that in their studies the increase in viscosity was due to the onset of vitrification, that is $\Delta T = (T_c - T_g)$ becoming small. However, it is essential that it is established that gelation occurs before vitrification for their interpretation to be generally applicable.

A more detailed examination of the change in visco-elastic properties of epoxy resins during cure is that of Adolf and coworkers (Martin, 1988; Adolf and Martin, 1990; Adolf *et al.*, 1990) who also used a percolation model and determined the critical viscosity exponent.

3.2.2 The glass transition

During cure as well as gelation the other critical phenomenon is vitrification, as illustrated in the TTT diagram (chapter 1, Figure 1.2). The glass transition in polymers has been discussed extensively and is treated in textbooks (e.g. Sperling, 1986; Ward, 1983). The review of Shen and Eisenberg (1966) is still of value. There is a universality of glass transition phenomena for all glass-forming systems, such as low molecular weight organic molecules, organic polymers or inorganic glasses, such as soda-lime-silica and many others (Scherer, 1986; Rekhson, 1986; Elliott, 1990).

The definition of a glass transition temperature T_g is related to its method of measurement and will also depend on the thermal history of the sample. For instance in determination of the volumetric glass transition temperature a sample is usually cooled from above T_g whilst the differential scanning calorimetric method (DSC) involves increasing the temperature of the sample at a prescribed rate from $T < T_g$, although other thermal histories may be prescribed. The previous thermal history of the sample and rate of cooling or heating affects the measured glass transition temperature. For a curing thermosetting resin, the decrease in molecular mobility with increasing extent of reaction causes the glass transition temperature of the sample to increase. Thus, it is with progress of the chemical reactions that the sample vitrifies and this process will be affected by the cure temperature, T_c, and the rate of cure which depends on the reactivity of the hardener with the resin.

Differential scanning calorimetry is a common method for the measurement of the glass transition temperature of curing epoxy resins, since such measurements can be combined with studies of the rate of cure (see section 3.3). However, many different methods may be used to measure glass transition temperatures (Sperling, 1986; Rabek, 1980a). Also, different methods of defining T_g have been used (Figure 3.4), such as

1. The temperature of the maximum in the mechanical tan δ peak.
2. The point of inflexion of the log E' vs T curve or the 'mid' point of the transition where E' is the storage modulus.
3. An intercept method.

With these alternative definitions there can be large differences between the

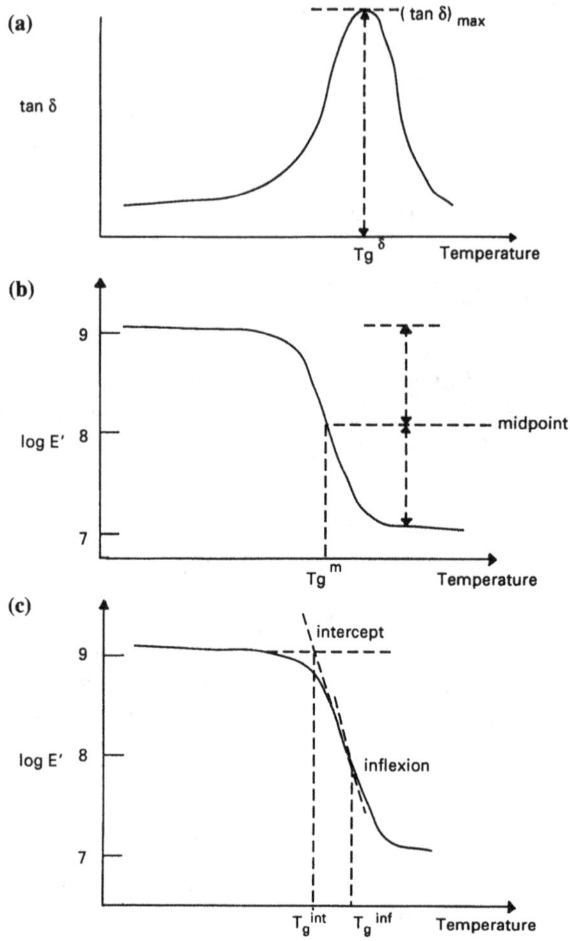

Figure 3.4 Definitions of the glass transition temperature, T_g. Measurement of dynamic mechanical properties using DMTA with standard rate of temperature rise. The differences between these estimates of T_g depend on the 'sharpness' of the transition. For example, BADGE type resin + DDM: cured short time (½ hr) at 100°C $T_g^{\delta} - T_g^{\mathrm{int}} \simeq 40$°C; post cured for longer times at 180°C $T_g^{\delta} - T_g^{\mathrm{int}} \simeq 10$°C (Bell *et al.*, 1992). The 'sharpness' of the transition increases with extent of cure. T_g^{δ} is the glass transition determined from the temperature of the $(\tan \delta)_{max}$ (method a). T_g^{int} is the glass transition determined from the temperature of the intercept (method c).

estimates of glass transition temperatures. These differences are greater the lower the extent of cure. Also, it should be appreciated that any method that involves increasing the temperature of the sample may allow cure reactions to proceed, thus the glass transition temperature increases whilst it is being measured. With care it is possible to obtain self-consistent sets of data, but comparisons between different sets must be treated with caution because of the range of operational parameters involved.

Prior to gelation the viscosity of the curing resin increases due to both an increase in molecular weight and also an increase in its glass transition temperature. The viscosity of glass-forming systems has a temperature dependence which is not Arrhenius and for polymeric materials it is often possible to 'fit' a set of viscosity measurements over a range of temperatures to the Williams, Landel and Ferry equation (see for example, Sperling, 1986 or Ward, 1983).

$$\log(\eta/\eta T_g) = \frac{-C_1(T - T_g)}{C_2 + (T - T_g)} \tag{3.8}$$

where the 'Universal' constants are $C_1 = 17.44$ and $C_2 = 51.6$ or alternatively $C_1 = 101.6$ when the reference temperature $T_s = T_g + 50°C$ is used in place of T_g. With an increase in the extent of cure T_g increases, ΔT ($= T_c - T_g$) decreases and the viscosity of the sample at T_c also increases. Similar non-Arrhenius viscosity temperature dependence has also been known for a long time for inorganic glasses. The equivalent temperature dependence is represented by the Tammann–Vogel–Fulcher equation (TVF), (Scherer, 1986; Jha and Parker, 1989)

$$\ln \eta = A + \frac{B}{T - T_o} \tag{3.9}$$

where $T_o < T_g$ may be regarded as an 'ideal' glass transition temperature but is often treated as an arbitrary constant.

Further discussion of the complexities of the glass transition would be out of place here. However, Stutz and coworkers (1990) have presented a generalized theory for cross-linked and uncrossed-linked polymers.

3.3 Techniques for monitoring cure

3.3.1 Monitoring cure

During the early stages of cure it is in principle possible to determine the concentration of epoxy or hardener reactive groups by 'wet' analysis or alternatively chromatographic methods, similar to those discussed for the characterization of epoxy prepolymers (chapter 1, section 1.4). However, there may be difficulties because of interference due to the presence of different types of reactive groups. When cure has reached the gel stage it is no longer possible to use methods which involve dissolution of the whole sample. With finely powdered resin, reactants diffuse only slowly into particles with a diameter of *c.* 5 μm (Bell, 1970a) and assay of finely comminuted resin is not a common method for monitoring cure. However, since cure involves the formation of a network it is not sufficient just to determine the extent of chemical reaction but changes in the physical and mechanical properties of the resin with cure treatment, time and temperature (t_c, T_c) are also important. The methods that are used to monitor cure as functions of T_c and t_c may be classified as follows.

1. Direct assay of the concentration of reaction groups present, usually the epoxy concentration.
2. Indirect estimation of the extent of chemical reaction.
3. Measurement of the changes in physical and mechanical properties.

For a full interpretation of the changes occurring during cure it is necessary to monitor both the concentration of reactive groups *and* other properties (Verchere *et al.*, 1990; Adolf *et al.*, 1990).

To obtain data amenable to a kinetic analysis of the rates of the competing consecutive reactions that occur during cure, the concentration of the reactive groups must be assayed as a function of cure time, t_c, at a specific isothermal cure temperature T_c. In some cases non-isothermal data have been analyzed but they are not so reliable as isothermal data. However, it should be noted that collection of isothermal rate data may involve considerable experimental difficulty. For instance it is not possible for samples to reach the cure temperature instantaneously, heat has to be supplied to the sample to raise its temperature, i.e. $T_{ambient} \rightarrow T_c$. Also, the resin is curing so that there may be temperature gradient across the sample and subsequently the heat of reaction may cause its centre to become much higher than the nominal cure temperature T_c. This may cause sampling problems since the extent of reaction may be inhomogeneous.

3.3.2 *Direct assay of the concentration of reactive groups*

Some of the methods that can be used for the assay of the concentrations of epoxy and other groups may be divided into two groups:

1. Wet analysis or chromatographic methods which involve dissolution of the partially cured resin.
2. Assay without dissolution where the whole sample is assayed using methods such as infrared spectroscopy.

As noted the former methods are only applicable prior to gelation, whilst the others can be used to the limit of their sensitivity. However, it should be noted that the sensitivity of both groups decreases as cure time increases. It is usually difficult to assay the extent of reaction for the 'final' few percent of the reactive groups. It is easier to detect the presence of residual epoxy groups but not possible to assay their concentration accurately. Thus, these methods are insensitive for the 'final' stages of cure when there are often very significant changes in the properties of the cured resins. Measurement of the glass transition temperature which has been correlated with an estimate of the extent of reaction has been proposed for monitoring cure since it is sensitive to the processes occurring in the final stages (Wisanrakkit and Gillham, 1990).

The 'wet' methods of analysis discussed for the characterization of uncured resins (chapter 1, section 1.4) have been adopted to the assay of reactive groups still present in partially-cured resins. Problems due to interference must be eliminated to obtain reliable results (Steinmann, 1989) and often more than one method of assay is necessary.

Chromatographic methods have been discussed in detail by Meston and Morris (1984) and both gel permeation chromatography and nuclear magnetic (NMR) spectroscopy have been used to determine the degree of branching in a partially-cured epoxy resin (Rogers, 1972). For partially soluble resins it is possible to determine their high resolution NMR spectra, usually proton chemical shifts, but the ^{13}C spectra can also be measured. For example, Sojka and Moniz (1976) have tabulated ^{13}C chemical shifts for DGEBA and the products of its reaction with piperidine. A general outline of the theory and experimental methods of NMR spectroscopy with its application to polymeric materials is given by Rabek (1980b), with also a description of 'magic angle' spinning, a technique which enables the chemical shifts in solid polymers to be measured. The application of this technique to the study of epoxy resins has been discussed in detail by Mertzel and Koenig (1986).

The integrated intensity of an infrared absorption band is directly related to the concentration of the absorbing species and two spectral regions have been used for the assay of epoxy groups. In the 4000–7000 cm^{-1} (2.5–15 μm) region there is an absorption band at 915 cm^{-1} (10.9 μm) assigned to the epoxy ring and in the overtone region 16 000–4000 (0.6 to 2.5 μm) the band at 4535 cm^{-1} (2.205 μm) has been used for the assay of epoxy group concentration. Listings of the infrared absorption spectra of uncured epoxy resins are given by Lee and Neville (1967) and Hadad (1988a) outlines the application of infrared spectroscopy to the analysis of epoxy resins. Rabek (1980c) gives a review of the theory of infrared spectroscopy and also the experimental techniques available for its application to the study of polymeric materials.

Quantitative analysis with a dispersive infrared spectrometer for the 4000–700 cm^{-1} region is now considered to be insufficiently accurate for the assay of epoxy groups due to the presence of adjacent absorption bands to the epoxy band and also the insensitivity of the method especially when the concentration of epoxy groups is low. Some of these problems for the measurement of 915 cm^{-1} epoxy band have been resolved by the application of **F**ourier-**T**ransform **I**nfra**R**ed spectroscopy (FT-IR) which has been reviewed by Mertzel and Koenig (1986).

Measurement of the near infrared absorption (16 000–4000 cm^{-1}) offers advantages since only strong bands have significant absorption in the overtone region and hence the spectra are less complicated so that problems due to the presence of overlapping bands are avoided. Also, the sample path length is longer, 1–10 mm, and hence sampling is more representative and it is possible to use a glass cell to confine a liquid sample. Goddu and Delker (1958) studied the overtone region and Dannenberg (1963) extended their work to establish calibrations which enable accurate assay of epoxy and amine group concentrations during cure. Recently (Ellis *et al.*, 1992) the method has been extended by the design of transparent soda-lime–silica glass cells so that the extent of cure could be studied from initiation until the epoxy absorption becomes too low to measure.

3.3.3 Thermal analysis

The basis for the application of differential scanning calorimetric (DSC) methods depends on the assumption that the measured thermograms can be directly related to the extent of chemical reaction, with usually a simple linear relationship between thermal response, e.g. exothermic heat evolved, and the consumption of epoxy groups. The technique is facile and requires only small samples and allows a determination of the glass transition temperature. An outline of DSC methods together with a brief discussion of sampling techniques with an extensive bibliography is given by Rabek (1980d). Prime (1981) discusses the detailed experimental procedures that are required for the accurate application of DSC methods for the monitoring of the cure of thermosetting resins, including epoxy resins which are also dealt with by Barton (1985) and Hadad (1988b). Fava (1968) has discussed methods of analysing DSC thermograms.

Recently, in a study of the cure of DGEBA (Dow-DER 332) with a stoichiometric concentration of trimethylene glycol di-*p*-aminobenzoate TMAB (Polaroid Corp. Polarcure 740M) Wisanrakkit and Gillham (1990) estimated the fractional conversion from

$$X = 1 - \frac{\Delta H_t}{\Delta H_T}$$

where ΔH_t and ΔH_T are the amounts of exothermic heat evolved in the time internal $t = 0$ to t_c and the total heat evolved. The method involved integrating the areas under exothermic peaks which requires the specification of a procedure for 'drawing' a baseline, which has been discussed by Barton (1985) and Prime (1981) cites a case where a dynamic DSC scan does not return to the baseline. However, Wisanrakkit and Gillham (1990) established a good correlation between the extent of reaction and the glass transition temperature of the curing resin samples.

Barton (1985) discussed, with reference to most of the early work, non-isothermal reaction kinetics when the temperature of cure is increased at a heating rate, $\beta = dT/dt$. Many mathematical approximations (Flynn and

Wall, 1966) have been applied to analyse the data, and some have been re-examined by Keenan (1987) who evaluated the errors involved in estimating activation energies for the curing of two epoxy resin adhesives. A recent suggestion for the transformation of dynamic DSC measurements into isothermal parameters for the cure of thermosetting resins is that of Khabenko and Dolmatov (1990).

3.3.4 Rheological changes during cure

During cure, an epoxy resin changes from a viscous liquid to a rigid glassy solid, as illustrated in chapter 1, Figure 1.1. Concomitant with the changes in structure, there are very large changes in its rheological properties, which are shown schematically in Figure 3.5. It can be appreciated that at $t_c = 0$ the real part of the complex shear modulus is zero, i.e. $G' = 0$, but will increase to about 3 GPa as cure progresses and essentially all of the epoxy groups are consumed. Ideally G^* and its components G' and G'' (or E^*, E' and E'') should be monitored throughout the cure process. However, it is difficult to design instruments which are capable of responding to the large changes in $G^* = f(T_c, t_c)$. Analogous measurements of the complex dielectric constant ε^* and its components as functions of both frequency and t_c over a range of cure temperatures, T_c, are relatively facile and are discussed in detail by Johari in chapter 7. Also, it is possible to monitor cure *in situ* using microdielectrometry sensors (Bidstrup and Sentura, 1989), but interpretation will involve an analysis such as that given in chapter 7 (see also Senturia and Sheppard, 1986).

Although initially the rate of reaction may be high, the size of the molecules increases slowly, as was discussed previously. Thus, the early stages of cure are often monitored by measurement of a shear viscosity. One of the most convenient instruments is a cone and plate viscometer (Whorlow, 1980), since the sample size is small and temperature control is good. Evaporation of a volatile hardener at high cure temperature may alter the epoxy/hardener stoichiometry and thus cure kinetics will be inaccurate. However, an advantage of this instrument is that the rate of shear is essential constant for the whole sample, and it is possible to test for the onset of non-Newtonian behaviour (Chen and Ellis, 1992a). Macosko (1985) gives an outline of the application of rheological measurements to monitor the cure of thermosetting resins including epoxies and the very large increase in viscosity at the gel point is discussed in detail by Malkin and Kulichikhin (1991). A fuller understanding of cure processes is attained by measuring other property changes as well as the shear viscosity. For instance Verchere *et al.* (1990) measured the shear viscosity with a concentric cylinder viscometer and also the extent of reaction by FT-IR spectra, size exclusion chromatography and DSC. Mijovic and Lee (1989) compared chemorheological models by combining viscosity and DSC measurements.

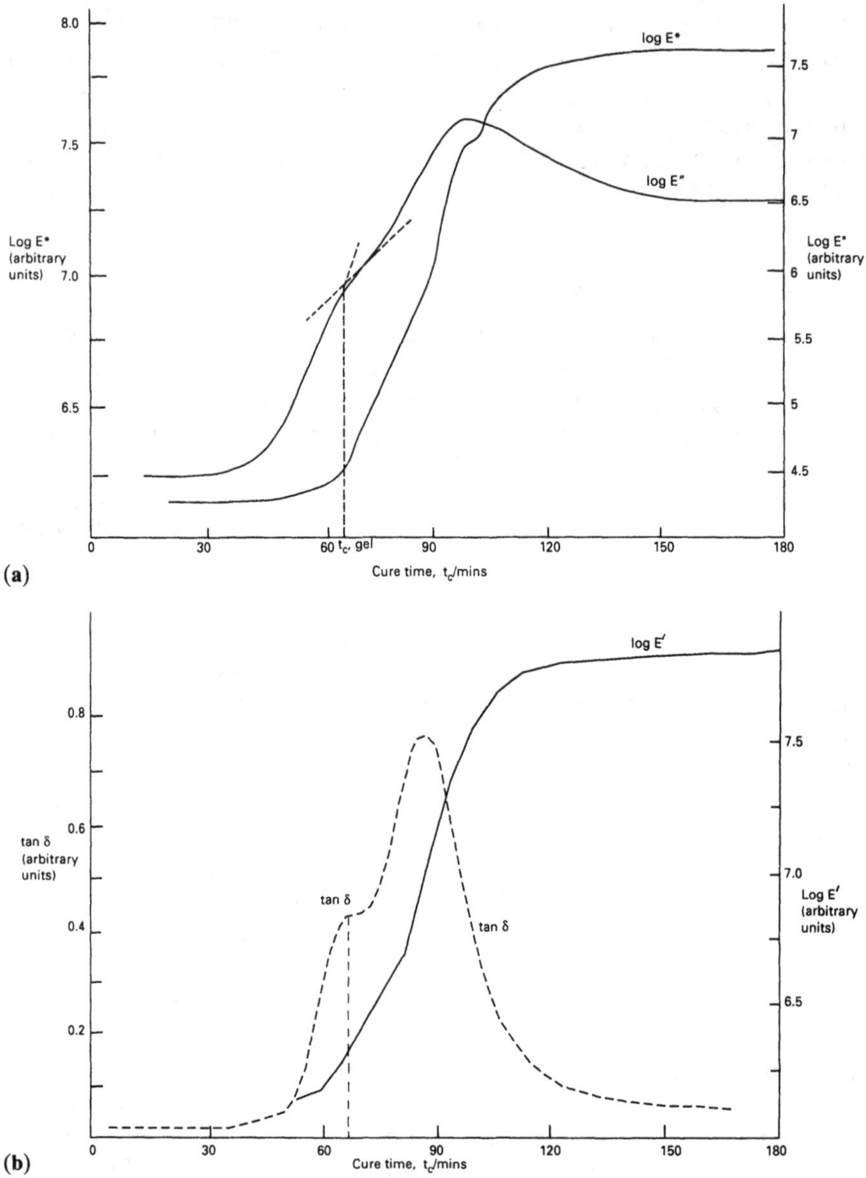

Figure 3.5 Dynamic mechanical properties as a function of cure time. (a) Log E^* and log E'' vs t_c. The gel time, $t_{c.gel}$, can be determined from the change in slope of log E'' vs t_c. (b) Log E' and tan δ vs t_c. The shoulder at lower cure times on the tan δ vs t_c curve is due to gelation.

The shear viscosity of a resin during cure may increase due to either gelation or the onset of vitrification, that is, $\Delta T = T_c - T_{g(t_c)}$ becomes small. Thus, the whole curve process cannot be monitored by measurement of a shear viscosity, it is necessary to measure a complex modulus and its components (e.g. G^*, G' and G''). Such measurements are difficult for a sample which is liquid initially and eventually solid (Figure 3.5). An early solution to this problem was the torsional braid technique introduced and developed by Gillham and coworkers (see Aronhime and Gillham (1986) for a review). An instrumented Torsional Brain Analyser is commercially available and there are several other instruments available for the measurement of the dynamic mechanical properties of polymers. The technique is known as dynamic mechanical thermal analysis, DMTA, or DMA (Flynn, 1989). The instruments are equipped with a computer and suitable programmes which enable measurement of the changes in the complex modulus and its components by factors of 10^3.

Adolf and Martin (1990) have measured the changes in G^* and its components as functions frequency and extent of reaction and interpreted their results using a percolation model. Harran and Landourd (1986) found that the slope of a plot of log G'' vs t_c decreases at the gel point, and Chen and Ellis (1992b) have confirmed that d (log E'')/dt_c decreases at the gel point. More recently Harran and coworkers (Serrano *et al.*, 1990) have applied a percolation model to analyse their complex shear modulus data, as mentioned previously.

3.4 Kinetics of cure

3.4.1 Introduction

The determination of the kinetics of cure of epoxy resins involves more than just measurement of the rates of reaction of epoxy groups with the hardener. This is because it is necessary to locate the gelation time, $t_{c(gel)}$, and also the rates of reaction become diffusion-controlled with the 'onset' of vitrification, i.e. when $\Delta T = T_c - T_{g(t_c)}$ becomes small. It is possible to inhibit reaction between hardener and epoxy groups by reducing the temperature so that the resin/hardener vitrifies. However, for elucidation of the mechanisms of the chemical reactions occurring during cure it is necessary to study the initial rates of reaction and also assay the extent of reaction between epoxy and reactive hardener groups. Usually the epoxy group conversion is determined but assay of the concentration of hardener reactive groups is also desirable, especially when their reactivity is unequal, as for example primary and secondary amine hydrogen atoms. Two approaches have been used:

1. Generalized empirical rate equations.
2. Rate equations derived from proposed chemical mechanisms.

Both of these approaches will be considered and also the effects of vitrification when the rates of the curing reactions become diffusion-controlled.

3.4.2 Amine-curing agents

3.4.2.1 Introduction. The importance of amine-curing agents is clear from chapter 2, and there have been many studies of the rates of cure of epoxy resins with amine hardeners. Barton (1985) tabulates 26 resin amine hardener systems, together with associated rate equations that have been determined using DSC methods.

The possible reactions that may be of importance in the cure of epoxy resins by polyamines are reaction of epoxy groups with primary amines, secondary amines, hydroxyl groups or other epoxy groups. Reaction of epoxy groups with primary amines

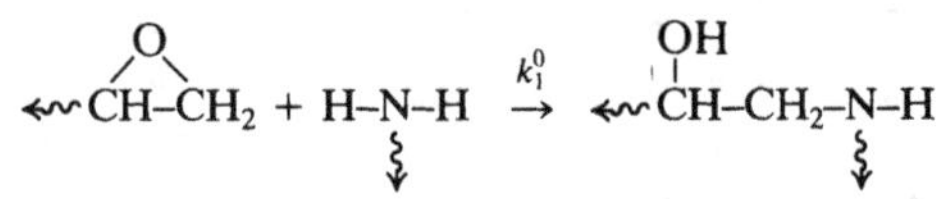

where k_1^0, is the rate constant for the uncatalysed reaction, but the reaction is often catalysed by hydroxyl or other groups and then the rate is faster, i.e. $k_1 > k_1^0$

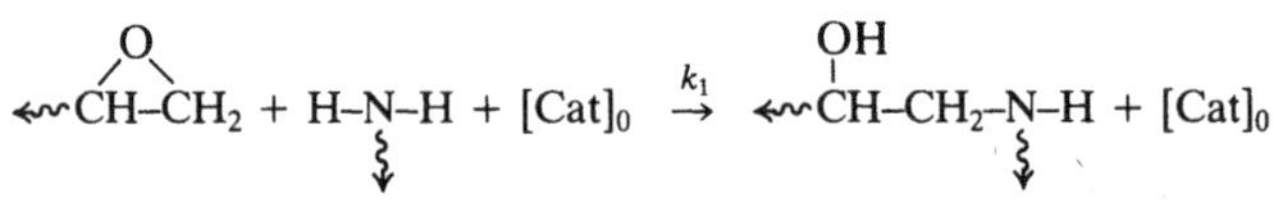

where $[\mathrm{Cat}]_0$ is the effective concentration of the adventitious catalyst present initially in either or both the resin and hardener, and its concentration is regarded as essentially constant. It will be noted that for each epoxy group that reacts a hydroxyl group is formed which can also catalyse the reaction between amine and epoxy groups. Thus the reaction will be autocatalytic and the rate equation will contain a rate constant k_{11} and concentration term, [P-OH], for the concentration of active hydroxyl groups formed, and its concentration will increase during cure. A similar set of equations can be written for the reactions of secondary amines with rate constants k_2^0, k_2 and k_{22}. Other reactions may also occur, such as that between epoxy and hydroxy groups:

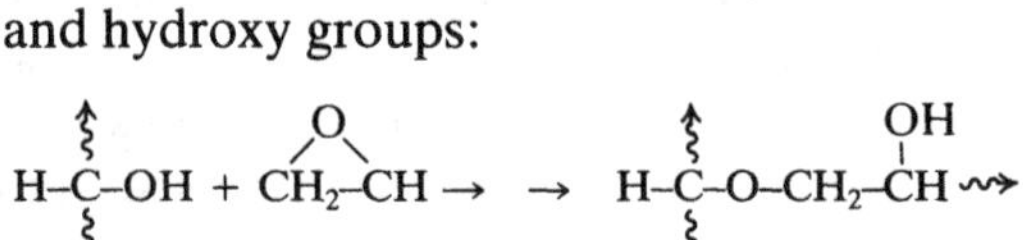

and also epoxy groups may react with each other when a suitable catalyst is present and the cure temperature is high enough.

3.4.2.2 Approximate kinetic analysis. A full kinetic scheme for the possible reactions that may occur during the cure of an epoxy resin with a primary amine would require the formulation and solution of a set of coupled differential equations. There would be a rate equation for each species that is involved in the curing process, with a term for every reaction of that species. Thus, the rate equation for the consumption of epoxy groups by reaction with primary amine hydrogen could be

$$-\frac{dE}{dt} = k_1^0\, EA_1 + k_1\, C_0\, EA_1 + k_{11}\, EA_1[\text{P-OH}]$$
$$+ k_2^0\, EA_2 + k_2\, C_0\, EA_2 + k_{22}\, EA_2[\text{P-OH}]$$
$$+ k_{OH}\, E[\text{P-OH}] + k_E\, E^2[\text{Cat}] \qquad (3.10)$$

where E, A_1, A_2, etc. represent concentrations and the ks are rate constants. The first set of terms is for the reaction of the primary amine, with concentration A_1, the second for the reaction of the secondary amine with concentration A_2, which of course will depend on the rate of consumption of primary amine. There are also terms for the reaction of epoxy groups with the hydroxyl group which is a product of the epoxy reactions and also for reaction with itself. In view of this complexity, various simplifications have been used and are often the basis for the empirical equations that have been proposed.

The kinetic treatment of Horie and co-workers (1970) has been the basis for many subsequent treatments of the kinetics of amine cure. They considered that only the catalysed reactions of epoxy groups were important, that is $k_1^0 = k_2^0 = k_{OH} = k_E = 0$. It is useful to define the relative rate, κ of the catalysed reaction of secondary to primary amine hydrogens, that is $\kappa = k_2/k_1$, and when $\kappa = 0.5$ the rate equation 3.10 can, using $x_e = E_0 - E$ and some algebraic manipulation be expressed as

$$\frac{dx_e}{dt} = (E_0 - x_e)\,(k_{11}\, x_e + k_1\, C_0)\,(A_1 + \kappa A_2) \qquad (3.11)$$

When κ is approximately a half then with use of $\kappa = 0.5 + \Delta\kappa$ and also using the stoichiometric relation for reacted epoxy and active hydrogen atoms $A_1 + A_2/2 = A_0 - x_e/2$, when $A_{20} = 0$, equation 3.11 can be re-arranged to

$$\frac{1}{(E_0 - x_e)\,(A_0 - x_e/2)} \cdot \frac{dx_e}{dt} = (k_1\, C_o + k_{11}\, x_e)\left[1 + \frac{2A_2\,.\,\Delta\kappa}{2A_1 + A_2}\right] \qquad (3.12)$$

A plot of the left hand side of equation 3.12, regarded as a reduced reaction rate, versus x_e will be initially linear when the conditions used to derive equation 3.12 apply. This is because the term $[2A_2\,.\,\Delta\kappa/(2A_1 + A_2)]$ is initially zero and small compared with unity up to maybe 50% consumption of epoxy groups. Horie *et al.* (1970) regarded their linear plots of reduced reaction rate versus conversion to be satisfactory for the initial stages of cure. Subsequently these plots have a maximum because the rate of conversion decreases as the curing reactions become more and more diffusion-controlled as $\Delta T = (T_c - T_g)$ becomes small.

Often the extent of reaction is expressed in terms of a normalised conversion $X_e = (E_0 - E)/E_0 = x_e/E_0$ and hence

$$\frac{dX_e}{dt} = \frac{1}{E_0}\left(\frac{dx_e}{dt}\right).$$

When the initial stoichiometry is exact and the concentration of active amine hydrogen atoms and epoxy groups are exactly equal, $A_0 = E_0/2$. Then using these substitutions equation 3.12 becomes:

$$\frac{1}{(1-X_e)^2}\cdot\frac{dX_e}{dt} = [K_1 + K_2 X_e]\,H_c \tag{3.13}$$

where $K_1 = \frac{1}{2}(k_1\, C_0\, E_0)$; $K_2 = \frac{1}{2}(k_{11}\, E_0^2)$ and $H_c = [1 + (2A_2 . \Delta\kappa / (2A_1 + A_2)]$. H_c is the Horie connection factor for the deviation of κ from 0.5. Thus, when a stoichiometric concentration of hardener is present a plot of $[1/(1 - X_e)^2(dX_e/dt)]$ vs X_e should be linear until H_c becomes significantly larger than unity and/or the rates of the curing reactions decrease because they become diffusion-controlled.

Another approach which has often been applied by others is derived from that of Sourour and Kamal (1976) who accept the postulates of Horie *et al.* (1980) except that they regard the relative reactivity of primary and secondary amines as equal, i.e. $\kappa = 1$ and the rate equation they propose is

$$\frac{dX_e}{dt} = \bar{K}_1(1-X_e)(R-X_e) + \bar{K}_2 X_e(1-X_e)(R-X_e) \tag{3.14}$$

where $\bar{K}_1 = \bar{k}_1\, C_0\, E_0$ and $\bar{K}_2 = \bar{k}_{11}\, E_0^2\, X_e$ are constants and $R = H_0/E_0$ the ratio of the initial concentrations of active hydrogen atoms to epoxy groups. Equation 3.14 can be derived for a catalysed reaction between active hydrogen atoms and epoxy groups and also a hydroxyl group autocatalysed reaction:

$$E + H + C_0 \xrightarrow{\bar{k}_1} A$$
$$E + H + [P\text{–}OH] \xrightarrow{\bar{k}_{11}} A$$

The rate equation is

$$-\frac{dE}{dt} = k_1\, C_0\, EH + k_{11}\, EH[P\text{–}OH] \tag{3.15}$$

which can be rearranged using $E = E_0\,(1 - X_e)$, $H = H_0 - (E_0 - E) = H_0 - E_0X_e = E_0(R - X_e)$ with $R = H_0/E_0$ and noting that $[P\text{–}OH] = E_0 - E = E_0X_e$. For a stoichiometric composition, $R = 1$, equation 3.14 can be put into the form

$$\frac{1}{(1-X_e)^2}\cdot\frac{dX_e}{dt} = k_1 + k_2 X_e \tag{3.16}$$

which has the same form as equation 3.13 but with $H_c = 1$ and the constants are different. Thus, the linearity of plots of the left hand sides of equations 3.13 or 3.16 vs X_e will not be sensitive to the value of κ if it is either 0.5 or 1.0. Thus, the value of the relative rates of reaction of primary and secondary amine hydrogen can only be found when the values of k_1, k_2 or $\bar{k}_1$ and $\bar{k}_2$ can be determined separately with sufficient accuracy.

A generalisation of equation 3.16 which has been used and regarded as 'quasi-theoretically' justified is:

$$\frac{dX_e}{dt}(\bar{K}_1 + \bar{K}_2 X_e^m)(1 - X_e)^m \tag{3.17}$$

However, it is difficult to relate this equation to the reaction steps that occur during an amine cure of epoxy resins. A simplified form of equation 3.17 with $K_1 = 0$ has also been used by Kamal and Sourour (1972, 1973):

$$\frac{dX_e}{dt} = KX_e^m(1 - X_e)^n \tag{3.18}$$

Macosko (1985) and Camargo *et al.* (1983) suggest that a general rate expression which is useful for fitting conversion data is

$$\frac{dX_e}{dt} = k(1 - X_e)^n \sum_{i=0}^{n} k_i X_e^i \tag{3.19}$$

where in some cases the second term can be regarded as an autocatalytic parameter. If the reactions are not autocatalytic $m = 0$ and hence equation 3.18 becomes

$$\frac{dX_e}{dt} = k(1 - X_e)^n \tag{3.20}$$

which is the rate equation for an *n*th order reaction (Wisanrakkit *et al.*, 1990).

Barton (1985) gives a list of the rate equations that have been fitted by about 20 different kinetic studies with the extent of cure measured by DSC methods. However, from the present discussion it can be appreciated that the kinetics of cure may be complicated, even before the rates of reaction become diffusion-controlled due to vitrification. The empirical equations yield only limited information on the mechanisms of the chemical reactions involved. However, they have utility by providing a rate function for the analysis of kinetic data, such as that obtained from DSC measurements. Thus they may be sufficient for specification of cure conditions for specific resin-hardener combinations. Also, they may be useful for quality control and the *in situ* monitoring of the cure process.

Mijovic and coworkers (Mijovic, 1990; Mijovic and Wang, 1989; Mijovic *et al.*, 1984) have been studying the cure of epoxy resins over a number of years and recently (Mijovic and Lee, 1989) used equation 3.17 with the 'overall' reaction order, $m + n = 2$, and solved it using a fourth order Rungge–Kutta numerical integration to obtain curves for the extent of reaction versus time of cure for several isothermal cure temperatures. Their system was DGEBA-cured with a mixture of 4,4 methylene-dianiline (MDA) and *m*-phenylene diamine (*m*-PDA) and they applied branching theory to model the increase in viscosity prior to the gel point.

Sheppard and Senturia (1986) as part of their programme of studies of the effects of cure on the dielectric properties of epoxy resins have also measured the extent of cure using DSC methods. They found that plots of

$$\frac{1}{(1-X_e)^2}\cdot\frac{dX_e}{dt} \text{ vs } X_e$$

(equation 3.16) were linear for the cure of DGEBA with DDS with cure temperatures in the range 410–460 K. They also compared the time required for 50% primary amine conversion estimated from the kinetic model and obtained satisfactory agreement with experimental determinations (see also chapter 7).

An even simpler rate equation is obtained when $m = 0$ in equation 3.18, an expression used by Acitelli and coworkers many years ago (1971) and recently Wisanrakkit and coworkers (1990) obtained a first order fit with $n = 1$. For cure of DGEBA (Epon 828) with PACM-20 (see chapter 2) plots of dX_e/dt vs $(1 - X_e)$ for cure at 61°C were linear. Keenan (1987) discusses in detail a method for determining the parameters κ_2, m and n with $\kappa_1 = 0$ when a simpler form of equation 3.17 is applicable. His method involved measurement of an exothermic peak and the procedure was satisfactory for representation of the cure of FM123-5 (American Cyanamid Co.), a nitrile modified-dicyandiamide-cured epoxy adhesive supported on a nylon carrier.

An interesting alternative approach is that of Pinsheng and Chune (1989) who apply the non-equilibrium thermodynamic fluctuation theory proposed by Hsich (1982). In this theory the change in a mechanical property is measured and the rate equation derived after some simplification is:

$$\frac{G_\infty - G(t)}{G_\infty - G_0} = \exp[-(t/\tau)^\beta]$$

where G_∞, $G(t)$ and G_0 are the final, time-dependent and initial values of the physical property, β is a constant which depends on the width of the relaxation spectrum and τ is the relaxation time. Pingsheng and Chune studied the cure of a commercial DGEBA resin E51 (Shanghai Resin Factory) with imidazole as the hardener and silica filler. They were able to model the change in the shear modulus as a function of both cure temperature and time.

3.4.2.3 Analysis of competitive consecutive reactions. The rate equation for the consumption of epoxy groups, equation 3.10, contains terms for the concentration of secondary amine and hence to specify the kinetics of cure, additional rate equations are required. When $k_1^0 = k_2^0 = 0$ the additional rate equations are

$$-\frac{dA_1}{dt} = k_1 A_1 EC_0 + k_{11} A_1 E[P\text{–}OH] \tag{3.21}$$

$$\frac{dA_2}{dt} = k_1 A_1 EC_0 + k_{11} A_1 E[P\text{–}OH] - k_2 A_2 EC_0 - k_{22} A_2 E[P\text{–}OH] \tag{3.22}$$

$$\frac{dA_3}{dt} = k_2 A_2 EC_0 + k_{22} A_2 E[P\text{–}OH] \tag{3.23}$$

and solution of these coupled equations is difficult, even when the autocatalytic terms are unimportant, which would be equivalent to the case discussed in detail by Frost and Pearson (1961) (see also Benson, 1960). However, it has been accepted, especially after the work of Smith (1961), that the reaction of amines with epoxy groups in epoxy resins is autocatalytic. An interesting approach to the solution of these coupled differential equations is that of Dobas *et al.* (1975). They regarded $k_1^0 = k_2^0 = 0 = k_1 = k_2 = k_{OH} = k_E$ so that equations 3.10 and 3.21–23 are reduced to

$$-\frac{dE}{dt} = k_{11} EA_1[P\text{–}OH] + k_{22} A_2 E[P\text{–}OH] \tag{3.24}$$

$$-\frac{dA_1}{dt} = k_{11} EA_1[P\text{–}OH] \tag{3.25}$$

$$\frac{dA_2}{dt} = k_{11} EA_1[P\text{–}OH] - k_{22} A_2 E[P\text{–}OH] \tag{3.26}$$

$$-\frac{dA_3}{dt} = k_{22} A_2 E[P\text{–}OH] \tag{3.27}$$

Thus, in this reaction scheme, only autocatalytic reactions are considered. Dobas *et al.* (1975) proceed by defining $Y = EA_1[P\text{–}OH]$ and $Z = EA_2[P\text{–}OH]$ so that from equation 3.25

$$k_{11a} = -\frac{dA_1}{dt} \cdot \frac{1}{Y} \tag{3.28}$$

and from equation 3.27

$$k_{22a} = -\frac{dA_3}{dt} \cdot \frac{1}{Z} \tag{3.29}$$

Other estimates of k_{12} can be obtained from three equations of the form

$$k_{11b} = \left[\frac{dA_2}{dt} - \frac{dE}{dt}\right] \cdot \frac{1}{2Y} \quad (3.30)$$

and a similar set of three equations for k_{22}

$$k_{22b} = -\left[\frac{dE}{dt} + \frac{dA_2}{dt}\right] \cdot \frac{1}{2Z} \quad (3.31)$$

Thus there are a total of eight equations which allow estimation of k_{11} and k_{22} from measured concentrations, E, A_1 and A_2 and also dE/dt, dA_1/dt and dA_2/dt. The concentrations were measured using near infrared spectroscopic assay and complete sets of all constants were determined for the reactions of seven diamines with *p*-tolylglycidyl ether at 80°C and 2,2-bis(4-(2,3-epoxypropyl) phenyl propane at several cure temperatures and related to the $pK_{a.1}$ values of the diamines.

It is possible to 'remove' times as a variable (Dušek *et al.*, 1975) by dividing equation 3.26 by 3.25 to obtain:

$$\frac{dA_2}{dA_1} = \frac{k_{22}}{k_{11}} \cdot \frac{A_2}{A_1} - 1 \quad (3.32)$$

which can be integrated using the initial condition that $A_1 = A_{10}$ when $A_2 = 0$, that is only primary amines are present initially. Charlesworth (1980) used this procedure to determine the value of κ, the ratio of the rate of reaction of primary and secondary amine hydrogens. He studied the reaction of phenyl glycidyl ether with amine at 100 and 150°C, $\kappa = 0.44$ at 100°C and $\kappa = 0.71$ at 150°C.

Williams and coworkers (Riccardi *et al.*, 1984; Riccardi and Williams, 1986a, 1986b) have published a number of studies of the reaction of amines with epoxy resins and in a recent paper (Verchère *et al.*, 1990) report a study of the reaction of a diamine with a monoepoxide, 4,4′diamino-3,3′ dimethyl dicyclohexylmethane (3DCM) and phenyl glycidyl ether. They distinguish the rates of reaction of a primary amine hydrogen on alternate amine groups in the diamine (see Figure 3.6). It should be noted that this hardener is 'hindered' (see chapter 2).

Thus they have a set of six coupled differential equations for the reaction of the four amine hydrogen atoms. These rate equations are formally autocatalytic so that for the reaction to start, they postulate that the concentration of hydroxyl groups is:

$$[OH] = C_0 + X_e E_0$$

where C_0 is the concentration of catalytic impurity. This is equivalent to regarding the rate constants for the two catalysed reactions of amine hydrogens with epoxide groups as essentially equal, that is, $k_1 = k_{12}$ and is equivalent to equation 3.25 but with [P–OH] replaced by [OH] =

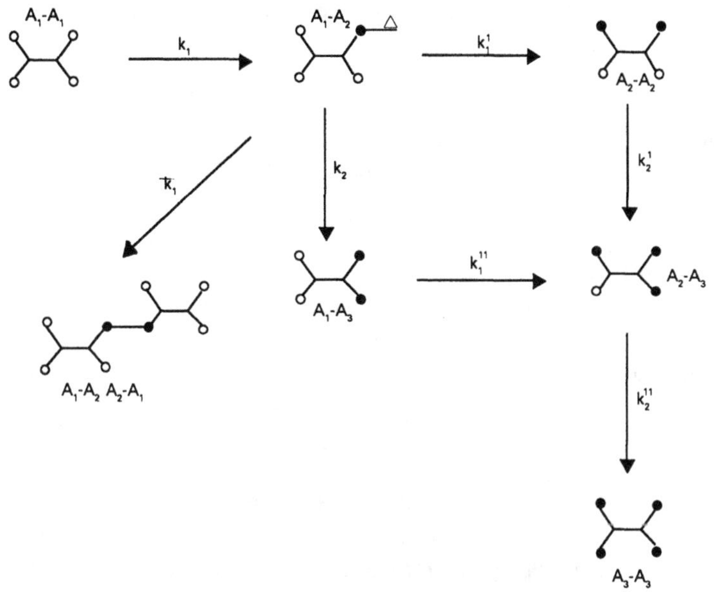

Figure 3.6 Reactions of diamines. A_1–A_1 unreacted primary diamine; A_1–A_2 unreacted primary amine and a secondary amine; A_3 reacted secondary amine, i.e. a tertiary amine. The epoxy prepolymer unit is omitted except for A_1–A_2. Verchère *et al.* (1990) measured the concentrations of A_2–A_2 plus A_1–A_3; A_2–A_3 and A_3–A_3 by size exclusion chromatography, and used a kinetic scheme with $k_1 = k_1'' = k_1' = k_1$ and $k_2'' = k_2' = k_2$.

$C_0 + X_e E_0$. The set of differential equations was solved by eliminating the dependence on time by division, as discussed previously, and then solved. The concentrations of PGE, $[A_2 - A_2]$ plus $[A_3 - A_1]$ and $[A_3 - A_2]$ plus $[A_3 - A_3]$ were measured as a function of reaction time at 50°C by size exclusion chromatography. They obtained agreement between experimental and calculated concentrations as a function of X_e with a value of $\kappa = 0.4$. Verchère and coworkers (1990) regarded their study of the reaction of 3DCM with PGE as a model for the cure of DGEBA with that hardener. They also treated this curing reaction as purely autocatalytic in the pre-gel stage, and measured the decrease in concentration of BADGE by SEC and were able to calculate $\bar{M}_n$ and $\bar{M}_w$ as a function of X_e, and compare experiment with branching theory. They found that vitrification does not affect the statistics of network formation.

In a study of the cure of BADGE with five aromatic amines, Grillet and coworkers (1989) regarded the predominant reaction mechanism to be non-catalytic for higher cure temperatures, that is $T_c > 100$°C. They also eliminated the time variable by dividing the set of differential equations by the one which contained only one concentration-dependent term, a similar procedure to that proposed by Dušek *et al.* (1975) and also used by Verchère (1990) referred to previously. From viscosity measurements and SEC they determined the reactivities of the diamines and found that they increased in the order:

$$4{,}4'\ \text{DDS} < 3{,}3'\ \text{DDS} < \text{BAPP} < \text{BAPS} < \text{DDM}$$

They also discuss methods for estimating the relative reactivities of secondary to primary amines, $\kappa = k_2/k_1$ and found that $\kappa < 1$, with $\kappa \sim 0.8$ to 0.9 for DDS diamines.

An interesting study that shows that the reaction of secondary amine hydrogens with an epoxy group is very much faster than either epoxy-hydroxyl or tertiary amine catalysed epoxy-epoxy reaction is that of Adolf and Martin (1990). They studied the reaction of Epon 828 with diethanolanine,

$$\text{H-N}\begin{matrix}\text{CH}_2\text{-CH}_2\text{-OH}\\ \text{CH}_2\text{-CH}_2\text{-OH}\end{matrix}$$

and the secondary amine reacts completely within 5 min at 90°C with about 1/5 of the epoxy groups to produce a secondary hydroxyl group. The reaction then proceeds by tertiary amine catalysed reaction of both the primary and secondary hydroxyl groups with equal probability. The gel point occurs after about 3 h at 90°C. Fully cured resins have a glass transition temperature of *c.* 70°C. This study confirms that in equation 3.10 $k_{OH} = k_E \simeq 0$ compared with the rates of reaction of primary and secondary amines with epoxy groups.

For the cure of BADGE with either TETA or diaminodiphenyl methane Chen and Ellis (1992a) have extended the approach given by Frost and Pearson (1961) so that both catalysis by impurities and also the product hydroxyl groups are included in the rate equations. For TETA it was found that the relative rates of reaction of secondary to primary amines increased with cure temperature, at 22°C $\kappa = 0.25$ and increased to $\kappa = 0.75$ at 50°C. Thus, the activation energy for the reaction of secondary amines was larger than that for primary.

3.4.3 Carboxylic acid anhydrides

Although acid anhydrides are used extensively as hardeners for epoxy resins (see chapter 2), studies of the kinetics of their curing reactions are far fewer than those for amine cures. Also, the kinetics appears to be more complicated and at the present time there is not a generally accepted rate equation for cure with these hardeners. The orders of the reaction that have been reported range from 0 to 4 and the statement of Mertzel and Koenig (1986) that 'a generally accepted reaction mechanism is not available' still applies.

Barton (1985) has reviewed the kinetic DSC studies of acid anhydride cures of epoxy resins, listing the kinetic parameters derived using an empirical rate equation

$$\frac{dX}{dt} = -\left.\frac{dC}{dt}\right|_{C_0} = k(1-X)^n \tag{3.33}$$

where C_0 is the initial concentration and k is an 'apparent' rate constant defined by $k = k_0\, C_0^{n-1}$, and has an Arrhenius temperature-dependence. The values of n in equation 3.33 ranged from 0.2 to 2 and the apparent activation energy from 56 to 159 kJ/mole, depending on the resin/hardener/catalyst combination studied. Although about half of the reactions are first order, $n = 1$, most of the others are fractional and it is not possible to relate them to a simple reaction mechanism.

Peyser and Bascom (1977) discuss the analysis of DSC data in some detail and found that for the cure of DGEBA with hexahydrophthalic anhydride with benzyldimethylamine as a catalyst, the cure was first order with an activation energy of 25 kcal/mole. Stevens and Richardson (1983) defined the extent of reaction, $P_{(t_c)}$, in terms of the time-dependence of the glass transition temperature.

$$P_{(t_c)} = \frac{T_{g(t_c)} - T_{g(0)}}{T_{g(\infty)} - T_{g(0)}} \tag{3.34}$$

and of course the use of such a parameter depends on the definition and experimental method used to determine $T_{g(0)}$, $T_{g(T_c)}$ and $T_{g(\infty)}$. Also, it is essential that the limited value $T_{g(\infty)}$ is determined unambiguously. They found that for cure of DGEBA type prepolymers, Ciba-Geigy CT200 and CY207, with phthalic anhydride and also a mixture of phthalic and tetrahydrophthalic anhydrides that the overall kinetics is neither first nor half-orders, but first order kinetics was satisfactory for 85–90% of the conversion.

Steinmann (1987, 1989) has considered many of the previous reports on anhydride cures and reports on her own work in which she studied the reactions of the several epoxy resins with hexahydrophthalic anhydride (HHPA) and tertiary amine catalysts.

3.4.4 *Diffusion control*

It is obvious that for any chemical reaction to occur the reactive centres must be in close proximity to each other and also be orientated so that a product is formed. Usually in liquids these processes may be regarded as relatively fast and their rate determines the observed reaction rate constant (North, 1964; Entelis and Tiger, 1976). For the cure of epoxy resins prior to vitrification or gelation the reactions occur in the 'liquid' state and the rate of cure is controlled by 'chemical kinetics'. However, when the viscosity increases the rate of cure then depends on diffusion of reactant to a reactive centre, that is the reactions are 'diffusion-controlled' (Eyring *et al.*, 1980). This is the factor which limits the extent of reaction at lower cure temperatures. The reaction between epoxy resin and hardener may be quenched by storing well below the glass transition temperature. Cure may be 'activated' by raising the temperature.

The onset of diffusion-controlled polymerizations and cure of thermosetting resins has been reviewed in considerable detail by Mita and Horie (1987). They discuss some epoxy resin systems and show that the effect of diffusion control of their rates of cure can be represented on Gillham's TTT diagram (Enns and Gillham, 1983; Aronhime and Gillham, 1986), see chapter 1, Figure 1.2. It is important to note that it may be difficult to determine the rate constants for the reaction of primary and secondary amines with epoxy groups without the onset of complicating processes, even when the product is not a covalently bonded cross-linked network. For instance, Rozenberg (1986) discusses the autoinhibition of the reaction between phenyl glycidyl ether with aniline. Even at longer reaction times the conversion is limited to about 0.8 due to effects of hydrogen bonding which reduces the rate of reorientation of the reaction product. This is a factor that may affect the ratio, $\kappa = k_2/k_1$, of the rate constants for the reaction of primary and secondary amines with epoxy groups in this and other systems.

Not only is it necessary for the mutual reorientation of potentially reactive groups to occur but a prior requirement is that they should diffuse together so that they are within a 'reaction zone' (Chiu *et al.*, 1983; Sharma and Soane, 1988). It is the diffusion coefficient of the smallest free molecule with a reactive group which will have the greatest effect on the overall rate of reaction. Most of the discussion of the factor affecting the onset of diffusion control of the rate of cure of epoxy resins has only considered the increase in the shear viscosity of the reacting mixture. However, it is a 'local molecular viscosity' which will determine the rate of diffusion of the smallest reactive molecule to a reactive site on an oligomeric molecule. The local viscosity has been determined for small molecules in some polymeric matrices. Guillet (1985) estimates that the microviscosity of polymeric systems is many orders of magnitude lower than their 'macroscopic' viscosity, that is their shear viscosity. He also observes 'that this fact has gone unrecognized in many studies of polymer reactions'.

At more advanced stages of cure, intramolecular reaction between epoxy and other reactive sites on the same molecule or in the gel become dominant. Thus, the rate of segmental reorientation dominates the cure process and has been discussed by Mita and Horie (1987) and also Rozenberg (1986) who outlines a model based on reptation theory (Larson, 1988). Rozenberg also discusses the 'topological reaction limit', since it is possible for potentially reactive groups to be isolated within the three-dimensional network. However, it appears that for a 'fully' cured resin, the concentrations of such unreacted groups is small, but because of the insensitivity of analytical methods, it is not possible to give a precise estimate of their concentration.

Dušek (1986) discusses diffusion control of the cure of epoxy resins in terms of a WLF temperature-dependent rate and concludes that branching theory can be applied to reactions occurring in the melt, rubbery and glassy states. In the vicinity of T_g, molecular mobility is a function of free volume (Dušek, 1985) so that log $k \propto \tau$, the segmental mobility (Lunak *et al.*, 1978) and epoxy curing reactions are quenched when $\Delta T \sim -30$ K (Dušek, 1985).

Others (Huguenin and Klein, 1985; Wisanrakkit and Gillham, 1990; Sandford and McCullough, 1990) have discussed diffusion-controlled curing reactions using free volume models coupled with the Rabinowitch theory (1937), a procedure which was also suggested by Havlicek and Dušek (1987). The overall rate constant, k_a, has both a temperature and extent of reaction dependence which can be formally represented by

$$\frac{1}{k_a(x, T_i)} = \frac{1}{k_T(T_c)} + \frac{1}{k_d(x, T_c)}$$

In the treatment of Wisanrakkit and Gillham (1990) k_T is the scaled rate constant with an Arrhenius temperature-dependence and k_d is the 'diffusion' rate constant which has a WLF type temperature-dependence. They give a discussion of a method for determining these two rate constants.

Matsuoka *et al.* (1989) have proposed a model based on a Kohlrausch–Williams–Watts (Williams and Watts, 1970) distribution of relaxation times, which is a function of the extent of cure. They applied their theory to the cross-linking of a commercial epoxy-novolak resin which is used for the encapsulation of integrated circuit chips. With a cure temperature of 140°C, the limiting glass transition temperature attained was 173°C, with very little increase in extent of cure for $1 < t_c < 10$ h. Thus, the cure reactions were quenched at about 90% conversion when $\Delta T \sim -30$°C as suggested by Dušek (1985).

Yu and van Meerwall (1990) report a very interesting application of NMR pulsed gradient spin echo methods for the measurement of the diffusivity of molecules during cure. They studied two epoxy resins, DGEBA and diglycidyl ether of 1,4 butanediol (DGEB), both cured with DDS at 120 and 140°C. The self-diffusion of both polymer and probe-molecules including free hardener were measured. There was no effect of gelation on the diffusivity but $D_{polymer}$ decreased by three orders of magnitude during cure. The change in D_{probe} was much smaller which indicates that it is the 'local' viscosity which is the important parameter determining the rate of cure as $\Delta T = T_c - T_g$ becomes smaller and smaller.

This discussion of diffusion control of the extent of cure when ΔT is small provides an explanation for the normal industrial practice of elevated temperature post-curing to attain optimum glass transition temperatures.

The relationship between extent of cure and the glass transition temperature is discussed in detail by Stutz *et al.* (1990). However, it may be difficult to determine the extent of cure as the concentrations of reactive groups approach zero. Even a DSC zero residual heat does not necessarily indicate 'full' cure (Stutz *et al.*, 1990; Stutz, 1992). Also, Stutz (1992) recommends control of the final stages of cure by measurement of the amount of solubles. Although this parameter may require experimental experience to obtain accurate measurements it can be used to calculate a degree of cure using branching theory (Macosko and Miller, 1976; see section 3.2).

3.5 Effect of cure on mechanical and related properties

3.5.1 Introduction

Epoxy resins find applications in areas such as coatings, adhesives or matrix materials for composites, as discussed in chapters 7, 8 and 9. One of the properties required for these applications is that the material should be essentially a rigid solid at the use temperature, T_{use}, although in some cases a degree of flexibility is required (see chapter 4). The use temperature in all these applications is below the glass transition temperature of the cured resin, $T_{use} < T_g$. For higher temperature applications there is continued development of epoxy-hardener systems with higher and higher glass transition temperatures and hence higher T_{use}. Of course it is important to ensure that chemical degradation, due to oxidative, photochemical or other processes, do not impair the properties of the cured resin. These relationships are represented schematically in the TTT diagram, Figure 1.2.

In many applications of epoxy resins it is their glassy state elastic moduli that are of paramount importance. Modulus maps for amorphous polymers have been proposed by Gilbert *et al.* (1986) who confined their discussion to thermoplastics, so-called 'linear' polymers, and considered poly(methyl methacrylate) and poly(styrene) in detail. There is a need to represent the structure–property relationships for thermosetting resins. The elastic properties will be a function of the structure of the network and the measurement temperature relative to the glass transition temperature of the cured resin. Thus, an elastic property $P = f(T,C)$ where T is the measurement temperature and C is a cure parameter. Choice of measurable quantities to define C presents several problems. Of course, C is related to extent of conversion, X_e, which is difficult to measure accurately when the concentration of reactive groups becomes very low, as discussed previously (see section 3.3.2). Also it is important to note that the relationship between T_g and X_e is non-linear (Wisanrakkit and Gillham, 1990; Wang and Gillham,

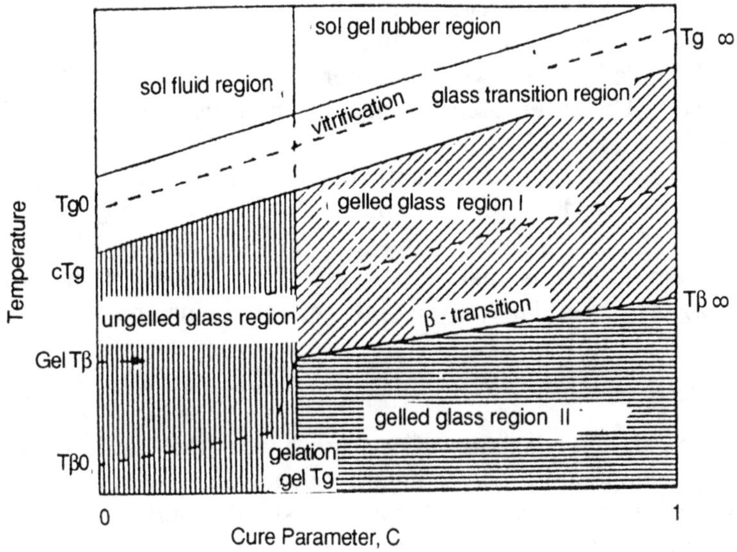

Figure 3.7 The CTP diagram: a modified version of a T_gTP diagram (Wang and Gillham, 1992)

$$C = \frac{T_g - T_{go}}{T_{g\infty} - T_{go}}$$

T_{go} is the initial glass transition temperature of the uncured resin; $T_{g\infty}$ is the glass transition temperature of the fully-cured resin; T_g is the glass transition temperature of the partially-cured resin.

1992), and would complicate any representation of the relationships between properties and extent of conversion. Wang and Gillham (1992) suggest that the glass transition temperature can be used as a measure of cure, C, and have presented a T_g – TP diagram. Their proposal has been slightly modified by replacing T_g by C in Figure 3.7, in which a normalized definition for the state of cure is used,

$$C = \frac{T_g - T_{go}}{T_{g\infty} - T_{go}} \tag{3.35}$$

where T_{go} is the initial and $T_{g\infty}$ is the 'fully cured' glass transition temperature. Such a definition requires that T_g is a measure of the structure of the cured resin regardless of the precise cure treatment, that is, cure temperature and cure time. With this definition C increases from zero to unity as cure progresses.

An earlier review of the effects of cross-linking on the properties of polymers is that of Nielsen (1969), which is still useful as a starting point for the study of the relationships between properties and cure of thermosetting systems. It should be noted that there has been more recent discussion of the relationship between the glass transition temperature and the structure of the cured resin (see for example Banks and Ellis (1982) and Stutz *et al.* (1990)). Kaelble (1973 and with co-authors, 1988) and Oleinik (1986) review epoxy resins. The latter discusses much of the original work published in Russian.

The change from a liquid to eventually a visco-elastic solid during cure of an epoxy resin was discussed in chapter 1 and illustrated in Figure 1.1. The importance of the T_gTP (or CTP) diagram (Figure 3.7) introduced by Wang and Gillham (1992) is that it provides a generalized framework for correlating the physical properties of thermosetting materials.

3.5.2 Glassy moduli

When visco-elastic relaxation effects are absent the elastic moduli will have a low temperature and/or high frequency limit, such as E_{go}, for the Young's modulus. The moduli that are measured experimentally at higher temperatures and lower rates of strain will be lower than E_{go}. When either the measurement temperature decreases or the rate of strain is increased the moduli increase. A major cause of this increase in modulus with decrease in temperature is due to a decrease in the volume of the resin. As the density increases the separation between adjacent chain segments decreases so that the van der Waals forces between intermolecular atoms increase, which of course leads to an increase in the elastic moduli. For thermoplastics, Gilbert *et al.* (1986) suggest a linear relationship

$$E = E_0(1 - \alpha_m \, . \, T/T_g) \tag{3.36}$$

where E_0 is the Young's modulus at 0 K, α_m is the temperature coefficient of modulus, which is regarded as constant in the range T to T_g. It should be noted that equation 3.36 is only applicable for the glassy modulus, that is $T < T_g$. Bondi (1968) has suggested an equation 3.37, which has a similar form to equation 3.36, for $T < 0.9\ T_g$.

$$E^*(T) = E_0^* \left[1 - \frac{b}{E_0^2} \cdot \frac{T}{T_g}\right] \tag{3.37}$$

where $E^* = EV_w/\Delta H_s$ is a dimensionless reduced modulus, and H_s is a lattice contribution to the enthalpy and V_w is the van der Waal's volume of a repeat unit. Bondi gives data for the calculation of molecular structure increments and a value for b, a dimensionless constant, of about 30. Morel *et al.* (1989) have measured Young's modulus for a series of epoxy resins and find that Bondi's method overestimates the elastic modulus.

The major factor that determines the elastic moduli of glassy polymers is the cohesive forces between molecules, or molecular segments. An estimate of these attractive forces is provided by the cohesive energy density of the material and Tobolsky (1960) suggested that the bulk modulus, B, of a polymeric glass can be estimated from

$$B = 8.04\ \delta^2 \tag{3.38}$$

where δ is the solubility parameter (Barton, 1983), which is the square root of the cohesive energy density. Kaelble (1973) and Kaelble *et al.* (1988), have used Tobolsky's method to estimate the elastic moduli of epoxy resins. Since they are interrelated by

$$E = 3B(1 - 2\nu) = 2G(1 + \nu) \tag{3.39}$$

where ν is Poisson's ratio and G is the shear modulus it is possible to calculate E from B when ν is known. Kaelble sets $\nu = 0.33$ to obtain

$$E = 2.67G = 8.04\ \delta^2 \tag{3.40}$$

Kamon and Furakawa (1986) used the same approach but considered the relationship unproven. Further, they noted that the elastic modulus of annealed samples was higher than for those that had been quenched, an effect that will be discussed later (see section 3.5.5). However, on annealing there is an increase in the ambient temperature density of an epoxy resin. Thus there will be an increase in the intermolecular forces and so also an increase in the elastic modulus.

A systematic examination of the relationship between the structure and elastic moduli of cured epoxy resins has been reported by Morel and coworkers (1989). They compared experimental determinations of elastic moduli for a range of resin-hardener systems with several methods of estimating the limiting glassy moduli, such as those discussed by van Krevelen and Hoftyzer (1976). Ultrasonic pulse measurements of both the longitudinal transverse wave velocities enable elastic constants to be calculated, from the relations

$$V_L = \frac{B}{\varrho} . \frac{3(1-\nu)}{(1+\nu)} \tag{3.41}$$

$$V_T = G/\varrho \tag{3.42}$$

where V_L and V_T are the velocities of the longitudinal and transverse waves, and ϱ is the density. Thus, the shear modulus can be calculated from equation 3.42 and also using equation 3.39 with equation 3.41 Young's modulus can be calculated.

Poisson's ratio may be estimated from its relationship with the longitudinal wave velocity by

$$V_L = \left[\frac{U}{V}\right]^6 \left[\frac{3(1-\nu)}{(1-\nu)}\right]^{1/2} \tag{3.43}$$

where U is the sum of molar group contributions, U_i, i.e. $U = \Sigma U_i$ and V is the molar volume. The U_is are known as Rao additivity increments and are listed by Morel *et al.* (1989) for their calculations for epoxy resins. A more complete listing is given by van Krevelen and Hoftyzer (1976). Use of this set of molar contributions U_i underestimates the longitudinal wave velocity and hence would lead to an underestimate of the elastic moduli.

Morel *et al.* (1989) also evaluated the applicability of equation 3.38 by plotting the bulk modulus B vs δ^2, the cohesive energy density. A straight line plot was obtained but with a slope of 11.5 compared with the predicted value of 8.04. They considered that their estimates of the solubility parameters neglected the hydrogen bonding that is well known to occur in epoxy resins. To estimate the degree of hydrogen bonding they measured the equilibrium water absorption, W_m, at 100°C and 95% relative humidity. There was a good linear fit between the experimental data and the water absorption and elastic modulus, equation 3.44.

$$M = M_0 + kW_m \tag{3.44}$$

where M is the modulus and the slope, k, depending on the contribution due to hydrogen bonding, and M_0 depending on the aromatic group concentration. Good linear correlations were obtained for all three moduli, B, E and G.

The moduli that are usually measured in the temperature range 50 to 150°C below the glass transition temperature of the resin are often considerably lower than the limiting low temperature high frequency moduli which have been discussed in this section. The stress-strain behaviour of epoxy resins will be considered next.

3.5.3 *Stress–strain curves and visco-elastic behaviour*

Stress–strain curves for epoxy resins are often non-linear, as illustrated schematically in Figure 3.8 for extension (Pink and Campbell, 1974). It is usual to determine Young's modulus from the limiting slope at low applied stresses and hence strains. This limiting slope increases slightly with increased rate of extension (Yamini and Young, 1980) and decreases with increase in temperature. Similar behaviour was also observed with conventional compressive tests. Yamini and Young (1980) also studied the effect of changes in the concentration of hardener and found that the tensile modulus decreased as the concentration of hardener increased (Shell 828 with TETA, 7.4 to 14.7 phr). The cure time and temperature obviously also affected the modulus. The change in relative modulus with T_g, as a measure of the state of cure, is illustrated in Figure 3.9, which shows that the modulus increases rapidly in the region if vitrification reaches a maximum and then decreases (Wang and Gillham, 1992). This maximum is not related to gelation, which has been suggested by some authors. Gelation was discussed previously (see section 3.3.4) and it was noted that there is a change in the slope of log E'' or log G'' vs cure time, t_c, at the gel point. However, it is

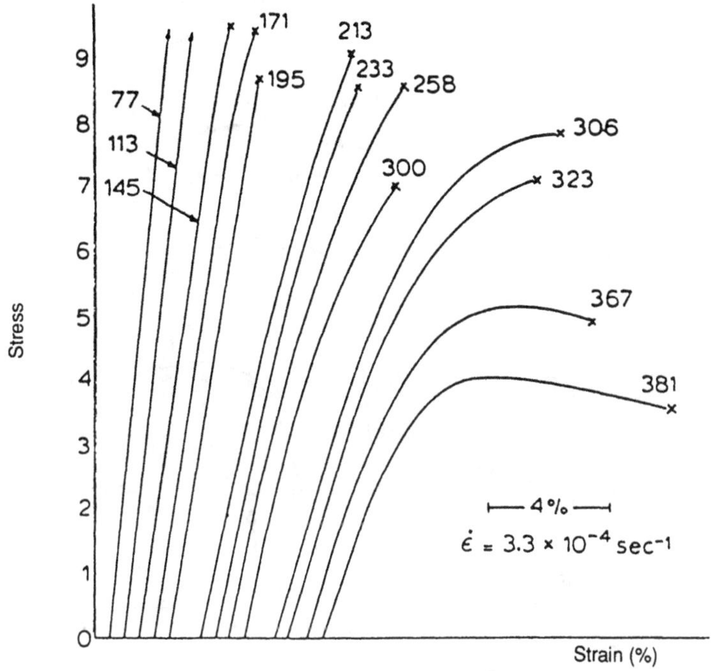

Figure 3.8 Stress-strain curves for an epoxy resin at temperatures in the range 77 to 381 K. Stress, kp/mm^2; strains, displaced for clarity, see gauge length (from Pink and Campbell, 1974).

pertinent to note that when the structure of a cured epoxy resin is altered by changing the concentration of hardener the Young's modulus has a minimum when there is an exact stoichiometric mixture of epoxy groups and reactive amine hydrogen atoms (Bell, 1970b; Selby and Miller, 1975; Kim *et al.*, 1978; Gupta *et al.*, 1985). It may be that the modulus is related to the density of the resin at the measurement temperature, which also passes through a maximum with cure time and hence extent of reaction (Bell *et al.*, 1992). The maximum in the relative modulus is found at all cure temperatures studied by Wang and Gillham (1992) with the relative modulus increasing with cure temperature and the maximum occurring at lower extents of reaction as the cure temperature increased (Figure 3.9).

From Figure 3.8 it can be seen that the curvature of the extensional stress-strain curves is more pronounced at higher test temperatures. At room temperature it has been found that for stresses lower that the yield stress the stress-strain curve fits a power law relationship $\sigma = M\varepsilon^n$ with a value of $n \simeq 0.93$ for a BADGE type resin cured with DDM (Bell *et al.* 1992). Williams (1979) suggests the use of the power law equation for the stress-strain relationship for polymeric materials which do not obey Hooke's law exactly with $n \simeq 0.9$. When strains are low, M is approximately equal to Young's for values of n not too different from unity.

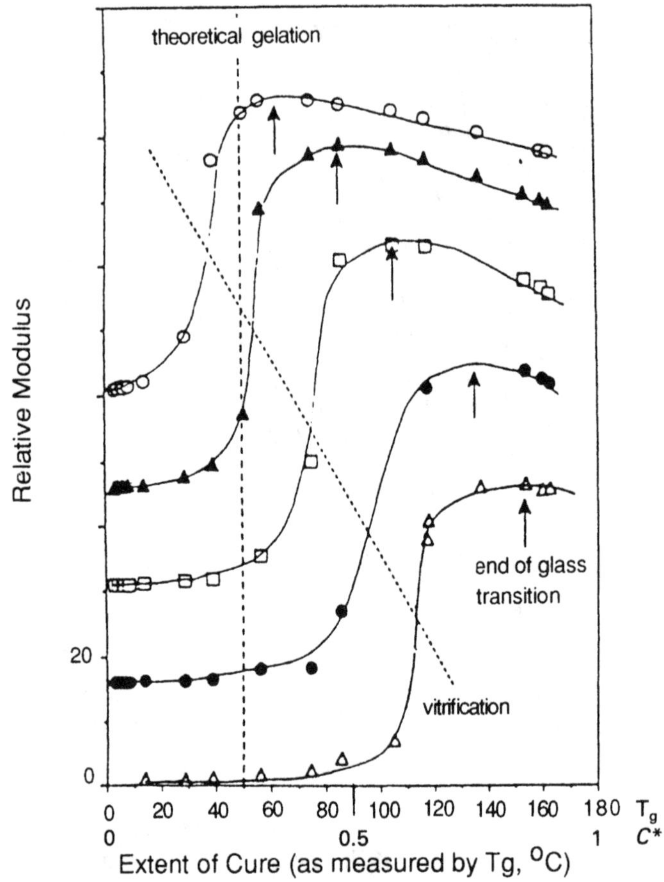

Figure 3.9 Relative isothermal elastic modulus versus T_g (extent of cure). Resin: DGEBA type (Dow, DER332); hardener: Trimethylene glycol di-*p*-aminobenzoate (TMAB, Polaroid Corp.) in a stoichiometric concentration. Measurement temperature (○) 30°C; (▲) 50°C; (□) 70°C; (●) 90°C; (△) 110°C (from Wang and Gillham, 1992). * *C* is the cure parameter as defined in Figure 3.7.

Although yield is observed in extension (Figure 3.8) especially at higher measurement temperatures, it is often more precisely measured in uniaxial compression (Moehlenpah *et al.*, 1969). Yamini and Young (1980) have studied the yield of a TETA-cured epoxy resin measuring yield stress as functions of resin/hardener concentration, post-cure and rate of straining. The most marked effects were due to systematic changes in the resin-hardener stoichiometry and the test temperature. The yield stress decreased as expected as the test temperature increased up to the glass transition temperature of the resin, which increased with increased hardener concentration for resin given the same cure and post-cure. They also found that both the yield theories of Argon (1973) or Bowden (1973) were satisfactory, but that Bowden's applied over a wider temperature range up to the glass transition temperature of the resin.

Above their glass transition, epoxy networks behave as elastomers with elastic moduli, E_r, very much lower than the glass moduli. Banks and Ellis (1982), compared $\Delta E = E_r/E_g$ for several cross-linked thermosetting resins and found that for epoxy resin, ΔE was in the range 7×10^{-3} to 1.3×10^{-2}, whereas for a very highly cross-linked phenol-formaldehyde resin ΔE was only 1.45×10^{-1}. Bell *et al.* (1992) have found that this ratio, ΔE, changes monotonically with epoxy conversion from 1.05×10^{-2} to 1.16×10^{-2} for a BADGE type resin cured with DDM, and is not much higher even after an extensive post cure (6 h) at 180°C.

The theory of rubber-like elasticity has been treated in considerable detail (Treloar, 1975), which it would be inappropriate to discuss here. The rubber-like behaviour of epoxy networks above their glass transition temperature have been discussed by Oleinik (1986) and Ilavsky *et al.* (1984). For an ideal rubbery material the moduli are related to the network structure by

$$E_r = 3G_r = \frac{3\varrho RT}{M_c} \tag{3.45}$$

where ϱ is the density, R is the gas constant, T is the absolute temperature and M_c is the average molecular weight of the network chains. From consideration of the ratio of the chain contour length to the root mean square distance between junction points Kaelble *et al.* (1988) discuss the calculation of the maximum extension ratio, $\lambda = l/l_0$, for epoxy resins and other network polymers. For epoxy resins it is 1.4 to 1.6 compared with 6.85 for a silicone rubber. Oleinik (1986) also concludes that the largest possible elongation of a cross-linked epoxy resin is about 50% (i.e. $\varepsilon \simeq 0.5 = \lambda - 1$, and $\lambda \simeq 1.5$). The highest measured elongations at break is somewhat lower, 35% (range 12–35%). Also, the stress-strain curve can be represented by

$$\sigma = G_r(\lambda - 1/\lambda^2) \tag{3.46}$$

for extension ratios of up to 1.2, where σ is the applied stress, equal to the applied force divided by the unstrained area of the sample. Thus, from this experimental value of G_r it is possible to calculate the equivalent molecular weight of the network chains. This is a direct method of determining a parameter which is a characteristic of the network structure. Other methods involve branching theory and either assumptions or direct determination of the relative reactivities of the reactive groups, such as the reactivity of primary and secondary amine groups.

3.5.4 *Visco-elastic properties*

The rate and temperature effects on the stress-strain curves, Figure 3.8, is a manifestation of the visco-elastic properties of epoxy resins. The change in the dynamic mechanical properties with cure were briefly discussed in section 3.3.4 (see Figure 3.5). As well as the glass transition temperature, known as the α process, there are other relaxations, β, γ, etc. (Pogany, 1969, 1970; Williams, 1979; Ochi *et al.*, 1985). These processes may be observed by measurement of dynamical mechanical properties as a function of temperature, and also creep or stress relaxation rates. These methods have been reviewed by Neilson (1969) and Kaelble (1973; Kaelble *et al.*, 1988) discuss the visco-elastic properties of epoxy resins and Kong (1986) extends the discussion to the properties of composites, as does Wolfe and Tod (1989).

Arridge and Speake (1972) report on the effects of cure on the low temperature γ relaxation. Crowson and Arridge (1979) constructed a time-temperature master curve from creep data, and determined the shift factors a_T. They found that the temperature dependence of the log a_T could not be represented by the WLF equation. A similar temperature dependence of the shift factors, log a_T, was found for master curves constructed from DMTA measurements (Bell *et al.*, 1992). Thus, further analysis of the temperature dependence of visco-elastic data is required.

Gupta and coworkers (1984–1985) have studied an epoxy resin and found as well as the α and β relaxation a third broad but not very intense relaxation situated between the α and β peaks which they call an α' transition. Dillman and Sefferis (1989) report their study in which they analysed both the cure processes and the dynamic mechanical behaviour, and pointed out its analogy with dielectric relaxation. Wang and Gillham (1992) have monitored the changes in rigidity and log (decrement) with cure for an epoxy resin. They located T_g and T_β from the maxima in the log (decrement) versus temperature plots. These data were used for the construction of their T_g-TP (CTP) diagram Figure 3.7. An interesting feature of their results was the 'jump' in the T_β vs T_g plot, Figure 3.10, which occurs prior to gelation. After gelation T_β increases linearly with T_g.

3.5.5 Physical ageing

When a glass forming material is quenched from above its glass transition temperature there is an 'instantaneous' contraction followed by a time-dependent reduction in volume. The rate of the time-dependent densification depends on the temperature to which the glass has been quenched. For temperatures only moderately below the glass transition temperature the rate of volume relaxation is relatively rapid but becomes slower and slower as the quench temperature is reduced. Such volume relaxations were studied many years ago and Kovacs (1958, 1977) carried out systematic studies on polymeric materials. Such volume relaxation processes can be treated in a similar way to a visco-elastic analysis of creep behaviour.

With a decrease in volume the free volume will also be reduced and hence

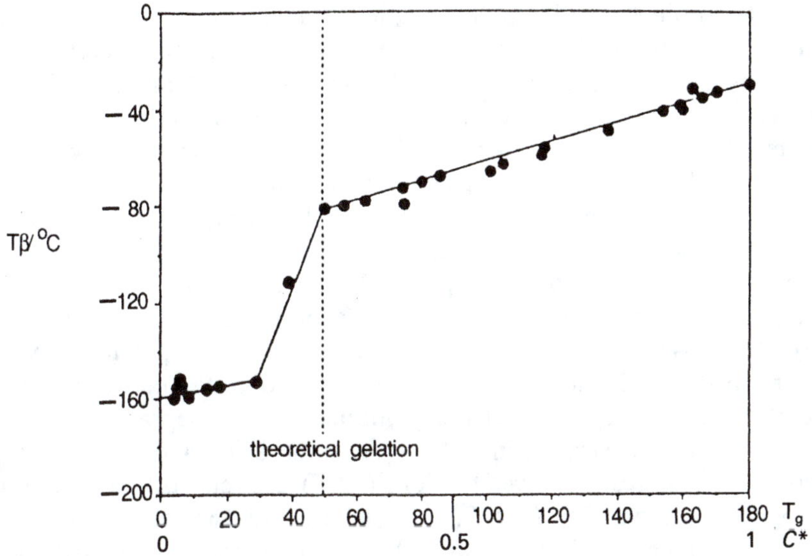

Figure 3.10 The β relaxation temperature T_β versus the glass transition temperature T_g. The resin-hardener system is the same as in Figure 3.9 (from Wang and Gillham, 1992). * C is the cure parameter as defined in Figure 3.7.

chain mobility which depends on molecular packing and is at least partially dependent on free volume will also be reduced. Thus, as the volume decreases so will molecular mobility and hence the rate of volume relaxation. Unfortunately it is still not possible to define free volume uniquely or unambiguously and hence these effects can only be explained qualitatively in terms of a change in free volume. However, the concept is very convenient as an aid to an understanding of the processes involved.

The properties of a polymeric glass will depend on its packing density which affects the distances between molecular sub-groups and hence the van der Waals forces between molecules. Generally, when a polymeric material is cooled through its glass transition temperature it will become stiffer and more brittle. Also, its visco-elastic properties change, the creep and stress relaxation rates decrease, as does the damping or tan δ. These properties also change with densification of the glass and these effects have been attributed to 'physical ageing', to distinguish the process from 'chemical ageing' such as photo-chemical or thermal degradation which are due to the breaking or formation of covalent chemical bonds. These chemical processes lead to permanent changes in the structure of the material. It is generally regarded that physical ageing effects are reversible and by increasing the temperature of the sample above its glass transition temperature its past history can be expunged.

Studies of physical ageing have been the subject of a monograph by Struik (1978) who observed that partially aged samples with identical thermal histories subjected to large deformations relaxed more rapidly than when subject to only small deformations. With the application of a tensile stress there will be an increase in volume of the sample. The magnitude of this increase in volume will depend on the value of Poisson's ratio for the material at the ageing temperature. Thus, with this increase in volume the rates of molecular relaxation processes will be more rapid. The term that has been used for this effect is 'erasure of ageing' but more recently 'rejuvenation' appears to be accepted.

Most of the initial studies of physical ageing were on thermoplastics, so called 'linear' macromolecules and it is only more recently that more attention has been directed to such processes in network polymers. A major problem with the study of cured resins is control of the network structure, since it is possible for either or both continued cure or degradative processes to obscure the physical ageing effects. A useful review and report of his own work on epoxy resins is given by Kong (1986). He reports that the lower the cooling rates from $T > T_g$ to $T_A < T_g$, where T_A is the ageing temperature, and/or with increased ageing time, the sample increases in density and this leads to lower impact strength, fracture energy, fracture strain, mechanical damping and lower creep and stress relaxation rates. With physical ageing there is a loss of ductility which is important in epoxy resin composites which are also discussed by Kim (Kim *et al.*, 1978).

McKenna and co-workers (see Lee and McKenna, 1988, 1990) have published a lot of detailed research work on the physical ageing of epoxy resins. In two recent papers (Santore *et al.*, 1991; McKenna *et al.*, 1991) they question the possibility of rejuvenation due to the applications of mechanical stresses. Specifically they find that the underlying volume recovery kinetics, which are much slower than the mechanical relaxation, remain unaltered in spite of the imposition of mechanical deformations. Thus, they report that their results support the argument that mechanical stimuli neither alter the underlying (non-equilibrium) thermodynamic state of the glass nor erase physical ageing.

Study of both curing and physical ageing processes have been carried out by Plazek and co-workers (Choy and Plazek, 1986; Plazek and Choy, 1989; Plazek and Frund, 1990). They have measured the change in volume of the resin during cure using a specially constructed pressurized volume dilatometer, and a frictionless magnetic-bearing torsion creep apparatus to measure creep and recovery in order to calculate viscosities and recoverable steady-state compliances of the cured resins. In their later study (Bero and Plazek, 1991) of volume-dependent rate processes in epoxy resins they report on four types of interrelated experiments, (i) specific volume with constant rates of cooling, (ii) time-temperature volume measurements, (iii) elongational creep compliances and (iv) shear creep compliances. A very interesting result they report is that the volume-temperature cooling cures are asymptotic to their glass lines irrespective of the rate of cooling. Thus, the lower temperature limit of the glass transition region is independent of the cooling rate. Hence, the molecular motions that occur in the glass transition region are inhibited at a specific temperature irrespective of the actual volume of the specimen. This observation is of considerable importance theoretically and practically and requires further study, as planned by Bero and Plazek (1991).

References

Acitelli, M.A., Prime, R.B. and Sacher, E. (1971) *Polymer* **12**, 335–343.

Adolf, D. and Martin, J.E. (1990) *Macromolecules* **23**, 3700–3704.

Adolf, D., Martin, J.E. and Wilcoxon, J.P. (1990) *Macromolecules* **23**, 527–531.

Amdouni, N., Sautereau, H., Gérard, J.-F. and Pascault, J.-P. (1990) *Polymer* **31**, 1245–1253. See Fig. 2 and Table 2, p. 1348.

Argon, A.S. (1973) *Phil. Mag.* **28**, 839.

Aronhime, A.T. and Gillham, J.K. (1986) *Adv. in Polym. Sci.* **78**, 83–113.

Arridge, R.G.C. and Speake, J.H. (1972) *Polymer* **13**, 443–454.

Banks, L. and Ellis, B. (1982) *Polymer* **23**, 1466–1472.

Barton, A.L.M. (1983) *Handbook of Solubility Parameters and other Cohesion Parameters*. CRC Press, Boca Raton, USA.

Barton, J.M. (1985) *Adv. in Polym. Sci.* **72**, 111–154.

Bell, J.P. (1970a) *J. Polym. Sci. A2.* **6**, 417–436.

Bell, J.P. (1970b) *J. Appl. Polym. Sci.* **32**, 1901–1906.

Bell, J., Ellis, B. and Found, M.S. (1992) Unpublished experimental measurements.

Benson, S.W. (1960) *The Foundations of Chemical Kinetics*. McGraw-Hill, New York, pp. 42–46.

Bero, C.A. and Plazek, D.T. (1991) *J. Polym. Sci., B. Polym. Phys.* **29**, 39–47.

Bidstrup, W.W. and Senturia, S.D. (1989) *Polym. Eng. Sci.* **29**, 290–294.

Billmeyer, F.W. (1984) *Textbook of Polymer Science* 3rd edn, Wiley, New York, pp. 40–47.

Bondi, A. (1968) *Physical Properties of Molecular Crystals, Liquids and Glasses*, Section 13.3.1, Wiley, New York, pp. 397–400.

Bowden, P.B. (1973) In *The Physics of Glassy Polymers* Ch. 5, ed. Haward, R.N. Applied Science Publishers, London, pp. 279–339.

Camargo, R.E., Gonzalez, V.M. and Macosko, C.W. (1983) *Rubber Chemistry and Technology* **56**, 774–778.

Carothers, W.H. (1936) *Trans. Far. Soc.* **32**, 39–53.

Case, L.C. (1957) *J. Polym. Sci.* **26**, 333–350.

Chen, X.M. and Ellis, B. (1992a) Paper submitted to *J. Appl. Polym. Sci.*

Chen, X.M. and Ellis, B. (1992b) Unpublished data, but see Chen, X.M. Ph.D thesis Sheffield 1988).

Charlesworth, J. (1980) *J. Polym. Sci. Polym. Chem. Ed.* **18**, 621–628.

Chiu, W.Y., Carratt, G.M. and Soong, D.S. (1983) see Figure 2 *Macromolecules* **16**, 348–357.

Choy, I.C. and Plazek, D.J. (1986) *J. Polym. Sci., B. Polym. Phys.* **24**, 1303–1320.

Crowson, R.J. and Arridge, R.G.C. (1979) *Polymer* **20**, 737–746.

Dannenberg, H. (1963) *S.P.E. Trans.* **3**, 78–88.

de Gennes, P.-G. (1979) *Scaling Concepts in Polymer Physics*, Ch. V, Cornell University Press, Ithaca, pp. 128–162.

Dillman, S.H. and Sefferis, J.C. (1989) *J. Macromol. Sci.-Chem.* **A26**(1), 227–247.

Dobas, I., Eichler, J. and Klaban, J. (1975) *Collection Czech. Chem. Comm.* **40**, 2989–2997.

Durand, D. and Bruneau, C.-M. (1983) *Polymer* **24**, 287–591 and 592–595.

Dušek, K. (1985) *Brit. Polym. J.* **17**, 185–189.

Dušek, K. (1986) *Adv. in Polym. Sci.* **75**, 1–59. See pp. 9–12 for discussion of the morphology of cured epoxy resins.

Dušek, K., Illavský, M. and Luňák, S.J. (1975) *J. Polym. Sci. Polym. Symp.* **53**, 29.

Elias, H.-G. (1976) *Macromolecules* Vol. 1 and 2. Plenum Press, New York, pp. 606–614.
Elliott, S.R. (1990) *Physics of Amorphous Materials* 2nd edn, Ch. 2, Longman Scientific and Technical, Harlow, pp. 29–69.
Ellis, B., Bell, J., Chen, X.M. and Found, M.S. (1992) Paper in preparation.
Enns, J.B. and Gillham, J.K. (1983) *J. Appl. Polym. Sci.* **28**, 2567–2591.
Entelis, S.G. and Tiger, R.P. (1976) *Reaction Kinetics in the Liquid Phase*, Ch. 1. Translated from Russian, John Wiley, Chichester, pp. 1–25.
Eyring, H., Lin, S.H. and Lin, S.M. (1980) *Basic Chemical Kinetics*, Ch. 9, John Wiley, New York, pp. 376–451.
Fava, R.A. (1968) *Polymer* **9**, 137–151.
Flory, P.J. (1941) *J. Amer. Chem. Soc.* **63**, 8083–3100.
Flory, P.J. (1953) *Principles of Polymer Chemistry*, Ch. IX Cornell University Press, Ithaca, pp. 347–398.
Flynn, J.H. (1989) *Enc. Polym. Sci. and Eng.* 2nd edn Suppl. pp. 715–723.
Flynn, J.H. and Wall, L.A. (1966) *J. Res. Natl. Bur. Stand.* **70A**, 487–523.
Frost, A.A. and Pearson, R.G. (1961) *Kinetics and Mechanism* 2nd edn, John Wiley, New York, pp. 178–184.
Ghaemy, M., Billingham, N.C. and Calvert, P.D. (1982) *J. Polym. Sci. Polym. Lett.* **20**, 439–443.
Gilbert, D.G., Ashby, M.F. and Beaumont, P.W.R. (1986) *J. Mat. Sci.* **21**, 3194–3210.
Goddu, R.F. and Delker, D.A. (1958) *Anal. Chem.* **30**, 2013–2016.
Gordon, M. and Ross-Murphy, S.B. (1975) *Pure and Applied Chem.* **43**, 1–26.
Grillet, A.C., Galy, J., Pascault, J.P. and Bardin, I. (1989) *Polymer* **30**, 2094–2103.
Grimmett, G. (1989) *Percolation* Ch. 1. Springer-Verlag, New York, pp. 1–24.
Guillett, J. (1985) *Polymer Photophysics and Photochemistry* see Ch. 3, pp. 52–70 and Ch. 6, pp. 114–135 especially p. 132. Cambridge University Press, Cambridge.
Gupta, V.B., Drzal, L.T., Lee, C.Y-C. and Rich, M.J. (1984–1985) *J. Macromol. Sci. B23*, 435–466.
Hadad, D.K. (1988a) In *Epoxy Resins, Chemistry and Technology* 2nd edn. ed. May, C.A. Marcel Dekker, New York, pp. 1111–1123.
Hadad, D.K. (1988b) In *Epoxy Resins, Chemistry and Technology*, 2nd edn. ed. May, C.A. Marcel Dekker, New York, pp. 1126–1136.
Harran, D. and Laudourd, A. (1986) *J. Appl. Polym. Sci.* **32**, 6043–6062.
Havlicek, I. and Dušek, K. (1987) In *Crosslinked Epoxies* ed. Sedlacek, B. and Kahovec, J. Walter de Gruyter, New York, pp. 417–424.
Horie, H., Hiura, H., Sawada, M., Mita, I. and Kambe, H. (1970) *J. Polym. Sci.* **A1.8**, 1357–1372.
Hsich, H.-Y. (1982) *J. Appl. Polym. Sci.* **27**, 3265–3277.
Huguenin, F.G.A.E. and Klein, M.T. (1985) *Ind. Eng. Chem. Prod. Res. Der.* **24**, 166–171.
Ilavsky, M., Bogdanova, L.M. and Dušek, K. (1984) *J. Polym. Sci. Polym. Phys. ed.* **22**, 265.
Jha, A. and Parker, J.M. (1989) *Physics and Chemistry of Glasses* **30**, 220–228.
Kaelble, D.H. (1973) In *Epoxy Resins, Chemistry and Technology*, Ch. 5. eds. May, C.A. and Tanaka, Y. Dekker, New York, pp. 327–371.
Kaelble, D.H., Moacanin, J. and Gupta, A. (1988) *Epoxy Resins, Chemistry and Technology* 2nd edn, Ch. 6. ed. May, C.A. Dekker, New York, pp. 603–652.
Kamal, M.R. and Sourour, S. (1972) *Society of Plastics Engineers* **18**, 93–98.
Kamal, M.R. and Sourour, S. (1973) *Polym. Sci. and Eng.* **13**, 59–64.
Kamon, T. and Furakawa, H. (1986) *Adv. in Polym. Sci.* **80**, 173–202.
Keenan, M.R. (1987) *J. Appl. Polym. Sci.* **33**, 1725–1734.
Khabenko, A.V. and Dolmatov, S.A. (1990) *J. Thermal. Anal.* **36**, 45–54.
Kim, S.L., Skibo, M.D., Manson, J.A., Hertzberg, R.W. and Janiszewski, J. (1978) *Polym. Eng. Sci.* **18**, 1093–1100.
Kong, E.S.W. (1986) *Adv. in Polym. Sci.* **80**, 126–171.
Kovacs, A.J. (1958) *J. Polym. Sci.* **30**, 131–147.
Kovacs, A.J., Hutchinson, J.M. and Aklonis, J.J. (1977) *The Structure of Non-Crystalline Materials* ed. Gaskell, P.H. Taylor and Francis, London, pp. 153–163.
Labana, S.S. (1986). *Enc. Polym. Sci. and Eng.* 2nd edn, Vol. 4, John Wiley, New York, pp. 350–360.

Larson, R.G. (1988) *Constitutive Equations for Polymer Melts and Solutions*, Ch. 4. Butterworth, Boston, pp. 93–127.
Lee, A. and McKenna, G.B. (1988) *Polymer* **29**, 1812–1817.
Lee, A. and McKenna, G.B. (1990) *Polymer* **31**, 423–430.
Lee, H. and Neville, K. (1976) *Handbook of Epoxy Resins* Appendix 4-1. McGraw Hill, New York, pp. 4-35–4-56.
Lunak, S., Vladyka, J. and Dušek, K. (1978) *Polymer* **19**, 931–933.
Macosko, C.W. and Miller, D.R. (1976) Macromolecules **9**, 199–206.
Macosko, C.W. (1985) *Brit. Polymer J.* **17**, 239–245.
Martin, J.E., Adolf, D. and Wilcoxon, J.P. (1988) *Phys. Rev. Lett.* **61**, 2620–2623.
Malkin, A.Yu and Kulichikhin, S.G. (1991) *Adv. in Polym. Sci.* **101**, 217–257.
Matsuoku, S., Quan, X., Bair, H.E. and Boyle, D.S. (1989) Macromolecules **22**, 4093–4098.
McKenna, G.B., Santore, M.M., Lee, A. and Durran, R.S. (1991) *J. Non-Crystalline Solids* **131–133**, 497–504.
Mertzel, E. and Koenig, J.L. (1986) *Adv. in Polym. Sci.* **75**, 73–112. For discussion of NMR see pp. 92–112 and FT-IR see pp. 73–92.
Mestan, S.A. and Morris, C.E.M. (1984) *J. Macromol. Sci. Rev. in Macromol. Chem. and Phys.* **C24**, 117–172.
Mijovic, J. (1990) *J. Appl. Polym. Sci.* **40**, 845–866.
Mijovic, J., Kim, J. and Slaby, J. (1984) *J. Appl. Polym. Sci.* **29**, 1449–1462.
Mijovic, J. and Lee, C.H. (1989) *J. Appl. Polym. Sci.* **38**, 2155–2170.
Mijovic, J. and Wang, H.T. (1989) *J. Appl. Polym. Sci.* **37**, 2661–2673.
Mita, I. and Horie, K. (1987) *J. Macromol. Sci. Rev. Macromol. Chem. and Phys.* **C27**, 91–169.
Moehlenpah, A.E., Ishai, O. and Di Benedetto, A. (1969) *J. Appl. Polym. Sci.* **13**, 1231–1245.
Morel, E., Bellenger, V., Bocquet, M. and Verdu, J. (1989) *J. Mat. Sci.* **24**, 69–75.
North, A.M. (1964) *The Collision Theory of Chemical Reactions in Liquids*. Methuen, London.
Nielsen, L.E. (1969) *J. Macromol. Sci., Rev. Macromol. Chem.* **C3**(1), 69–103.
Ochi, M., Iesako, H. and Shimbo, M. (1985) *Polymer* **26**, 457–461.
Oleinik, E.F. (1986) *Adv. in Polym. Sci.* **80**, 50–99.
Peyser, P. and Bascom, W.D. (1977) *J. Appl. Polym. Sci.* **21**, 2359–2373.
Pingsheng, H.E. and Chune, Li (1989) *J. Mat. Sci.* **24**, 2951–2956.
Pink, E. and Campbell, J.D. (1974) *Mat. Sci. Eng.* **15**, 187–194.
Pinner, S.H. (1956) *J. Polym. Sci.* **21**, 153–157.
Plazek, D.J. and Choy, I.C. (1989) *J. Polym. Sci., B. Polym. Phys.* **27**, 307–324.
Plazek, D.J. and Frund, Z.N. (1990) *J. Polym. Sci., B. Polym. Phys.* **28**, 431–448.
Pogany, G.A. (1969) *J. Mat. Sci.* **4**, 405–409.
Pogany, G.A. (1970) *Polymer* **11**, 66–78.
Prime, R.B. (1981) In *Thermal Characterization of Polymeric Materials* Ch. 5. ed. Turi, E. pp. 435–569 see p. 445ff.
Rabek, J.F. (1980a) *Experimental Methods in Polymer Science* John Wiley, Chichester, Ch. 32, pp. 527–542. (1980b) Ch. 20, pp. 299–331. (1980c) Ch. 15, pp. 221–253. (1980d) Ch. 34, pp. 549–581.
Rabinowich, E. (1937) *Trans. Far. Soc.* **33**, 1225–1233.
Rashid, H.U. (1978) *Ph.D Thesis*, University of Sheffield.
Rekhson, S.M. (1986) In *Glass: Science and Technology*, Vol. 3, Ch. 1, ed. Uhlmann, D.R. and Kreidl, N.J. Academic Press, Inc, Orlando, pp. 1–117.
Riccardi, C.C., Adahbo, H.E. and Williams, R.J.J. (1984) *J. Appl. Polym. Sci.* **29**, 2481–2492.
Riccardi, C.C. and Williams, R.J.J. (1986a) *J. Appl. Polym. Sci.* **32**, 3445–3456.
Riccardi, C.C. and Williams, R.J.J. (1986b) *Polymer* **27**, 913–920.
Rogers, M.G. (1972) *J. Appl. Polym. Sci.* **16**, 1953–1958.
Rozenberg, B.A. (1986) *Adv. in Polym. Sci.* **75**, 113–165.
Sandford, W.M. and McCullough, R.L. (1990) *J. Polym. Sci. (Pt. B) Polym. Phys.* **28**, 973–1000.
Santore, M.M., Duran, R.S. and McKenna, G.B. (1991) *Polymer* **31**, 2377–2381.
Scherer, G.W. (1986) *Relaxation in Glass and Composites*, Ch. 12, John Wiley & Sons, New York, pp. 162–174.
Selby, K. and Miller, L.E. (1975) *J. Mat. Sci.* **10**, 12–24.
Senturia, S.D. and Sheppard, N.F. (1986) *Adv. in Polym. Sci.* **80**, 1–47.

Serrano, D., Peyrelasse, J., Borred, C., Harran, D. and Monge, P. (1989) *Macromol. Chem. Macromol. Symp.* **25**, 55–61.

Serrano, D., Peyrelasse, J., Borred, C., Harran, D. and Monge, P. (1990) *J. Appl. Polym. Sci.* **39**, 679–693.

Sharma, D.K. and Soane, D.S. (1988) *Macromolecules* **21**, 700–710.

Shen, M.C. and Eisenberg (1966) *Prog. Solid State Chem.* **3**, 407–481. See also *Rubber Chemistry and Technology* **43**, 95–170.

Sheppard, N.F. and Senturia, S.D. (1986) *Polym. Sci. and Eng.* **26**, 354–357.

Smith, T.T. (1961) *Polymer* **2**, 95–108.

Sojka, S.A. and Moniz, W.B. (1976) *J. Appl. Polym. Sci.* **20**, 1977–1982.

Sourour, S. and Kamal, M.R. (1976) *Thermochimica Acta* **14**, 41–59.

Sperling, L.H. (1986) *Introduction to Physical Polymer Science*, Ch. 6, John Wiley, New York, pp. 224–300.

Stanley, H.E. (1986) *Enc. Polym. Sci. and Eng.* 2nd edn, Vol 4, pp. 323–349.

Stauffer, D. (1985) *Introduction to Percolation Theory* Taylor and Francis, London (A 2nd edn is scheduled for 1992).

Stauffer, D., Coniglio, A. and Adam, M. (1982) *Adv. in Polym. Sci.* **44**, 103–158.

Steinmann, B. (1987) In *Cross-Linked Epoxies* eds Sedláček, B. and Kahovec, J. Walter de Gruyter & Co., Berlin, pp. 117–130.

Steinmann, B. (1989) *J. Appl. Polym. Sci.* **37**, 1753–1776.

Stevens, G.C. and Richardson, M.J. (1983) *Polymer* **24**, 851–858.

Struik, L.C.E. (1978) *Physical Aging in Amorphous and Other Materials.* Elsevier, Amsterdam.

Stutz, H., Illers, K.-H. and Mertes, J. (1990) *J. Polym. Sci. (Pt. B) Polym. Phys.* **28**, 1483–1498.

Stutz, H. (1992) Private communication.

Tanaka, Y. and Kakiushi, H. (1965) *J. Polym. Sci.* **3A**, 3279–3299.

Tanaka, Y. and Mika, T.F. (1973) In *Epoxy Resins, Chemistry and Technology* Ch. 3, ed. May, C.A. and Tanaka, Y. section IV, pp. 212–238.

Tobolsky, A.V. (1960) *Properties and Structure of Polymer* Wiley, New York, pp. 1–12.

Treloar, L.R.G. (1975) *The Physics of Rubber Elasticity*, 3rd edn, Oxford University Press, London.

van Krevelen, D.W. and Hoftyzer, P.J. (1976) In *Properties of Polymers, their estimation and correlation with chemical structure*, 2nd edn, Ch. 3, Elsevier, Amsterdam (see also van Krevelen, D.W. (1990), 3rd edn).

Verchère, D., Sautereau, H., Pascault, J.P., Riccardi, C.C., Moschier, S.M. and Williams, R.J.J. (1990) *J. Appl. Polym. Sci.* **38**, 725–731.

Wang, X. and Gillham, J.K. (1992) *J. Coatings Tech.* **64**, 37–45.

Ward, I.M. (1983) *Mechanical Properties of Solid Polymers* 2nd edn, Ch. 8. John Wiley, Chichester, pp. 166–174.

Whorlow, R.W. (1980) *Rheological Techniques* Ellis Horwood Ltd, Chichester, pp. 123–191, Appendix A1, p. 422.

Williams, G. and Watts, D.C. (1970) *Trans. Far. Soc.* **66**, 80–85.

Williams, J.G. (1979) *J. Appl. Polym. Sci.* **23**, 3433–3444.

Williams, J.G. (1980) *Stress Analysis of Polymers* 2nd edn. Ellis Horwood, Chichester, pp. 142–143.

Wisanrakkit, G. and Gillham, J.K. (1990) *J. Appl. Polym. Sci.* **41**, 2885–2929.

Wisanrakkit, G., Gillham, J.K. and Enns, J.B. (1990) *J. Appl. Polym. Sci.* **41**, 1895–1912.

Wolfe, S.V. and Tod, D.A. (1989) *J. Macromol. Sci. Chem.* **A26(1)**, 249–272.

Yamini, S. and Young, R.J. (1980) *J. Mat. Sci.* **15**, 1814–1822.

Yu, W. and von Meerwall, D. (1990) *Macromolecules* **23**, 882–889.

Ziman, J.M. (1979) *Models of Disorder* Ch. 7, Cambridge University Press, Cambridge, pp. 246–285.

4 Additives and modifiers for epoxy resins*

S.J. SHAW

4.1 Introduction

In addition to the two main ingredients of an epoxy formulation, i.e. resin and curing agent, numerous other formulatory materials are available and have frequently been employed to modify the properties and characteristics of epoxies, both uncured and in their cured form. In this chapter each of the main types of modifying material will be discussed with examples given of the most common types that have been employed together with the potential benefits they impart.

Although this discussion will be approached via a series of convenient headings it is important to recognise that many ingredients employed in epoxy formulations could legitimately be discussed under more than one heading. Thus, for example, some of the materials described in this text as flexibilisers could also be considered as reactive diluents and *vice versa*. The author apologises for any confusion this may cause.

4.2 Diluents

Diluents are usually employed as a means of reducing viscosity in order to aid general processability as well as allowing for greater incorporation of other formulatory ingredients such as fillers and as a means of improving characteristics such as wetting and the incorporation of resins into various fibrous reinforcements.

In addition to viscosity, it is important to recognise that other properties will also be modified, the properties in question and the magnitude of modification being dependent on both the type and quantity of diluent employed. In addition many of the materials capable of exerting diluent effects possess adverse toxicological characteristics, in many cases being more severe than those found with both resin and curative. Great care should therefore be taken in both selection and use.

Diluents that have been considered for use in epoxy resin technology can be divided into two broad classes comprising those which can be considered as either non-reactive or reactive.

4.2.1 Non-reactive diluents

A number of materials which can be regarded as non-reactive have been considered and used to provide viscosity reduction in epoxies. These include aromatic hydrocarbons such as toluene or xylene which are capable of reducing viscosity in a simple diglycidylether of bisphenol A (DGEBA) resin system sufficient to permit a room temperature viscosity approximately 20% of its initial value with a diluent loading of about 5%. When employed at this level, studies have shown that the level of impact on physical and mechanical properties are not usually dramatic, indeed in some cases the presence of the diluent has been shown to increase the extent of cure in the formulation, presumably through the reduction in viscosity providing a reaction medium more conducive to molecular mobility and thus reactivity. At higher concentrations however, the properties of the cured epoxy can deteriorate markedly and, due to their non-reactivity, diluents of this type can lead to the presence of voids in the cured polymer, primarily if the formulation is subjected to elevated temperature cure sufficient to allow vaporisation of the diluent.

Other non-reactive materials which have been considered as potential diluents include dibutyl phthalate, styrene and various phenolic compounds.

In addition to the drawbacks mentioned above, since the compounds are not chemically bound into the cross-linked network, cured formulations containing these materials usually exhibit inferior chemical resistance. Consequently, attempts at viscosity reduction with epoxies have generally made use of materials capable of chemical reaction with the epoxy resin, these generally being referred to as reactive diluents.

4.2.2 Reactive diluents

Two major classes of reactive diluent can and have been utilised commercially, these being either epoxy-containing diluents or compounds which derive their reactivity from functional groups other than epoxide.

4.2.2.1 Epoxy-based reactive diluents Two types of diluent can be considered under this classification, i.e. (i) mono-epoxy compounds and (ii) di/polyfunctional epoxies. With both classes the presence of the epoxide group(s) allows the diluent to participate, with the resin and curing agent, in the polymerisation and cross-linking reaction thus permitting the diluent to become chemically bound into the cross-linked network. Thus, to a degree, the use of reactive diluents can alleviate some of the debilitating effects found with the non-reactive compounds.

Figure 4.1 shows some mono epoxy reactive diluents which have been considered and indeed in some cases used to modify epoxy resin viscosity.

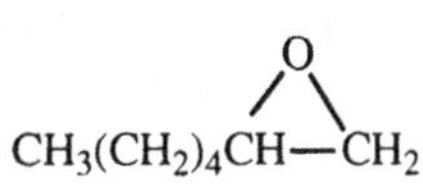

Octylene oxide

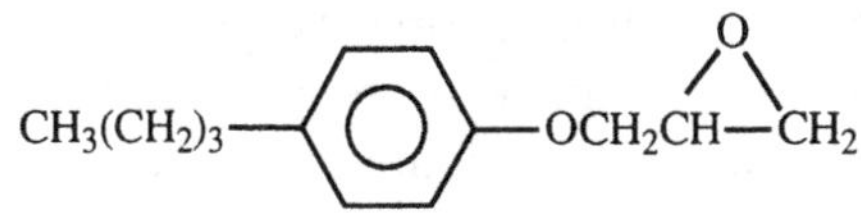

p-Butyl phenol glycidyl ether

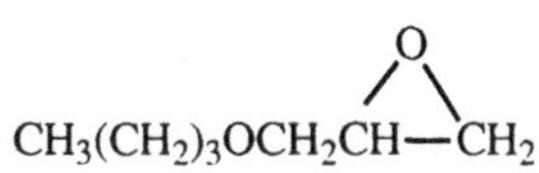

Butyl glycidyl ether

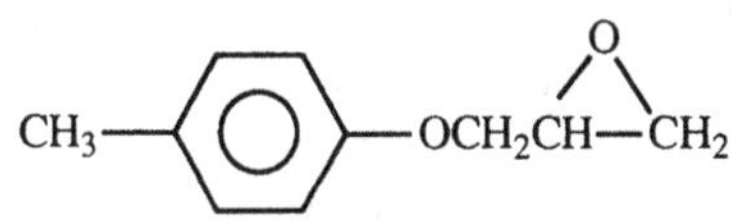

Cresyl glycidyl ether

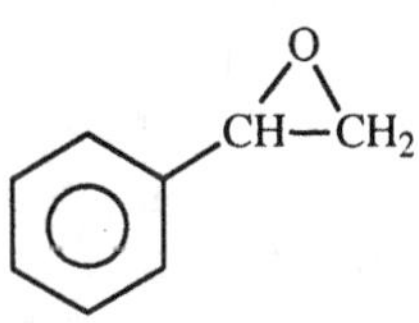

Styrene oxide

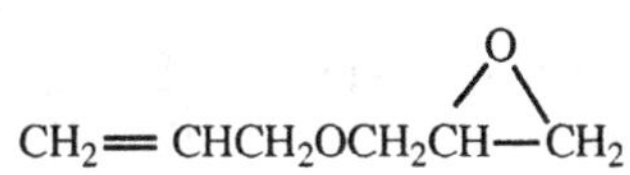

Allyl glycidyl ether

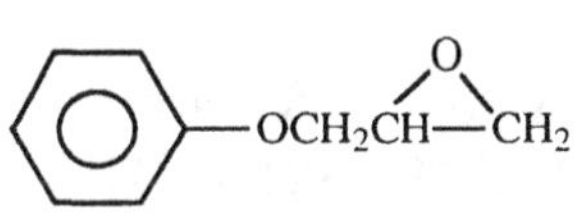

Phenyl glycidyl ether

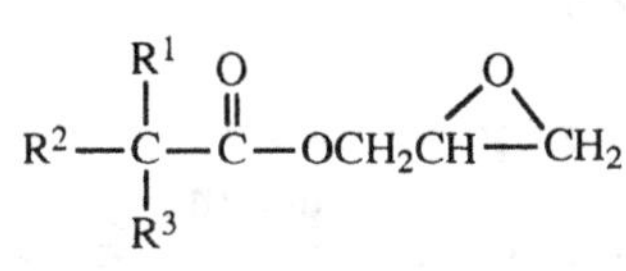

Glycidyl ester of *tert*-carboxylic acid

Figure 4.1 Monoepoxy reactive diluents.

Although capable of exerting a substantial viscosity-reducing effect, these materials essentially reduce the functionality of the reaction system leading to a decrease in the cross-link density of the cured formulation. Consequently important physical and mechanical properties are frequently affected, particularly the ability to exert a 'high temperature capability'.

The extent to which both cured properties and initial viscosity are reduced will be a function of both diluent type and concentration. Thus choice of both parameters will clearly be crucial in order to derive the greatest reduction of viscosity with minimal cost to other properties.

In an attempt to preserve physical/mechanical properties at elevated temperatures, the use of polyfunctional epoxy diluents has been considered. Although to a degree they can be regarded as epoxy resins in their own right, their use in more conventional formulations, e.g. those employing a DGEBA as the basic resin component, has demonstrated their ability to provide viscosity reduction. Typical of the polyfunctional epoxy compounds which have been suggested as reactive diluents are those shown in Figure 4.2. In most cases, since the funtionality of the system as a whole is not greatly affected by the incorporation of these materials, cross-link density, and therefore properties associated with it are not usually affected detrimentally. Indeed some property enhancement has been shown to be possible.

4.2.2.2 Non-epoxy based reactive diluents. Although epoxide-based reactive diluents are employed in the majority of cases where viscosity reduction is considered desirable, a number of compounds which derive reactivity via non-epoxide groups have been considered. Most notable amongst these have been triphenyl phosphite and lactone compounds such as butyrolactone, the structures of which are shown in Figure 4.3. Both materials have been shown to have effective viscosity-reducing characteristics, although frequently at the expense of an 'elevated temperature capability'.

4.3 Fillers

Apart from the resin and curative, fillers are possibly the most common formulatory ingredient employed in the majority of epoxy formulations. Literally hundreds of different filler types can and have been used to modify, in one form or another, the properties and characteristics of epoxies, in addition to reducing cost. Table 4.1 indicates in terms of advantages and disadvantages, the various effects that filler incorporation can have. Although fillers can be considered beneficial for many applications, disadvantageous characteristics such as an increased density (and hence weight) together with an increase in viscosity which is likely to influence the processing behaviour of the formulation, obviously require serious consideration.

Table 4.2 lists some of the more important particulate fillers which have been employed in epoxy formulations. Although the list cannot be regarded as exhaustive, it does indicate the range of characteristic modifications which the various filler types shown can provide. It is of interest to briefly consider some of these property modifications.

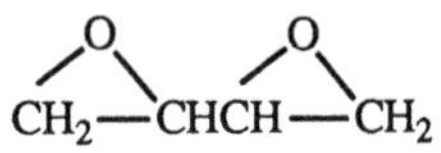

Butadiene dioxide

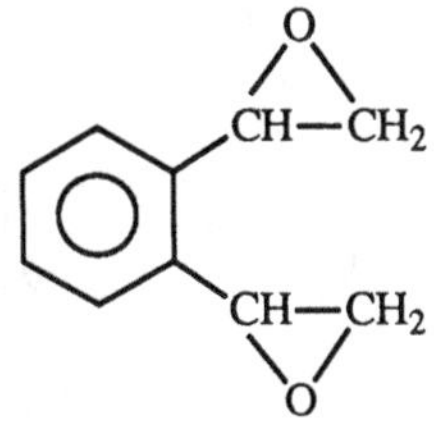

Divinylbenzene dioxide

Diglycidyl ether

Butanediol diglycidyl ether

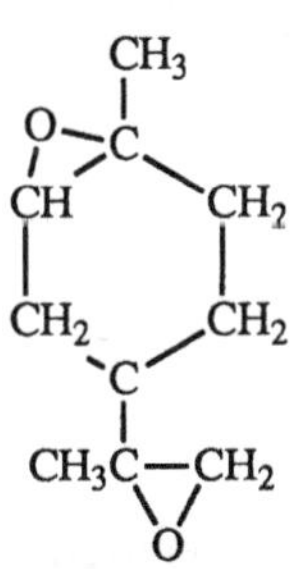

Limonene dioxide

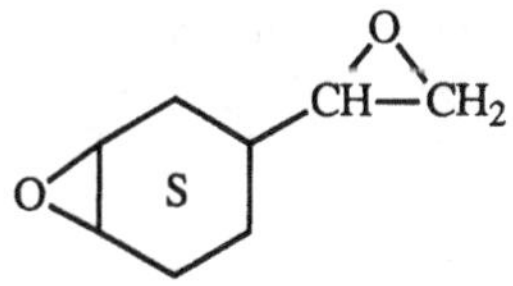

Vinylcyclohexene dioxide

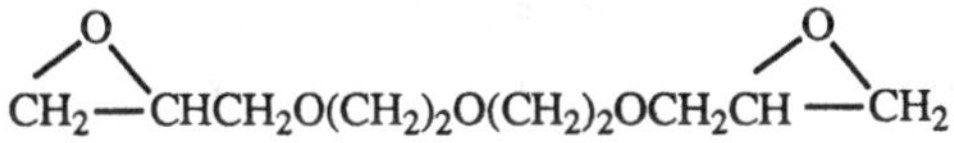

Diethylene glycol diglycidyl ether

Figure 4.2 Polyfunctional epoxy reactive diluents.

4.3.1 Physical/mechanical properties

The addition of particulate fillers generally results in reductions in strength characteristics such as tensile and flexural strength. Fibrous fillers, however, when employed at loading levels greater than a critical value, frequently provide improvements in these strength parameters. As would be expected, incorporation of fillers, particulate or fibrous, invariably produces

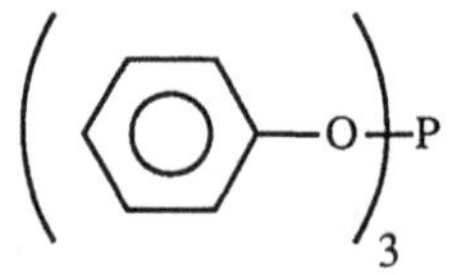

Triphenyl phosphite

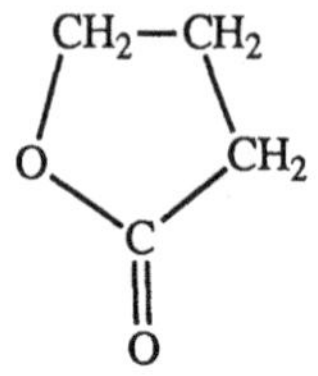

Butyrolactone

Figure 4.3 Non-epoxy based reactive diluents.

Table 4.1 Advantages and disadvantages of filler incorporation

Advantages	Disadvantages
Reduced formulation cost	Increased weight
Reduced shrinkage	Increased viscosity
Improved toughness	Machining difficulties
Improved abrasion resistance	Increased dielectric constant
Reduced water absorption	
Increased heat deflection temperature	
Decreased exotherm	
Increased thermal conductivity	
Reduced thermal expansion coefficient	

substantial increases in modulus, the magnitude being dependent on filler type and loading level.

One specific application where fillers have been shown to impart substantial strength improvements is with structural adhesive formulations where the addition of fillers such as aluminium and alumina is common practice (Lewis, 1988).

Fillers do not usually provide any significant enhancement of glass transition temperature, T_g, or other measures of high temperature distortion.

4.3.2 Thermal characteristics

Fillers allow the possibility of modifying various thermal characteristics of epoxy resins including properties such as exothermic behaviour, thermal conductivity and thermal expansion coefficient.

The exothermic characteristics of many epoxy formulations can give rise to serious processing difficulties, particularly when the production of bulky items is being considered. Since epoxies generally exhibit good thermal insulating characteristics, extremely high temperatures are possible in the middle of particularly large castings to the extent that component volatilisation and even charring can result. The incorporation of fillers can

Table 4.2 Filler types and potential property modifications

Filler	Property modification
Aluminium	Machinability, impact resistance, thermal conductivity, mechanical properties, dimensional stability
Alumina	Abrasion resistance, electrical resistivity, dimensional stability, toughness, thermal conductivity
Aluminium silicate	Extender, pigmentation, dimensional stability, chemical resistance
Aluminium trioxide	Flame retardation
Arsenic pentoxide	Thermal resistance
Barium sulphate	Extender
Beryllium oxide	Thermal conductivity
Calcium carbonate	Extender, pigmentation, dimensional stability, machinability, mechanical properties
Calcium sulphate	Extender, dimensional stability
Calcium silicate	Mechanical properties
Carbon black	Reinforcement, pigmentation, thermal conductivity, electrical conductivity, thermal resistance
Copper	Electrical conductivity, thermal conductivity, mechanical properties
Colloidal silica	Thixotropy
Fibrous glass	Impact strength
Graphite	Lubricity, pigmentation, thermal conductivity, electrical conductivity, abrasion resistance
Glass microballoons	Density reduction
Kaolin clay	Extender
Lithium aluminium silicate	Thermal expansion coefficient
Mica	Electrical resistance, dielectric properties, chemical resistance, toughness, moisture resistance, lubricity
Molybdenum disulphide	Lubricity
Quartz	Electrical properties, dimensional stability, extender
Sand	Abrasion, thermal conductivity
Silica	Abrasion resistance, electrical properties, extender, dimensional stability, thermal conductivity, moisture resistance
Silver	Electrical conductivity, thermal conductivity
Titanium dioxide	Pigmentation, dielectric properties, extender
Talc	Extender
Zirconium silicate	Arc resistance

substantially diminish exothermic heat generation by reducing the quantity of resin in the formulation and increasing thermal conductivity so as to allow more efficient removal of the exothermic heat. Indeed thermal conductivity can be increased by a factor of approximately five with the addition of fillers such as aluminium, alumina and copper. Generally an increase in filler concentration increases thermal conductivity, the upper loading limit usually being associated with the maximum permissable processing viscosity.

Fillers, when incorporated in sufficient concentration, can also be successful in reducing the thermal expansion coefficient of epoxies down to similar levels to that obtained with many metals. This can be advantageous in electronic and structural adhesive bonding applications.

4.3.3 Shrinkage

When epoxies undergo polymerisation and cross-linking, shrinkage occurs throughout the cure process which can be damaging in a number of applications ranging from electronics to adhesive bonding. Although generally regarded as superior to other thermosetting polymers such as phenolics and polyesters, further reductions in shrinkage beyond those found for most unformulated epoxies is frequently considered necessary. This can often be achieved through the use of various fillers and flexibilising ingredients. Filler incorporation reduces shrinkage by simple bulk replacement of resin with an inert compound which does not participate in the cross-linking process.

4.3.4 Electrical conductivity

For many electronic and electrical applications, epoxy resin formulations exhibiting electrical conductivity are required. Unformulated epoxies, like the majority of polymeric materials, exhibit high levels of electrical resistivity. The judicious use of fillers can, however, increase the levels of electrical conductivity quite significantly (Lee, 1988). The filler material most commonly employed to meet this requirement is silver, employed either in flake or powdered form. Occasionally other fillers such as copper, carbon black or graphite have also been considered, primarily as lower cost but reduced efficiency alternatives. The popularity of silver stems primarily from the absence of an oxide layer; such layers generally imparting electrically insulating characteristics. For this reason metallic fillers such as aluminium are seldom considered for this application.

4.3.5 Viscosity

The incorporation of fillers into an epoxy invariably results in an increased viscosity, the maximum loading of filler for most applications often being restricted by the maximum permissible working viscosity. Generally fibrous fillers exert greater viscosity-enhancing effects, on an equivalent weight basis than particulates. With the latter filler, particle size usually exerts a dominating effect with fillers of small particle size tending to increase viscosity to a greater degree than corresponding fillers of greater particle size. This can be attributed to the greater surface area of the former (Potter, 1970).

Some fillers, in particular various types of silica, are capable of exerting a thixotropic effect which has been put to good use in structural adhesives and formulations requiring anti-sag characteristics (Lewis, 1988).

4.3.6 *Toughness*

The usefulness of epoxy resins for many applications is often limited by their comparatively brittle nature and susceptibility to catastrophic failure. Consequently numerous investigations have been conducted over many years in an attempt to provide processes and procedures which would allow significant improvements in toughness, achieved at minimum expense to other important properties and parameters such as modulus and T_g. Probably the most successful means by which this has been achieved is by a process known as rubber-modification and the subject will be discussed in more detail later (section 4.6).

A number of investigations have shown that the incorporation of particulate fillers such as silica, glass microspheres and alumina trihydrate can increase the toughness of various epoxy formulations (Moloney *et al.*, 1983, 1984, 1987; Spanoudakis and Young, 1984). Perhaps of major advantage here is that this can be achieved with an improvement in modulus; although rubber modification generally provides a more efficient means of toughness enhancement, this is usually achieved at the expense of other important properties such as modulus.

Variables such as particle size, particle size distribution, particle surface chemistry and particulate volume fraction have been studied by various researchers which have indicated the most important variables necessary for optimum toughness enhancement.

A mechanism based upon the concept of 'crack pinning' (Figure 4.4) has been proposed to explain the toughness improvements brought about by particulate reinforcement (Lange, 1970; Evans, 1972; Green *et al.*, 1979).

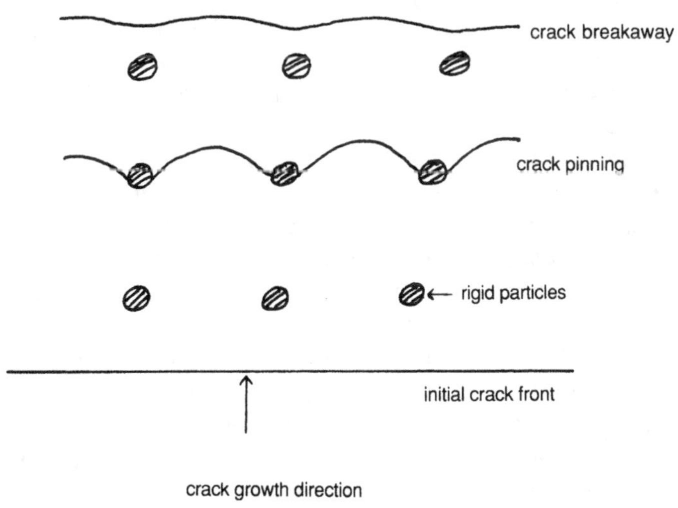

Figure 4.4 Crack pinning mechanism for toughness improvements.

It proposes quite simply that a propagating crack front, when encountering an inhomogeneity, becomes temporarily pinned at that point. An increase in load increases the degree of bowing between pinning points resulting in both new fracture surface and an increase in length of crack front. These processes lead to an absorption of energy and therefore an improvement in toughness (chapter 5, section 5.8.1).

4.4 Resinous modifiers

A wide variety of what could possibly be described as resinous materials have been considered for use in combination with epoxies either as a means of cost reduction or so as to impart property modifications. Typical examples include combinations of epoxies with nylons, polysulphides, polyvinyl formal and butyral, polyurethanes, styrene–butadiene copolymers, chlorosulphonated polyethylene, fluorinated polymers, silicones, isocyanates, furfural resins and acrylics.

One particular critical aspect of epoxies mentioned on numerous occasions in this chapter concerns brittleness, with the use of particulate fillers, rubber modification and thermoplastics in particular having been shown as fairly effective routes to toughness enhancement. In addition to these approaches, 'alloying' epoxies with other materials of a resinous nature has also been shown to be a particularly effective means of enhancing toughness. Combinations of epoxies with, in particular, nylon, polysulphides and polyvinyl formal/butyral have been the more traditional approach to toughness improvement (Lewis, 1988). In particular with nylon-epoxies, the levels of toughness achieved have been sufficient to allow their use as structural adhesives exhibiting high lap-shear and peel strength values at ambient and slightly elevated temperatures. However, due to the presence of the extremely hygroscopic nylon constituent, these systems have been shown to be vulnerable to the effects of atmospheric moisture which has greatly limited their use (chapter 7, section 7.7.2).

The combination of epoxies with phenolics has been shown to produce resins having a higher temperature capability than those obtained with epoxies alone. Indeed structural adhesives systems based upon an epoxy-phenolic chemistry have been shown to exhibit extremely attractive short-term high temperature capabilities (Wake, 1982).

Shaw and Tod (1990) have recently combined epoxy resins with bismaleimide prepolymers in an attempt to produce structural adhesives exhibiting the prime attributes of the two systems, i.e. the high temperature capability of the bismaleimide with the excellent processability of the epoxy. Reasonable success was achieved. Indeed resin systems supposedly based on epoxy-bismaleimide chemistry have recently appeared commercially as structural adhesives and as the basis of composite prepregs (Landman, 1984).

Since epoxies are capable of absorbing, in some cases, considerable quantities of water which can have a severely debilitating effect on various important mechanical properties, many investigations have been conducted in an attempt to identify the various molecular and structural factors which are of importance. As a result it is now possible to tailor epoxies with a lower propensity for water absorption together with a lesser sensitivity to the effects of absorbed moisture.

Many of the formulatory ingredients described in this chapter will exert an influence on water absorption characteristics. Studies conducted by various workers have shown that the incorporation of halogens into the network structure of various epoxy systems can improve hydrophobic characteristics considerably. In particular Griffith (1982) has developed fluorinated epoxy resins together with specifically designed curing agents having the structures shown in Figure 4.5. Investigations conducted by Shaw and co-workers (1988) have demonstrated the pronounced hydrophobicity which these

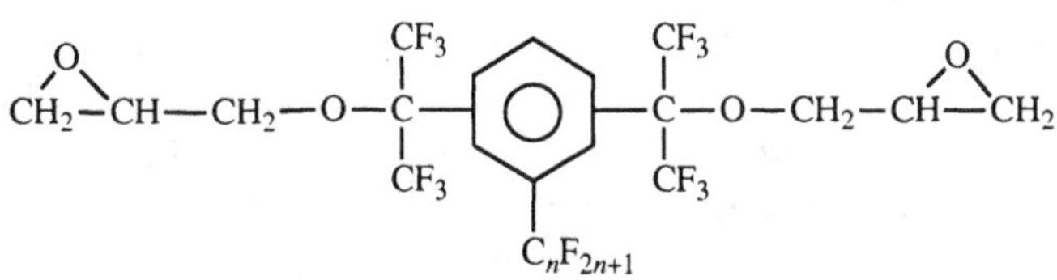

Fluoroepoxy resin

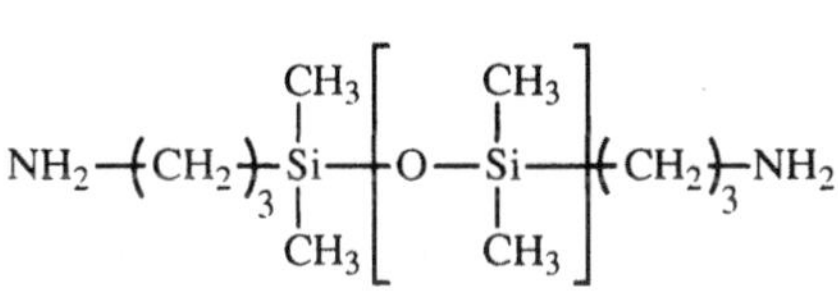

Siliconeamine curing agent

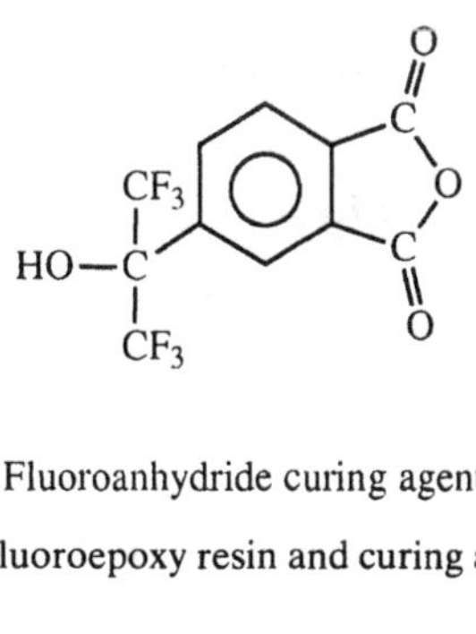

Fluoroanhydride curing agent

Figure 4.5 Fluoroepoxy resin and curing agent structures.

systems are capable of in comparison to the more traditional non-halogenated epoxies. Although such studies have demonstrated that these systems are capable of use on their own, Twardowski and Geil (1991) have suggested that, due to their relatively low T_g values, blending of fluoroepoxies with, for example, DGEBA resins would provide a means of both incorporating hydrophobicity whilst maintaining reasonable levels of mechanical performance. Similarly, studies conducted by Goobich and Marom (1982) have shown that the incorporation of a brominated epoxy additive to both tetraglycidyldiaminodiphenylmethane (TGDDM) and DGEBA systems improves hydrophobicity whilst essentially maintaining the level of important mechanical properties.

4.5 Flexibilising/plasticising additives

In addition to filler or elastomer incorporation, a further means by which the inherently hard and brittle characteristics of epoxies can be alleviated is by making use of plasticising/flexibilising additives. The former approach traditionally makes use of long-chain, non-reactive molecules which do not become incorporated into the cross-linked epoxy structure. It is here that the distinction is usually made between plasticising and flexibilising materials, plasticisers acting in a non-reactive manner whilst flexibilisers is the term usually applied to materials that react with the epoxy system during cure. Consequently they become incorporated into the backbone structure which generally results in them achieving a greater effectiveness than the plasticising ingredients.

4.5.1 Plasticisers

Plasticisers of the type conventionally employed in vinyl type polymers such as PVC have not achieved wide usage in epoxy technology. In addition to being essentially incompatable with epoxies they also suffer from the traditional disadvantage exhibited by inert plasticising materials in being prone to separation from the base epoxy. Even in circumstances where these disadvantages are not manifested, the extent of flexibilisation achieved is generally slight in comparison to the more efficient reactive systems.

4.5.2 Reactive flexibilising additives

In addition to reducing the basic hardness and rigidity of epoxies, flexibilising ingredients are also able to confer improvements in terms of reduced exotherm and, in certain circumstances, reducing shrinkage (Lee, 1988). The flexibilised nature of the cross-linked network can result in a more strain tolerant system, better able to relieve internal stresses which can develop in cured networks, particularly when employed in high volume castings. They can improve adhesive joint properties such as lap shear and particularly peel strength in addition to improvements in impact strength and low temperature crack resistance. On the negative side, reactive flexibilisers can yield cured systems that are less strong mechanically and have reduced electrical and, in particular, chemical and solvent resistance in comparison to their non-flexibilised counterparts.

Some of the reactive flexibilisers which have been considered for use in epoxies will now be briefly discussed.

4.5.2.1 Polyamides. The polyamides typically employed in many epoxy formulations are produced by a Diels–Alder reaction of the 9,12 and 9,11 isomers of linoleic acid to form the dimer, which, on further condensation with an aliphatic polyamine, results in a polyamide. A typical structure is shown in Figure 4.6. As indicated, the molecule contains not only amide groups but also primary and secondary amine groups; the latter groups being capable of participating in cross-linking reactions with the epoxy. They are therefore capable of exhibiting flexibilising characteristics as well as being curing agents in their own right. With DGEBA resins, polyamide concentrations of approximately 25 to 50 phr are sufficient to provide cured products exhibiting both toughness and impact resistance. Although higher loading levels will result in further increases in flexibility, this will be accompanied by reductions in, for example, T_g and properties associated with it together with products exhibiting a weak, 'cheesy' consistency.

4.5.2.2 Carboxyl-terminated polymers. Possibly the most commonly employed carboxyl-terminated polymers which have been employed in epoxy resin technology are the carboxyl-terminated butadiene-acrylonitrile rubbers. When employed in concentrations of less than 20 phr they usually exert a toughening rather than a flexibilising effect (see section 4.6).

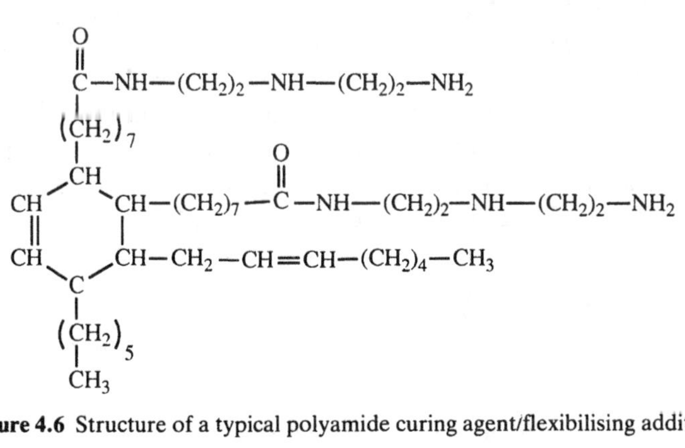

Figure 4.6 Structure of a typical polyamide curing agent/flexibilising additive.

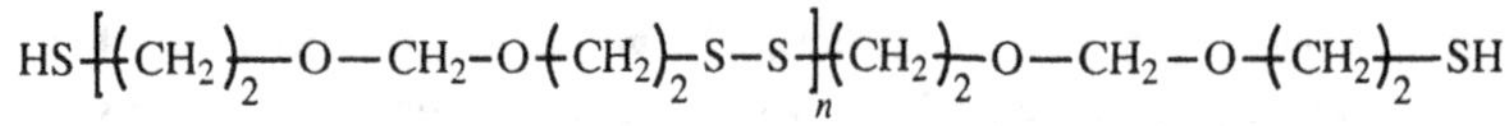

Figure 4.7 Polysulphide structure.

However, by judicious choice of elastomer concentration, the curing nature of the fomulation (type of curative and cure conditions) as well as the chemical nature of the elastomer, it is possible to introduce a high level of flexibility into an epoxy using these materials. Improvements in flexibility have also been achieved by the use of carboxyl-terminated polymers such as carboxyl terminated polyisobutylenes (Lee, 1988).

4.5.2.3 Polysulphides. Polysulphides are essentially mercaptan-terminated elastomeric materials having the structure shown in Figure 4.7. The value of n in the structure can vary from about 2 up to 26, yielding a wide range of properties and flexibilising capabilities, although the higher molecular weight materials can impose serious processing difficulties due to their high viscosities. The polysulphides are usually used at concentrations of between 25 and 100 phr in order to provide fairly tough, impact-resistant products. At even higher concentrations (>100 phr) formulations can exhibit poor tear resistance together with a decrease in other important characteristics such as hardness, shrinkage and electrical resistivity with increased polysulphide loading. As with most attempts at flexibilisation, the incorporation of polysulphides will generally result in substantial reductions in T_g and properties associated with it. In addition they also exhibit an objectionable odour which, fortunately, usually disappears after cure.

4.5.2.4 Polyglycol diepoxides. The general structure of the polyglycol diepoxides is as shown in Figure 4.8. These materials are generally produced by reacting the terminal hydroxyl groups of the appropriate glycol with epichlorhydrin so as to obtain the diglycidyl ether. They are generally used as blends with epoxies of the DGEBA or novolac-epoxy types, typically at concentrations of between 10 and 30 phr. At these levels, properties such as elongation, impact strength and occasionally tensile strength are increased with, as is usually typical, reductions in T_g and heat distortion temperature (HDT). Chemical resistance and electrical properties also tend to suffer as the concentration of flexibiliser is increased (Lee, 1988).

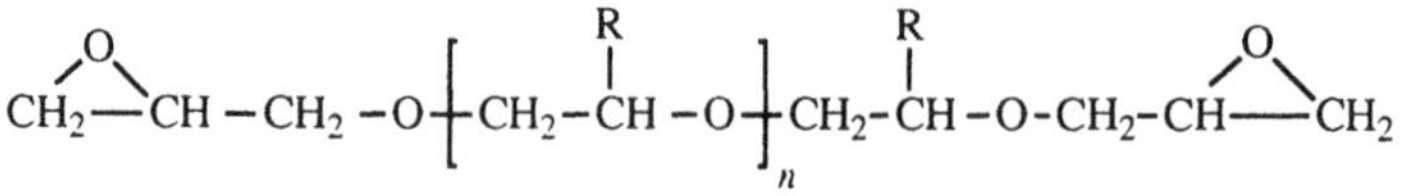

Figure 4.8 Structure of a polyglycol diepoxide.

4.6 Elastomeric modification

Of all the approaches that have been considered and adopted in an attempt to alleviate the brittle characteristics of epoxy resins, elastomeric modification has possibly been the most successful, where quite dramatic improvements in toughness have been achieved with, in many cases, only modest reductions in other important properties. Due to the technological importance with which elastomeric modification is now viewed, it will be of interest and indeed importance to consider this subject in some detail. In this account the various factors which have been shown to be of importance for successful toughness enhancement, including factors such as elastomer structure and chemistry, compatibility and morphology, will be discussed together with an account of the main mechanism which has been proposed to account for the effects of elastomeric modification.

4.6.1 Types of elastomeric modifiers

The various types of elastomeric materials which have been considered for the rubber modification of epoxies can be catagorised thus:

1. Reactive butadiene-acrylonitrile rubbers
2. Polysiloxanes (Yorkgitis *et al.*, 1984)
3. Fluoroelastomers (Mijovic *et al.*, 1984)
4. Acrylate elastomers (Kirshenbaum *et al.*, 1984).

Of these the first, the butadiene-acrylonitrile rubbers have been by far the most widely studied and employed in the rubber-modified systems which have achieved the most commercial success. Thus the majority of this section will be concerned primarily with these materials.

Butadiene-acrylontrile rubbers, as the name implies, are composed of a relatively low molecular weight backbone consisting of butadiene and acrylonitrile groups with reactive groups in the terminal positions of the molecule as indicated:

$$X\!\left[\!\left(CH_2\text{–}CH{=}CH\text{–}CH_2\right)_x\!\left(CH_2\text{–}\underset{\displaystyle CN}{\underset{|}{C}}H\right)_y\right]_z\!X$$

Numerous variants of these systems have been investigated with the main variables being (a) acrylonitrile content (Rowe *et al.*, 1970), (b) molecular weight (McGarry and Wilner, 1968) and (c) nature of the terminal group (Siebert and Riew, 1971).

The acrylonitrile portion of the molecule imposes a degree of polarity on an otherwise non-polar structure. As will be explained later a compatibility/incompatibility balance dictates to a large degree the success or otherwise of attempts at toughness enhancement. This is largely controlled by the relative polarities of the resin and rubber components, the latter being influenced by the acrylonitrile content. Various morphological characteristics and indeed mechanical properties have been related to the rubber acrylonitrile level. For reasons also related to compatibility and morphology, elastomer molecular weight has also been shown to be an important variable.

Experience has shown that terminal reactivity is an equally important requirement for effective toughness enhancement. It is now generally recognised that good adhesion between both the elastomeric and matrix components of a rubber-modified thermoset is of vital importance. This can usually be achieved by ensuring a means by which the liquid rubber can react with the thermosetting polymer, in this case an epoxy. The type of elastomer chosen therefore, at least in terms of terminal functionality, will depend upon the chemical characteristics of the matrix polymer. With epoxies a number of terminal functionalities have been studied including epoxy, hydroxyl, phenol, mercaptan, vinyl, amine and carboxyl with at least reasonable success being achieved in the majority of cases (Riew *et al.*, 1976). However, possibly the greatest benefits have been obtained from elastomers containing the carboxyl functionality (CTBNs).

Other elastomer types such as the polysiloxanes and polyacrylates are usually only considered when application conditions are of such an extreme nature so as to prevent employment of the butadiene-acrylonitrile systems. For example, the polysiloxanes exhibit important advantages in terms of both low temperature flexibility (due to a low T_g of about −120°C relative to −35°C for butadience-acrylonitrile systems) and excellent weatherability and moisture resistance. In addition the unsaturated structure of butadiene-acrylonitrile systems leaves them prone to thermal instability and thus unsuitable for long-term use at elevated temperatures. The polysiloxanes, acrylates and fluoroelastomers can be regarded as having superior thermal resistance and would thus be considered more suitable for such conditions. Although exhibiting these distinct advantages, work to date has shown the clear superiority of the butadiene-acrylonitrile rubbers in promoting the greatest toughness enhancement in epoxy resins.

4.6.2 Compatibility and morphology

It is now universally recognised that in order to promote a substantially enhanced toughness phase separation and the development of a two-phase morphology is critically important whilst maintaining or at least minimising the deterioration in other important properties/parameters such as modulus and T_g. For significant toughness enhancement the rubber must initially dissolve and become dispersed on a molecular level in the resin but precipitate when the epoxy begins to cross-link and thus form the required two phase morphology with rubbery particles dispersed in and bonded to the cross-linked epoxy matrix. Thus elements of both initial compatibility and eventual incompatibility must be regarded as vital elements of the system. This approach can be viewed in a more theoretical sense by considering the thermodynamics of mixtures in general and solubility parameters in particular.

The compatibility/incompatibility of any mixture of two substances is controlled fundamentally by the Gibbs free energy of mixing, ΔG_m given by

$$\Delta G_m = \Delta H_m - T\Delta S_m \tag{4.1}$$

where ΔH_m and ΔS_m are the molar enthalpy and entropy of mixing respectively.

By considering a cohesive energy density approach to molecular interaction, compatibility can, to a degree, be characterised by the solubility parameter, δ, the value of which for a particular material determines the tendency of its molecules to attract its own species in preference to those of a dissimilar species. This approach can be expressed quantitatively in the equation,

$$\Delta H_m = V(\delta_1 - \delta_2)^2 \phi_1 \phi_2 \tag{4.2}$$

where V is the molar volume of the mixture, δ_1 and δ_2 are solubility parameters for phases 1 and 2 and ϕ_1 and ϕ_2 are the respective volume fractions of the two phases. Molecules having similar chemical structures are generally found to have similar solubility parameters resulting, in turn, in a reduced tendency to demix. However since this approach only considers molecular interactions resulting from dispersion forces, anomalies do occur, particularly in systems where polar interactions and hydrogen bonding play a prominent inter-molecular role.

Combining equations 4.1 and 4.2 yields the following equation providing a more detailed view of the factors influencing ΔG_m and hence miscibility

$$\Delta G_m = V(\delta_1 - \delta_2)^2 \phi_1 \phi_2 - T\Delta S_m \tag{4.3}$$

Although, as mentioned above, lacking in certain respects, equation 4.3 provides a clear insight into the factors that will play a prominent role in both phase separation and morphology development. In order to develop the required morphology (i.e. small elastomeric particles present in the cross-linked resin matrix) good compatibility between epoxy and elastomer is required upon mixing of the two phases. Thus an initial requirement is that ΔG_m should be negative. As shown clearly in equation 4.3 this would be promoted by small differences in the δ parameters together with a low molar volume (i.e. low molecular weight). However, in order to avoid the continued intimate dispersion of the elastomeric component and hence allow flexibilisation of the cross-linked polymer, at some stage in the cure process, ΔG_m must become positive thereby leading to both phase separation and two-phase morphology development. From a molecular standpoint this could be achieved by changes in molecular structure resulting in diverging values of δ or by an increase in molecular weight or a combination of both occurring during curing. With most, if not all, rubber-modified epoxy formulations, it is the molecular weight build-up during cure (resulting in an increased magnitude of V) that is believed to result in a change in ΔG_m from negative to positive and thus phase separation. Thus equation 4.3 in a somewhat simplistic sense, indicates possibly the two main molecular factors that will control the critically important process of phase separation and morphology development, i.e. factors which will play a prominent role in determining the properties of the cross-linked polymer, these being chemical structure and molecular weight.

With the more commonly studied DGEBA/CTBN formulations, various factors have been shown to influence morphology. The most notable of these include rubber concentration (Riew *et al.*, 1976), type and extent of curing agent employed and cure conditions (Chan *et al.*, 1984; Shaw and Tod, 1989).

As would be expected, increasing the quantity of elastomer in the system results in an increase in the volume fraction of the elastomeric phase and, as has been shown in various studies, also produces an increase in the size of elastomeric domains. Elastomeric concentrations in excess of about 20% by weight have generally been shown to exhibit a process known as phase inversion where a reversal in morphological characteristics occurs resulting in cured epoxy particles embedded in an essentially elastomeric matrix (Bascom *et al.*, 1975). Although many studies have shown a positive relationship between rubber volume fraction and the extent of toughness enhancement (Chan *et al.*, 1984) as would be expected, mechanical properties generally deteriorate rapidly at elastomer concentrations in excess of this value (Bascom *et al.*, 1975).

The molecular weight and acrylonitrile content of the rubber have also been extensively studied. Generally, molecular weight reductions and enhanced acrylonitrile content have shown similar effects namely reductions in both elastomer particle size and volume fraction together with increased tendencies towards entrapment of elastomer in the matrix and thus flexibilisation.

The type of curing agent employed will play a major role in determining the mechanical properties of a rubber-modified epoxy. Not only will it strongly influence the structure of the cross-linked epoxy phase and therefore its mechanical properties, it will also be capable of exerting a substantial influence on morphology. This it can do by firstly altering the compatibility/incompatibility balance between the elastomer and epoxy. This effect can be particularly severe with curing agents which, by necessity, need incorporation in large quantities e.g. anhydrides and certain modified amine-based curatives. Perhaps of greatest significance, however, is the curing agent reactivity. Substantial evidence is now available which indicates that curing agents of a highly reactive nature are relatively inefficient in promoting substantial toughness improvements. This is generally attributed to what could be described as 'morphological constraint', where the highly reactive nature of the curative agent results in a relatively rapid onset of gelation which thus interfere with and prevents full phase separation. Increased curative reactivity, as a result, is generally found to produce systems exhibiting both low volume fractions of the rubber phase and relatively small particle size. Since a high volume fraction is of major importance for optimum toughness enhancement, the use of highly reactive curing agents should, if possible, be avoided. Since many commercial applications, driven by economic considerations, require the rapid cure times provided by highly reactive curatives, this can cause some shortfall in properties.

Recent research has shown that cure conditions can have as significant an effect on the mechanical properties of a rubber-toughened epoxy as the formulation variables highlighted above (Shaw and Tod, 1989). Once again the influence of this parameter, particularly cure temperature, on morphology is generally considered responsible. Although the effects of formulation variables on morphology are now well recognised, the influence of cure temperature is somewhat less predictable, due to its ability to influence several factors responsible for affecting phase separation. For example, an increase in cure temperature will promote both an improvement in epoxy/rubber compatibility and an increase in the cure rate; factors which will tend to retard the phase separation process leading to relatively small values of both rubber volume fraction and particle size. Conversely an increase in temperature would also result in a reduction in viscosity which would enhance the phase separation process thereby having the reverse effect on the morphological characteristics described above. The final influence of cure temperature, therefore, will be dependent on the relative importance of compatibility, viscosity and cure rate on the phase separation process. Although some conflicting trends have been observed, particularly with regard to rubber volume fraction, an increase in cure temperature has been found to usually result in an increased rubber particle size (Shaw and Tod, 1989).

At this point it is perhaps beneficial to briefly discuss the morphological factors that are generally believed to control the toughness of rubber-modified epoxies. Although this has been, and to a degree still is a somewhat controversial area, evidence accumulated in recent years has focused upon what would now seem to be the most critical parameters. Essentially the debate is centred upon the relative importance of two morphological characteristics, namely rubber volume fraction and particle size/particle size distribution. From their early pioneering research, Sultan and McGarry (1973) suggested that the key morphological factor responsible for optimum toughening was particle size, with relatively large particles ($> 1\ \mu m$

diameter) promoting crazing within the epoxy matrix and smaller particles (approximately 0.01 μm) leading to shear deformation. Craze generation, in their view, was seen to be the most efficient means of toughness enhancement and they therefore proposed that the formation of relatively large rubber particles was the key to maximum toughness. Crazing is now, of course, almost universally recognised as not being a significant deformation mechanism with epoxies. So in this respect alone this theory can be regarded as flawed. However, in one major respect involving the formation of particularly tough rubber-modified epoxies, the role of particle size and, in particular, particle size distribution has been and indeed continues to be emphasised. In the 1970s Riew *et al.*(1976) found that by incorporating bisphenol A into a DGEBA/CTBN/piperidine system, substantial improvements in toughness could be achieved over and above that obtained from the non-bisphenol A formulation. They noted in particular that this highly beneficial effect was accompanied by a bimodal particle size morphology, which they regarded as responsible for the enhanced mechanical performance. Since this early work, similar observations have been found by other workers employing the same approach and indeed the bimodal particle size concept has been utilised in commercial rubber-modified epoxy formulations, where the incorporation of both low and high molecular weight CTBN rubbers have been found to produce the required morphology together with impressive fracture performance data. Although such evidence would appear to strongly support the particle size protagonists, it is important to recognise that the parameters of interest, volume fraction and particle size, are extremely difficult to vary in magnitude independently of each other. Thus, for many formulations, although bimodal particle size distributions may exist and indeed exert an influence on the properties of the epoxy, it is likely that volume fraction could also vary in a significant manner.

Although a relatively small number of researchers would still appear to view particle size and size distribution as providing a not insignificant contribution to toughness enhancement, most now recognise that the critical morphological factor is rubber volume fraction. Generally the greater the volume fraction of the phase-separated elastomeric component the greater the improvement in toughness. Although this will partly depend on the amount of rubber initially added to the formulation, the other factors described previously, e.g. rubber molecular weight and acrylonitrile concentration, curing agent reactivity etc. will also have an important influence. It is particularly worth while noting that numerous studies have shown that rubber phase volume fraction usually exceeds the volume fraction of added rubber thus providing an insight into the composite nature of the rubber domains. An anomalous effect such as this can only be resolved if it is assumed that the rubber particles have a not insignificant concentration of epoxy, either on the molecular level or as sub-inclusions dispersed in the elastomeric phase.

4.6.3 *Toughening mechanisms*

Several mechanisms have been proposed to account for the toughness enhancements generated by the incorporation of elastomeric particles into glassy polymers such as epoxies. These include theories based upon contributions arising from rubber tear, multiple crazing and shear yielding.

A toughening model developed independently by Kinloch *et al.* (1983) and by Pearson and Yee (1983) based upon both cavitation and matrix shear yielding is now generally recognised as the most consistent in terms of the wealth of experimental data generated in recent years. Due to its importance it is of interest to discuss this model in some detail.

Many studies have indicated a positive correlation between fracture toughness and the extent of plastic deformation found on fracture surfaces. This, together with only very limited evidence for crazing processes have suggested a mechanism based upon yielding and plastic shear flow of the matrix as the primary source of energy dissipation during fracture. Since greatly enhanced plastic deformation has been almost universally associated with rubber particle incorporation it is clearly necessary to focus attention on the rubber particles themselves and in particular the distribution of stress which exists around them when located in the vicinity of a loaded crack tip. This is best achieved by consideration of the work of Goodier (1933) who studied the case of an isolated spherical particle embedded in an isotropic elastic matrix which is subjected to a uniform uniaxial tensile stress at points remote from the particle. He derived equations which indicated that for a rubbery particle, which possesses a considerably lower shear modulus than the matrix, the maximum stress concentration occurred in the equatorial regions of the particle with a value of approximately 1.9. However, allowing for complicating factors such as the large volume fraction of rubber particles which would invariably exist, together with the triaxial as opposed to the uniaxial stress state which particles would encounter, a stress concentration factor of approximately 1.6 would exist.

The interactions between a triaxial stress field in the vicinity of a loaded crack tip and that existing around rubber particles in close proximity could be envisaged resulting in the initiation of two important processes.

First, the development of a triaxial stress would lead to dilatation of the matrix. This, in combination with triaxial stresses inherent in the rubber particles due to differential thermal contraction effects resulting from the cure process, would cause cavitation of the particles. Indeed it is commonly found that elastomeric materials undergo cavitation quite readily under the action of a triaxial stress field. This cavitation process can be viewed as responsible for the stress whitening effects observed in the numerous studies conducted on rubber-modified epoxies.

The second process concerns the initiation and growth of shear yield deformation in the matrix. During loading the stress concentrations developing at the rubber particle equators would act as sites for the initiation of shear deformation. Due to the large number of particles present, the degree of yielding generated would be substantially greater than would otherwise occur in an unmodified material. However, particles would also act as sites for shear yield termination which would therefore keep the yielding localised.

It is possible and indeed likely that both of the above processes would occur during the early stages of loading, but with rubber particle cavitation probably being the most dominant. However, once initiated, cavitation within the particles would lead to more pronounced void formation which in turn would greatly enhance further shear yielding in the matrix. This would be partly attributable to an increased stress concentration which would be likely to accompany cavitation and void formation. However, of greater importance, void formation would result in reduced constraint on the matrix adjacent to voided particles, thus relieving the degree of triaxiality. This in turn would lower the yield stress in the inter-particle regions and thus promote further extensive shear yielding. This would cause extensive crack blunting ultimately resulting in the increasing development of the plastic zone at the crack tip. It is the enhanced development of this plastic zone which results in the greatly improved fracture toughness exhibited by rubber-modified epoxies in relation to their unmodified counterparts.

4.6.4 The hybrid modification approach

Although some fairly modest improvements in toughness have been observed with particulate reinforcement (section 4.3.6) these have not generally been of the same magnitude as that obtained with elastomeric modification. However, as previously described, the substantial modulus increase which generally accompanies toughness enhancement with particulate fillers can be regarded a major advantage. As a result several studies have considered a hybrid approach to toughening; making use of both particulate reinforcement and elastomeric modification (Maxwell *et al.*, 1984; Kinloch *et al.*, 1985). Much of this work has shown significant benefits in terms of both substantial improvements in toughness and modulus, which can be achieved by a combination of this type. It remains to be seen whether these attractive features result in any significant commercial usage.

4.7 Thermoplastic modification

The inclusion of thermoplastics into epoxy formulations can be regarded as a more recent means of imposing property variations into cross-linked epoxies and recent studies have shown that once again toughness enhancement is the major driving force.

Attempts at imparting significant improvements in toughness with high cross-link density epoxies such as those based upon the traditional aerospace epoxy, TGDDM, by use of elastomeric modifiers has generally resulted in failure. This has been attributed to the inability of the cross-linked epoxy matrix to undergo any significant shear yielding in response to the presence of the rubber particles. A somewhat different approach, using for example thermoplastics, has therefore been seriously considered, particularly for the more difficult to toughen multifunctional systems.

Attempts at thermoplastic modification have been conducted using a fairly limited range of thermoplastic polymers including, for example, polyethersulphone (Bucknall and Partridge, 1983) and polyetherimide (Bucknall and Gilbert, 1989), the basic structures of which are shown in Figure 4.9.

Bucknall and Partridge (1983) attempted to study the relationship between the structure and mechanical properties of two types of polyethersulphone-modified epoxy, one being trifunctional and the other based upon the TGDDM resin. They found that morphology, studied using both dynamic mechanical spectroscopy and scanning electron microscopy, was dependent upon both the type and concentration of resin and the curing agent employed. In certain circumstances two-phase morphological characteristics were observed with a polyethersulphone-rich phase dispersed in a cross-linked epoxy matrix. Because of both high modulus and T_g values of the polyethersulphone employed (2.8 GPa and 210°C respectively), large concentrations of thermoplastic generally had a minor effect on these particularly important parameters. This, in fact, represents a significant advantage over rubber-modification where reductions in these parameters, particularly modulus, can occur. Unfortunately, in this study,

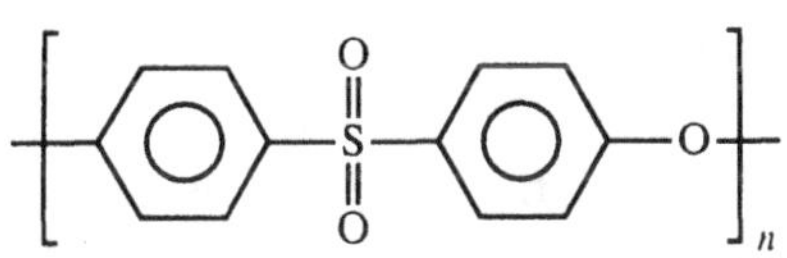

polyethersulphone

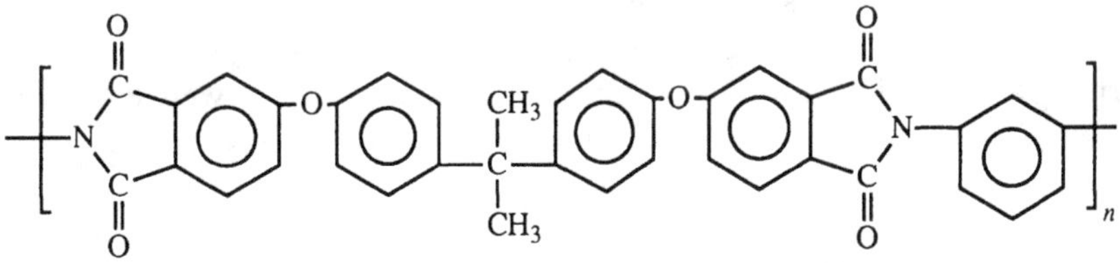

polyetherimide

Figure 4.9 Basic structures of polyethersulphone and polyetherimide.

polyethersulphone addition was found to have only a minor beneficial effect on fracture energy with thermoplastic additions of up to 40 phr resulting in less than a doubling in fracture energy.

Somewhat similar studies have been conducted by others including Diamont and Moulton using polyetherimide (Diamont and Moulton, 1984) and McGrath and co-workers (Hedrick *et al.*, 1985; Cerere *et al.*, 1986) who employed hydroxyl- and amine-terminated sulphones as modifiers for DGEBA epoxies. Although in both cases increases in toughness were obtained, as in the work of Bucknall and Partridge, the magnitude of improvement was small in comparison to that which could be expected with elastomeric modification.

More recent work by Bucknall and Gilbert (1989) using polyetherimide modification, showed that the incorporation of 30 phr of thermoplastic into an anhydride-cured TGDDM resin resulted in an approximately threefold increase in fracture toughness, K_{IC}, with modulus remaining virtually constant.

Although the use of tough, thermoplastic polymers provides for improvements in epoxy toughness, in comparison to rubber-modification the level of improvement is often generally poor, particularly with relatively low T_g systems. With high cross-link epoxies where rubber-modification would probably provide a similar, relatively poor, level of toughness enhancement as that provided by thermoplastics, the comparatively high modulus and T_g values of the latter would be regarded as highly advantageous. Thus in the majority of cases it would seem likely that thermoplastic modification would only be considered worth while for relatively high cross-link density epoxies under consideration for high temperature applications.

4.8 Miscellaneous additives

In addition to the main formulation types discussed in this chapter, countless other miscellaneous additives have been either considered or used in various epoxy systems. Since space clearly prevents a detailed account of all these ingredients, it is of interest to consider a few that have either been employed or which recent studies have indicated have considerable potential.

The use of coupling agents in epoxy resin technology, particularly in structural adhesive bonding applications, is now fairly well established. Most of the published literature within this context has been concerned primarily with organosilanes as pre-bond primers. Studies devoted principally to one specific organosilane, γ-glycidoxypropyltrimethoxy-silane, have shown that this material can dramatically improve the moisture resistance of epoxy adhesive bonded metallic joints (Kinloch *et al.*, 1975; Gledhill *et al.*, 1990). Unfortunately, the use of a primer as a coating prior to application of an adhesive adds an extra step to the bonding process. As a result, the possibility of adding a coupling agent to the adhesive has been considered and indeed is common commercial practice with some epoxy adhesive manufacturers. Since coupling agents such as organosilanes are generally considered 'surface active', i.e. they exert an influence on the interfacial zone between the substrate and epoxy, a mechanism of migration and interface activity would clearly need to apply with an approach of this kind.

As well as adhesive bonding, organosilanes have also been employed to enhance the properties of filled epoxy systems, particularly where resistance to moisture is required.

Since most epoxies are capable of burning, for many applications flame retardency is considered important. This can be achieved by incorporating into the formulation halogen (e.g. bromine)-containing resin or curative species (Tanaka, 1988), flame retarding fillers such as antimony trioxide (Potter, 1970) or somewhat more esoteric additives such as amino alkyl phosphates or halogenated β-lactones (Mika and Bauer, 1988).

Although the presence of a solvent in an epoxy formulation would be regarded as a substantial handicap in many applications, they are an important formulatory ingredient for one specific application, namely surface coatings (Tess, 1988). Solvents which have been considered for this purpose include methyl ethyl ketone, 2-ethoxyethanol, xylene, 2-ethoxy ethylacetate and toluene either alone or more generally in blends of various solvents where greater solvating power has been demonstrated. Numerous factors will require consideration in the appropriate choice of solvent(s). These include solvent power and evaporation rate which can strongly influence the final properties of a surface coating together with viscosity, cost and of course the highly important subject of health and safety.

In previous sections in this chapter, the general question of resin shrinkage upon cure has been briefly discussed, with particular emphasis on the beneficial effects of ingredients such as fillers and flexibilisers. In recent years, a somewhat novel approach, making use of the so-called 'expanding monomer' concept has yielded beneficial effects. Research conducted by Piggott and co-workers (1985) has shown that, by copolymerising with epoxies compounds such as dinorbornene spiro ortho carbonate, the levels of shrinkage encountered upon cure can be greatly diminished resulting in substantial improvements in properties which would be affected by the internal stresses resulting from shrinkage. In particular, improvements in the toughness of epoxy-carbon composites have been shown to result from such modifications.

Studies recently published have demonstrated a novel approach to improving the moisture resistance of epoxies. Fisher *et al.* (1985), Hu *et al.* (1987) and Kelly and co-workers (1988) showed that the amount of water absorbed by a TGDDM epoxy cured with diaminodiphenylsulphone (DDS) could be lowered significantly by so-called blocking agents when absorbed into thin films of the cured epoxy. The agents employed were isocyanate compounds and were believed to function by reacting with polar functional groups (such as hydroxyls and secondary and tertiary amines formed during cure) contained within the cross-linked structure. It is generally accepted that such functional groups are responsible for water absorption via hydrogen bonding mechanisms. The fact that this approach relied upon absorption of the blocking agent into the *cured* epoxy essentially meant that the modification could only be applied to very thin films. More recent work by Lonikar *et al.* (1990) has attempted to overcome this problem by incorporating the reagents into the epoxy formulation prior to cure. This was achieved by employing 'masked' isocyanates, the reactivity of which is suppressed at temperatures well above room temperature. At suitably elevated temperatures the isocyanate loses its latency allowing it to react with polar functional groups formed during cure of the epoxy, resulting in significant decreases in equilibrium moisture absorption. As far as the author is aware, blocking agents of the type described have not been employed in commercial practice as a means of promoting enhanced moisture resistance in epoxies.

References

Bascom, W.D., Cottington, R.L., Jones, R.L. and Peyser, P. (1975) *J. Appl. Polym. Sci.* **19**, 2545.

Bucknall, C.B. and Partridge, I.V. (1983) *Polymer* **24**, 639.

Bucknall, C.B. and Gilbert, A.H. (1989) *Polymer* **30**, 213.

Cerere, J.A., Hedrick, J.L. and McGrath, J.E. (1986) *31st SAMPE Symp.* 583.

Chan, L.C., Gillham, J.K., Kinloch, A.J. and Shaw, S.J. (1984) In *Rubber-modified Thermoset Resins*, eds. Riew, C.K. and Gillham, J.K., Adv. Chem. Ser. 208, Am. Chem. Soc., Washington.

Diamont, J. and Moulton, R.S. (1984) *29th Nat. SAMPE Symp.* **29**, 422.

Evans, A.G. (1972) *Phil. Mag.* **26**, 1327.

Fisher, C.M., Gilbert, R.D., Fornes, R.E. and Memory, J.D. (1985) *J. Polym. Sci., Polym. Chem. ed.* **23**, 2931.

Gledhill, R.A., Shaw, S.J. and Tod, D.A. (1990) *Int. J. Adhesion and Adhesives* **10**, 192.

Goobich, J. and Marom, G. (1982) *Polym. Eng. Sci.* **22**, 1052.

Goodier, J.N. (1933) *J. Appl. Mech.* **55**, 39.

Green, D.J., Nicholson, P.S. and Emberg, J.D. (1979). *J. Mat. Sci.* **14**, 1657.

Griffiths, J.R. (1982) *Chemtech* 290.

Hedrick, J.L., Yilgor, I., Wilkes, G.L. and McGrath, J.E. (1985) *Polym. Bull.* **13**, 201.

Hu, H.P., Gilbert, R.D. and Fornes, R.E. (1987) *J. Polym. Sci. Polym. Chem. ed.* **25**, 1235.

Kelly, B.K., Gilbert, R.D., Fornes, R.E. and Stannett, V.T. (1988) *J. Polym. Sci., B: Polym. Phys.* **26**, 1261.

Kinloch, A.J., Gledhill, R.A. and Dukes, W.A. (1975) In *Adhesion Science and Technology*, ed. Lee L.H., Plenum, New York.

Kinloch, A.J., Shaw, S.J. and Hunston, D.L. (1983) *Polymer* **24**, 1355.

Kinloch, A.J., Maxwell, D. and Young, R.J. (1985) *J. Mat. Sci.* **20**, 4169.

Kirshenbaum, S.L., Gazit, S. and Bell, J.P. (1984) In *Rubber-modified Thermoset Resins*, eds. Riew, C.K. and Gillham, J.K. Adv. Chem. Ser. 208, Am. Chem. Soc., Washington.

Landman, D. (1984) *Polyimides. Synthesis, Characterisation, and Applications,* Vol. 1. ed. Mittal K.L., Plenum, New York.
Lange, F.F. (1970) *Phil Mag.* **22**, 983.
Lee, S.M. (1988) *Epoxy Resins, Chemistry and Technology*, 2nd edn, Ch. 9. ed. May C.A., Dekker, New York, p. 783.
Lewis, A.F. (1988) *Epoxy Resins, Chemistry and Technology*, 2nd edn, Ch. 7., ed. May C.A., Dekker, New York.
Lonikar, S.V., Rungsimuntakul, N., Gilbert, R.D. and Fornes, R.E. (1990) *J. Polym. Sci. A. Polym. Chem.* **28**, 759.
Maxwell, D., Young, R.J. and Kinloch, A.J. (1984) *J. Mats. Sci. Letts.* **3**, 9.
McGarry, F.J. and Willner, A.M. (1968) *Report R68-8*, MIT Cambridge, USA.
Mertz, E.H., Claver, G.C. and Baer, M. (1956) *J. Polym. Sci.* **22**, 325.
Mijovic, J., Pearce, E.M. and Foun, C.C. (1984) *Rubber-Modified Thermoset Resins.* eds. Riew, C.K. and Gillham, J.K. Adv. Chem. Serv. 208, Am. Chem. Soc., Washington.
Mika, T.F. and Bauer, R.S. (1988) *Epoxy Resins. Chemistry and Technology*, 2nd Edn, Ch. 4. ed. May C.A., Dekker, New York, pp. 465–550.
Moloney, A.C., Kausch, H.H. and Stieger, H.R. (1983) *J. Mat. Sci.,* **18**, 208.
Moloney, A.C., Kausch, H.H. and Stieger, H.R. (1984) *J. Mat. Sci.* **19**, 1125.
Moloney, A.C., Kausch, H.H., Kaiser, T. and Beer, H.R. (1987) *J. Mat. Sci.* **22**, 381.
Pearson, R.A. and Yee, A.F. (1983) *Polym. Mat. Sci. Eng.* **49**, 316.
Piggott, M.R., Lam, P.W.K., Lim, J.T. and Woo, M.S. (1985) *Comp. Sci. Tech.* **23**, 247.
Potter, W.G. (1970) *Epoxide Resins*, Iliffe, London.
Riew, C.K., Rowe, E.H. and Siebert, A.R. (1976) *Toughness and Brittleness of Plastics*. Adv. Chem. Ser. 154, Am. Chem. Soc., Washington.
Rowe, E.H., Siebert, A.R. and Drake, R.S. (1970) *Mod. Plast.* **47**, 110.
Shaw, S.J., Tod, D.A. and Griffith, J.R. (1988) *Adhesive Sealants and Coatings for Space and Harsh Environments.* ed. Lee L.H., Plenum, New York, p. 45.
Shaw, S.J. and Tod, D.A. (1989) *J. Adhesion* **28**, 231.
Shaw, S.J. and Tod, D.A. (1990) *Adhesion '90,* Plastics and Rubber Institute, London, 12/1.
Siebert, R.A. and Riew, C.K. (1971) *Org. Coat. Plast. Chem.* **31**, 552.
Spanoudakis, J. and Young, R.J. (1984) *J. Mat. Sci.* **19**, 473.
Sultan, J.N. and McGarry, F.J. (1973) *J. Polym. Eng. Sci.* **13**, 29.
Tanaka, Y. (1988) *Epoxy Resins Chemistry and Technology* 2nd edn, Ch. 2. ed. C.A. May, Dekker, New York, pp. 9–283.
Tess, R.W. (1988) *Epoxy Resins Chemistry and Technology*, 2nd edn, Ch. 8. ed. May C.A., Dekker, New York, pp. 719–782.
Twardowski, T.E. and Geil, P.H. (1991) *J. Appl. Polym. Sci.* **42**, 69.
Wake, W.C. (1982) *Adhesion and the Formulation of Adhesives* , 2nd edn, Applied Science, London.
Yorkgitis, E.M., Trau, C., Eiss, N.S., Hut, T.Y., Yilgor, I., Wilkes, G.L. and McGrath, J.E. (1984) *Rubber Modified Thermoset Resins.* eds Riew, C.K. and Gillham, J.K., Adv. Chem. Ser. 208, Am. Chem. Soc. Washington.

5 Fracture behaviour of epoxy resins

W.J. CANTWELL and H.H. KAUSCH

5.1 Introduction

As a result of their superior mechanical properties, low shrinkage and thermal stability, epoxy resins are finding increasing use in a wide range of engineering applications. Over a twenty year period between 1965 and 1985, epoxy resin consumption in Western Europe increased by over 500% to reach approximately 132 000 tonnes by the mid-1980s (Feuerhahn, 1986). Since epoxies cost between three and fiteen times more than other thermosets such as isophthalic polyesters, they tend to be used in more advanced engineering applications where the increased cost of the product is offset by their overall superior load-bearing capacity and/or chemical resistance. In Western Europe the coatings industry represents the principal user of epoxy resins followed by the electrical and construction industries (Feuerhahn, 1986). Although much of the research published in the literature relates to epoxies for use in advanced composites, this sector represented only 4% of the total market in this year. In almost all of these applications, properties such as high strength, stiffness and toughness are fundamental requirements. The consequences of defects or local stress raisers in these systems can vary from inconvenient, for example the case of a somewhat unaesthetic damaged surface coating, to catastrophic such as the case of a heavily delaminated load-bearing aircraft structure. A detailed understanding of the deformation and failure mechanisms that occur in these materials is therefore clearly necessary.

During the last thirty years a large amount of work has been undertaken on epoxy resins. Although many of these studies have been both systematic and comprehensive, many questions remain unanswered and the influence of varying many basic material parameters is still poorly understood. A brief glimpse of the extensive literature highlights the large number of widely differing epoxy systems currently available. This is shown in Figure 5.1 where previously reported data from the literature are presented. Here, the mode 1 fracture energy (see section 5.2.1) of the various epoxies is plotted as a function of their glass transition temperatures.

Clearly, a range of systems is available varying from relatively tough, low temperature epoxies for use in the construction industry for example, to high temperature, although somewhat brittle, epoxies such as those destined for use in the aerospace sector. The figure also highlights one of the

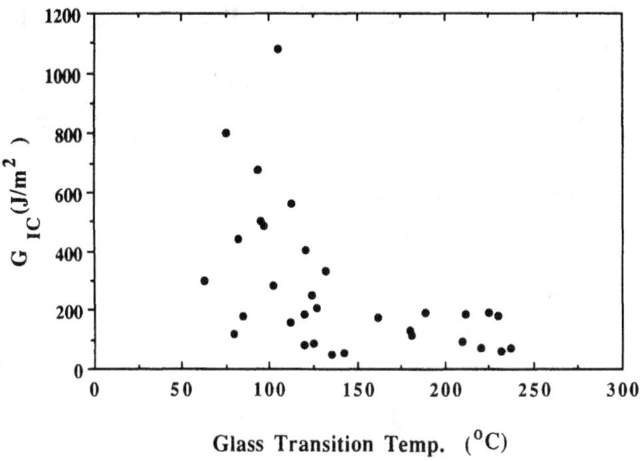

Figure 5.1 The wide range of epoxy properties and associated glass transition temperatures presently available.

great challenges to epoxy chemists, that is, a tough system with a good high temperature capability.

In this chapter the fracture bahaviour of epoxy resins will be considered in some detail. Since most of the recent advances in the understanding of the fracture behaviour of these materials have been achieved as a result of the application of fracture mechanics principles, most particularly linear elastic fracture mechanics (LEFM) considerable use of this approach will be made here. Attention will be given to assessing the influence of structure (cross-link density etc.) as well as the test conditions (temperature, rate, etc.) on the basic fracture properties of epoxy resins. In the later part of the chapter the various methods currently employed to enhance the relatively brittle nature of epoxies will be presented.

The following section will present a brief summary of some of the more important LEFM principles necessary for characterizing the behaviour of these materials.

5.2 Linear elastic fracture mechanics (LEFM)

Theoretical considerations have shown that the stress to cause cleavage fracture in a brittle solid should be approximately one tenth of the Young's modulus of the material. Assuming that the typical modulus of an epoxy resin is of the order of 3 GPa one would expect to obtain a failure strength of 300 MPa. Experimental evidence has clearly shown this not to be the case with the fracture strength of epoxy resins rarely exceeding one third of this value. This disparity was first recognised by Griffith (1920) who identified the significant influence of small defects on the load-bearing properties of a brittle isotropic material.

Modern-day linear elastic fracture mechanics is based on two inter-relatable conditions for fracture, the energy analysis (G approach) and localized stress field approach (K approach). The former supposes that fracture will occur when sufficient energy is released by crack growth to supply the requirement for creation of new fracture surfaces. LEFM is applied to materials which obey Hooke's law where the measured strain is proportional to the applied stress. Although all epoxies exhibit some inelastic deformation around the crack tip, the fact that the bulk of the remaining material behaves elastically permits the successful application of this analysis.

5.2.1 The G *approach*

By applying an energy balance argument to a linearly elastic cracked body of thickness B subjected to an applied load, it is possible to determine the critical strain energy release rate G_c for fracture to occur

$$G_c = \frac{P_c^2}{2B} \frac{\partial C}{\partial a}$$

where C is the compliance of the cracked body for a given crack length a and P_c the critical load for the onset of crack propagation. The above equation represents the basis for most G_c calculations with the basic requirement being a knowledge of the variation of specimen compliance with crack length. The compliance-crack length dependency can be determined either analytically or experimentally depending upon the complexity of the test specimen.

It is important to note that all modes of loading encountered in operational service can be decomposed into the three basic modes denoted I, II and III presented in Figure 5.2. Mode I corresponds to simple cleavage

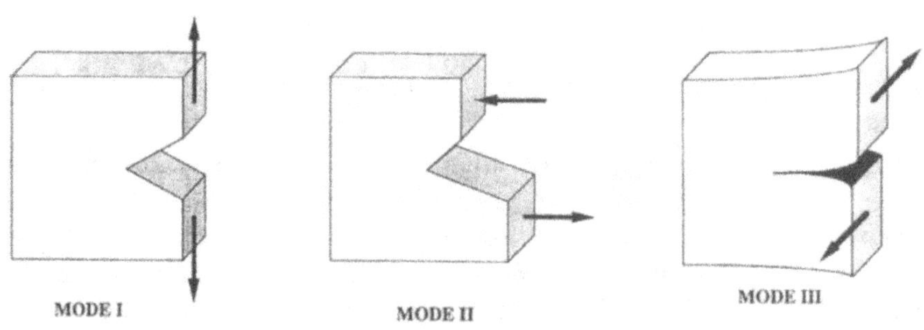

Figure 5.2 Modes of loading: mode I opening mode, mode II inplane shear, and mode III antiplane shear.

or tensile opening, Mode II to inplane shear and Mode III to antiplane shear. The fracture energies associated with I, II and III modes of deformation are denoted G_{Ic}, G_{IIc} and G_{IIIc} respectively. Mode I deformation is technically the most important of the three since it represents the most commonly encountered loading condition and generally has the lowest associated fracture energy. Mode I-type loading is also much more readily simulated in laboratory conditions. Typical mode I test specimens are presented in Figure 5.3. Of these, the single edge notch (SEN) and single

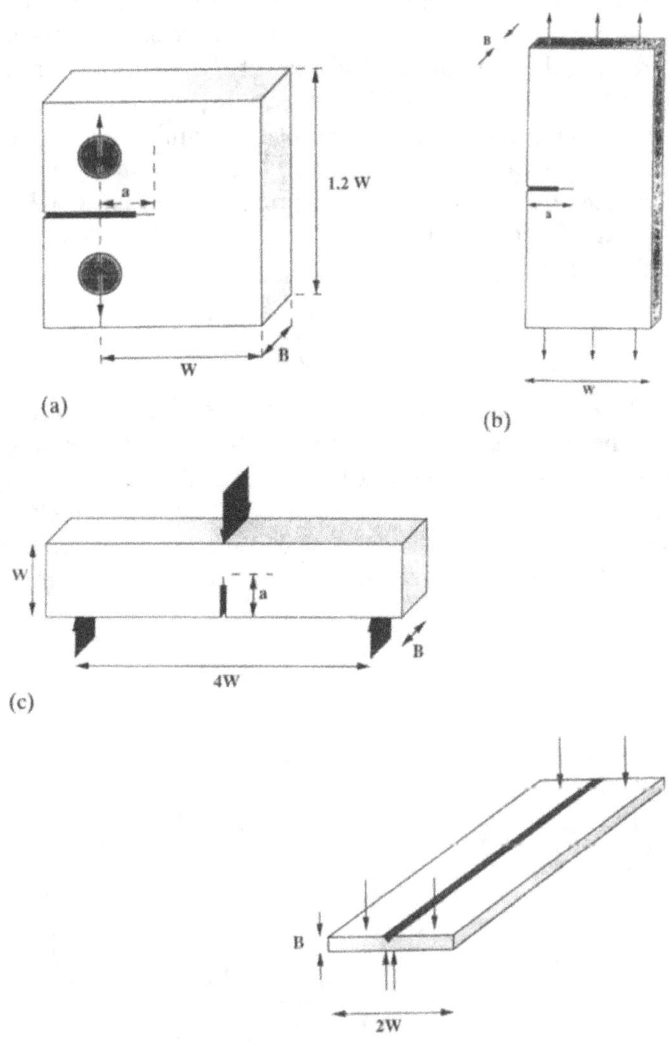

Figure 5.3 Typical test geometries used for characterizing the mode I fracture toughness of epoxies. (a) compact tension; (b) single edge notch; (c) single edge notch bend; (d) double torsion.

edge notch bending (SENB) are particularly unstable geometries whereas stable crack propagation can, under certain conditions, be achieved in the double torsion (DT) and compact tension (CT) specimens.

In selecting the precise dimensions for the chosen specimen geometries, a number of size criteria have to be first respected. Perhaps the most important of these concerns the specimen thickness B. Such a thickness-dependency relates to the state of stress at the crack tip. This can vary from plane stress conditions in a thin plate to plane strain conditions in a much thicker specimen. Since the stress at which a material yields is greater in a plane strain field than in a plane stress field, a smaller plastic zone size is observed under plane strain conditions. The effect of this smaller plastic zone is to reduce the measured toughness of the specimen. In designing components for use in load-bearing applications, it is therefore of utmost importance to ensure that the minimum (plane strain) value is employed.

Solutions enabling the direct determination of G_c exist for the commonly used specimen geometries such as those presented in Figure 5.3. Williams (1984) adapted the strain energy analysis for the SENB specimen and developed the following expression for G_{Ic}:

$$G_{Ic} = \frac{U}{BD\Phi}$$

where U is the energy under the load-displacement curve, B the specimen thickness, D the specimen height and Φ a calibration factor taking account of the specimen geometry and crack length. The filled circles in Figure 5.4 represent a typical set of experimental data for Araldite B. It is clear that the level of scatter is low and that the data can be fully characterized using the G analysis. The slope of the curve gives a value for G_{Ic} of 151 J/m^2, a typical

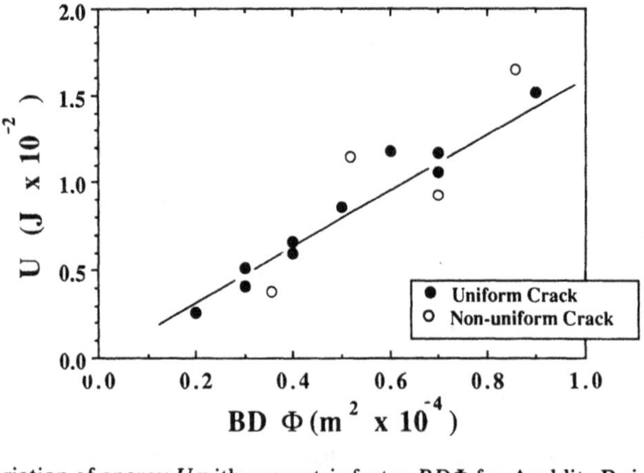

Figure 5.4 Variation of energy U with geometric factor $BD\Phi$ for Araldite B single edge notch bend specimens. The influence of a non-uniform crack front is apparent.

value for a relatively brittle epoxy of this nature. The figure also highlights one important point, that is that most epoxy resins are well suited to the application of linear elastic fracture mechanics. However, when undertaking fracture mechanics tests, it is very important that a straight uniform pre-crack be obtained. Generally speaking, it is not difficult to introduce shart pre-cracks into epoxy specimens. Difficulty is often experienced, however, in assuring a straight and uniform crack front. The open circles in Figure 5.4 correspond to tests on specimens in which the crack front was not straight. From the figure it is clear that the non-uniform cracks result in greater levels of scatter and an artificially high value of G_{Ic}. When pre-cracking epoxy samples with a razor blade and hammer care should be taken to ensure that the crack propagates as a result of one single blow rather than a series which may result in a non-straight crack front.

5.2.2 The K *approach*

The second approach frequently used to characterize the fracture resistance of epoxy resins is the critical stress intensity factor analysis. Here, the stress field around a sharp crack in a linear elastic material can be defined by a stress intensity factor K_I (for the case of mode I loading). The general form of the stress intensity factor is given by:

$$K_I = Y\sigma\sqrt{a}$$

where Y is a geometrical factor taking account of the specimen size, a the crack length and σ the applied nominal stress.

It is assumed that fracture occurs when K_I reaches a critical value termed K_{Ic}, a material property for a given test condition (temperature, rate etc.).

K_{Ic} for a finite plate is given as:

$$K_{Ic} = Y\sigma_c\sqrt{a}$$

where σ_c is the applied stress at the onset of fracture. In order to determine K_{Ic} it is therefore necessary to know the geometrical factor Y for the particular geometry employed. Such factors have been established for a wide range of geometries containing cracks and are listed in a number of handbooks (Rooke and Cartwright, 1976; Sih, 1973; Tada *et al.*, 1973).

If the data presented in Figure 5.4 are analysed using the K approach, a value for the fracture toughness of 0.64 MPa $m^{0.5}$ is achieved, this again being a typical value for a relatively brittle epoxy.

For linear elastic fracture mechanics a simple relationship exists between K and G. For Mode I fracture this relationship takes the form of

$$G_{Ic} = \frac{K_{Ic}^2}{E} \qquad \text{for plane stress conditions}$$

and

$$G_{Ic} = \left(\frac{K_{Ic}^2}{E}\right)(1 - \nu^2) \quad \text{for plane strain conditions}$$

where E is the Young's modulus and ν the Poisson's ratio.

5.2.3 *Crack opening displacement*

Several workers (Gledhill and Kinloch, 1976; Marshall *et al.*, 1974) have used the crack opening displacement criterion in order to characterize failure in a number of different thermoplastic and thermosetting polymers. The tensile stress applied to the material containing the crack causes a displacement of the two crack surfaces resulting in a crack opening displacement δ. Failure is assumed to occur when the crack opening displacement reaches a critical value δ_c, where:

$$\delta_c = \frac{K_{Ic}^2}{\sigma_y E} \quad \text{for plane stress}$$

Under conditions of plane strain, this value is multiplied by the factor $(1 - \nu^2)$.

Quoted values for δ_c vary from 1 to 20 μm depending on the test temperature and crack velocity (Gledhill *et al.*, 1978).

5.3 Deformation mechanisms

Much work has been undertaken in order to characterize the mechanical behaviour of epoxy polymers (Lilley and Holloway, 1973; Gledhill and Kinloch, 1976; Morgan *et al.*, 1979; Morgan, 1980, 1985; Yamini and Young, 1980; Young, 1980; Narisawa *et al.*, 1982; Bradley, 1989; Smith, 1989; Barbezat, 1990; Bradley *et al.*, 1992; Fischer, 1992). Simple tests using uni-axial tension, compression and flexure specimens have been used extensively to quantify the effects of material variables, test temperature, environmental conditions and loading rate on the mechanisms of deformation and fracture in these materials.

A number of workers have reported evidence for crazing around stress concentrations in cross-linked epoxies (van den Boogaart, 1966; Lilley and Holloway, 1973; Morgan and O'Neal, 1977). van den Boogaart (1966) reported seeing crazes around crack tips in Epikote 828 that was not completely cured. Subsequent work by Lilley and Holloway (1973) on a number of fully cured epoxies highlighted craze-like features ahead of the crack tip. Morgan and O'Neal (1977) undertook a number of *in situ* fracture tests on epoxy films in the scanning electron microscope. They concluded that when loaded in tension, their DGEBA-DETA epoxy deformed and failed through a crazing process. The evidence presented is not convincing since the supposed craze-like structures are much less well defined than those observed in thermoplastics. A more recent detailed study by Glad (1986), again undertaken using thin epoxy films, failed to identify any evidence for crazing in cross-linked epoxies. He noted that all thermoplastics with strand densities greater than $8 \times 10^{25}\ m^{-3}$, deformed through a shear yielding mechanism. The typical strand density of a cross-linked epoxy is approximately $150 \times 10^{25}\ m^{-3}$. He suggested that the 'crazes' observed by Morgan were artefacts resulting from the formation of voids around dust particles or other defects. It is therefore generally accepted that crazing mechanisms of the type observed in many thermoplastics do not occur in cross-linked epoxy resins. Instead, shear yielding represents the principal micro-deformation process in these materials. Figure 5.5a shows shear bands in a hydantoin-based epoxy tensile specimen tested at 85°C. The shear bands initiated at a very small defect located close to or at the surface of the sample. The presence of the highly deformed zones served to redistribute the local stress field and reduce the stress-concentrating effect of the defect. When a crack did initiate in this region, it propagated in a slow stable manner as can be seen on the fracture surface (Figure 5.5b).

The presence of localized shear-yielding in areas of high stress concentration is therefore beneficial since it results in crack tip blunting, a re-distribution of the local stress field and an increase in the measured fracture toughness. Several workers have attempted to measure the local plastic deformation at the tip of a sharp crack (Glad, 1986; Hibbs and Bradley, 1987). Optical density measurements by Glad (1986) indicated that

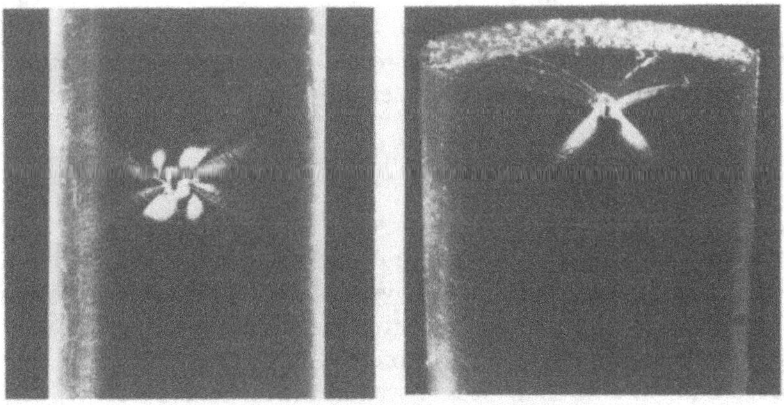

(a) (b)

Figure 5.5 Shear bands in a hydantoin-based epoxy tested in tension at 85°C. (a) Just prior to crack propagation and (b) after failure (Cantwell *et al.*, 1988).

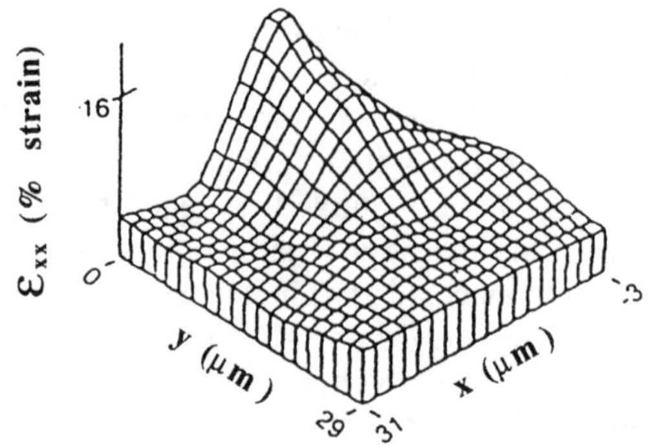

Figure 5.6 Strain field at the tip of a sharp crack in 3501-6 epoxy resin determined by *in situ* strain measurements in a scanning electron microscope (from Bradley, 1989).

the maximum extension ratio λ in the damage zone of a moderately cross-linked DGEBA epoxy could be as high as 1.4. Hibbs and Bradley (1987) developed a novel technique to quantify the crack tip strain field. Here, the beam of a scanning electron microscope (SEM) was used to burn a series of dots around a pre-crack in a compact-tension specimen. The specimen was then loaded using an SEM *in situ* stage and the relative movement of the dot pattern monitored using a continuous image processing system. The local strain field was then determined by numerically differentiating the displacement field data. A typical strain field for 3501-6 epoxy resin (a brittle epoxy for aerospace applications) is shown in Figure 5.6. As would be expected, the strain concentration immediate to the crack tip is significant. It is interesting to note, however, that the maximum value of strain is approximately 15% whereas the elongation at break of this resin when measured in a standard tensile test is typically only 1.5%. This significant difference almost certainly relates to the presence of small defects or local variations in the cross-link density which are clearly going to be more numerous in a larger volume. The results yielded by simple tensile tests should therefore be interpreted with caution.

5.4 Modes of crack propagation

The inelastic deformation mechanisms that take place at the tip of a sharp crack in an epoxy resin will clearly have a significant effect upon the manner in which the crack subsequently propagates. Many workers have made considerable use of the double torsion (DT) test geometry in an attempt to characterize both the toughness as well as the mode and stability of crack propagation in epoxy resins (Young and Beaumont, 1976; Yamini and Young, 1977; Phillips *et al.*, 1978; Scott *et al.*, 1980). Extensive testing of a number of pure and rubber-modified epoxies has highlighted three basic modes of crack propagation, these being:

(i) Stable brittle propagation
(ii) Unstable brittle propagation
(iii) Stable ductile propagation.

5.4.1 *Stable brittle propagation*

The first of these, stable brittle propagation, results in a flat, featureless fracture surface and a smooth continuous double-torsion load-displacement curve (Figure 5.7a). This type of behaviour is thought to be typical of classical brittle fracture (Kinloch, 1985). Gledhill *et al.* (1978) have shown that this mode of failure can be successfully characterized by a unique value of crack tip opening displacement CTOD. For a DGEBA/9.8 phr TETA epoxy this value was approximately 0.9 μm.

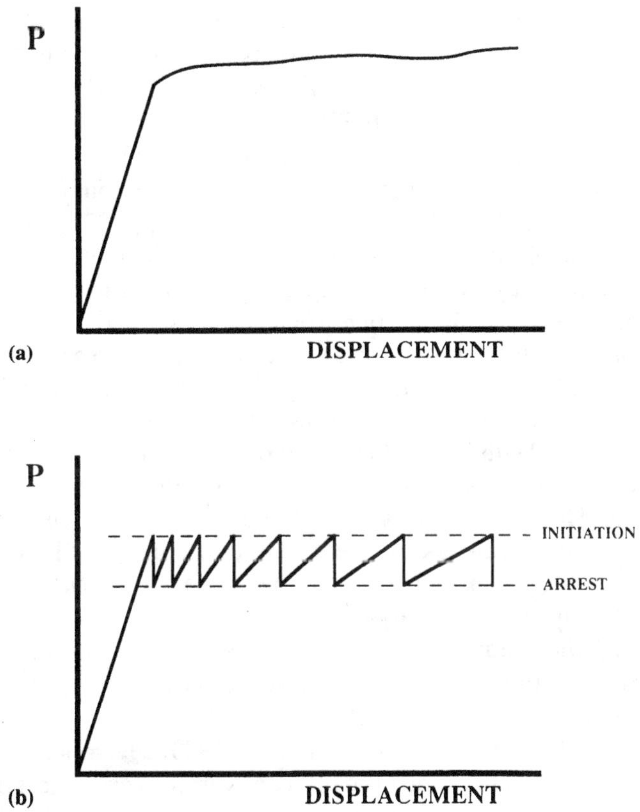

Figure 5.7 Schematic representations of (a) stable crack and (b) unstable crack growth in a double-torsion test specimen.

5.4.2 *Unstable brittle propagation*

Unstable brittle propagation results in the crack advancing in a jump-like fashion often referred to as stick-slip behaviour (Figure 5.7b). Since the load at which fracture initiates and arrests in the DT specimen are not the same, many workers quote initiation and arrest values of $K(G)$, that is K_{Ii} (G_{Ii}) and $K_{Ia}(G_{Ia})$.

The transitions at which crack propagation changes from stable brittle to unstable brittle or from unstable brittle to stable ductile depend upon a large number of both material and test parameters (Young and Beaumont, 1976; Gledhill *et al.*, 1978; Yamini and Young, 1979; Scott *et al.*, 1980; Young, 1980; Cherry and Thomson, 1981). Crack instability in these materials may occur for a number of reasons. Firstly, certain test geometries are such that the strain energy release rate increases with crack growth, that is, dG/da is positive. A second possibility is that the fracture energy of the material decreases with increasing crack length, i.e. dG_I/da is negative. This might occur, for example, as a result of deleterious environmental effects (Cherry and Thomson, 1981). Another possibility is that the polymer's fracture energy decreases with increasing crack velocity yielding a negative value for dG_I/da. Yamini and Young (1977) have suggested that the latter might explain the tendency for unstable crack propagation in moisture-conditioned amine-cured Epikote 828.

Many workers have identified the fundamental role of plastic flow and subsequent crack-tip blunting around stress concentrations in epoxies resins (Young and Beaumont, 1976; Yamini and Young, 1977; Phillips *et al.*, 1978). Localized plastic deformation is clearly a fundamental mechanism in this respect and many of the observations presented in the literature have been explained at least quantitatively in terms of this plastic flow process. In general, stable brittle failure occurs at lower temperatures and higher rates of loading. Under these conditions, the yield stress of the material is likely to be high and the extent of plastic flow and therefore crack blunting low (Barbezat, 1990). With increasing temperature and decreasing rate, failure tends to occur in the stick-slip manner associated with unstable brittle propagation. Here, the yield stress of the material is lower and greater plastic deformation can take place at the crack tip. The latter therefore blunts out and takes a more rounded appearance (Scott *et al.*, 1980). When the crack finally does re-sharpen, the release of energy is significantly greater than that required to create a fracture surface and the crack accelerates, sometimes up to several hundreds of metres per second, and may eventually bifurcate (Cantwell *et al.*, 1988b).

Williams (1973) and Kinloch and Williams (1980) developed and applied a crack blunting model to predict the effect of varying certain test parameters on the mode I fracture toughness of certain epoxy systems. Here, it is assumed that fracture occurs when a critical stress σ_{tc} is attained at a certain distance c ahead of the crack tip. Then, by combining the analysis for the local crack-tip stress field at a distance r ahead of the crack tip (this being equal to c):

$$\sigma_{yy} = \frac{\sigma_0 \sqrt{a}}{\sqrt{(2r)}} \frac{1 + \frac{\varrho}{r}}{\left(1 + \frac{\varrho}{2r}\right)^{3/2}}$$

where ϱ is the crack tip radius and σ_0 is the applied stress with the classical fracture mechanics equation

$$K_{\mathrm{Ics}} = \sigma_{\mathrm{tc}} \sqrt{(2\pi c)}$$

it was shown that the critical intensity factor K_{Ic} could be related to the value for a sharp crack K_{Ics} by the relationship:

$$\frac{K_{\mathrm{Ic}}}{K_{\mathrm{Ics}}} = \frac{\left(1 + \frac{\varrho}{2c}\right)^{3/2}}{1 + \frac{\varrho}{c}}$$

Figure 5.8 shows the variation of the relative fracture toughness parameter $K_{\mathrm{Ic}}/K_{\mathrm{Ics}}$ with the square root of the crack tip radius for a series of specimens containing natural cracks (open circles) and machined blunt cracks (closed circles) (Kinloch *et al.*, 1983). In the former, the crack tip radius was taken to be equivalent to the calculated crack tip opening displacement.

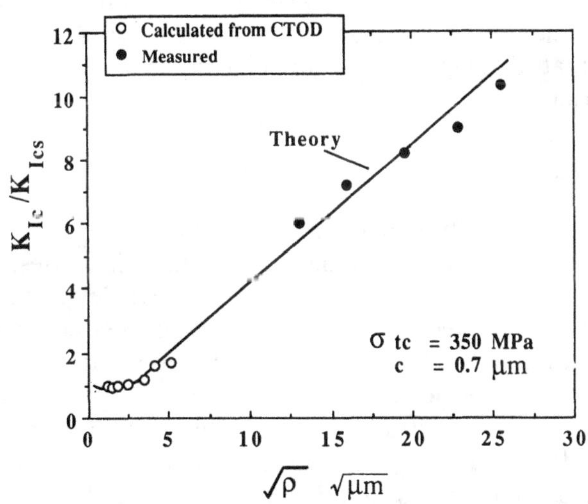

Figure 5.8 Variation of $K_{\mathrm{Ic}}/K_{\mathrm{Ics}}$ as a function of the square root of the notch tip radius, ϱ, for a DGEBA resin cured with 5 phr piperidine. The calculated values of the critical stress δ_{tc} and critical distance c are 350 MPa and 0.7 microns, respectively (after Kinloch *et al.*, 1984).

This crack blunting analysis implies that a stress of up to five times the value of the material's yield stress is attained at the crack tip. Kinloch (1985) suggested that this critical stress, σ_{tc}, could be interpreted as a constrained yield stress.

5.4.3 *Stable ductile propagation*

The third mode of failure, stable ductile propagation yields a smooth load-displacement DT plot similar to that observed under stable brittle conditions (Figure 5.7a). This type of failure is synonymous with high energy dissipation, a relatively rough fracture surface and, as a result, higher toughness.

At high temperatures and very low strain rates, cracks frequently advance through a tearing process in which much energy is dissipated in plastically deforming the material and creating multiple finger-like markings. Kinloch *et al.* (1986) proposed a model for this type of crack growth based on a meniscus instability analysis. The original model considers the advance of a meniscus between a less-dense and a more-dense fluid. By developing the model and applying it to the specific test conditions in question, Kinloch *et al.* (1986) calculated a value of 60 μm for the critical wavelength of the finger-like instabilities, this correlating well with the value of 68 $\pm$ 24 μm observed on the fracture surfaces.

5.5 Effect of test conditions

Epoxies, like all polymers, exhibit a viscoelastic response when tested under both short- and long-term loading conditions. Two fundamental test parameters that highlight this pronounced viscoelastic behaviour are test temperature and loading rate.

5.5.1 *Temperature*

Tests undertaken at different temperatures have shown that the mode I fracture toughness of untoughened epoxies increases, sometimes dramatically, with increasing test temperature (Bradley *et al.*, 1992). This increase is clearly related to the deformation process taking place at the crack tip. Increasing the temperature has the effect of reducing the yield stress of the polymer which in turn results in a larger plastic zone size and greater crack blunting. Bradley (1991) suggested that the degree of temperature-dependency depends upon the relative difference of the test temperature, T, and the glass transition temperature, T_g, of the polymer in question. Tests undertaken at temperatures considerably below the T_g of the polymer should always, therefore, result in brittle failure. By plotting the fracture toughness, values G_{Ic} (or G_q if the plain strain requirements were not satisfied) of his DGEBA epoxies as a function of $T - T_g$ he obtained a relatively unique curve with all the data falling within a narrow band (Figure 5.9). In all cases, the initial increase in toughness was found to occur at approximately 125°C below T_g with a dramatic increase occurring at about 60°C below T_g. Unfortunately, no such relationship was obtained for the rubber-toughened counterparts of these resins.

An interesting and, as yet, unexplained temperature-related effect was observed by Scott *et al.* (1980) following low temperature tests on a series of amine-cured epoxies. Whilst undertaking DT tests in a nitrogen gas environment they observed that at temperatures below −100°C crack propagation became unstable with the value of G_{Ii} rising rapidly with further decreasing temperature. The authors concluded that this change in failure mode could not be attributed to a simple crack blunting argument but might, instead, result from increased energy absorption by molecular relaxation as well as an interaction of thc crack tip with the gaseous envionment near its liquefaction temperature. Thomson (1981) suggested that any frozen water present on the surface of the specimen might be able to permeate and plasticize the resin around the crack tip. Young (1980) pointed out that since N_2 has been shown to influence the failure mechanisms occurring in thermoplastics, it could have a similar effect on epoxy resins. He suggested that a critical test would be to undertake such tests in a helium gas environment, a medium which does not have any observable effect on thermoplastic polymers.

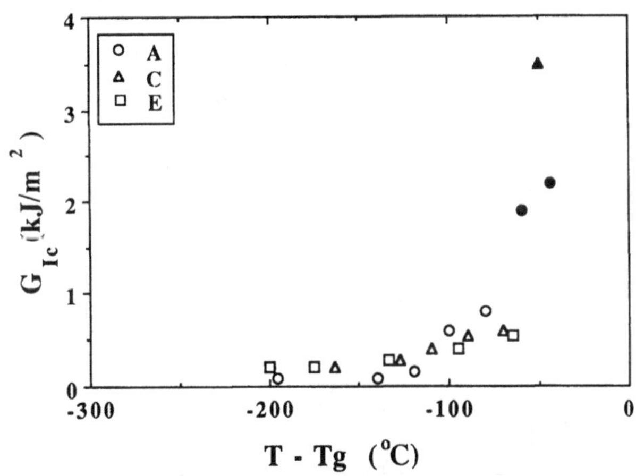

Figure 5.9 Mode I fracture toughness versus $T - T_g$ (T = test temperature, T_g = glass transition temperature) for three DGEBA epoxies. The T_gs of epoxies A (○), C (△) and E (□) are 200, 189 and 225°C respectively. The filled symbols represent lower bound G_q values which do not satisfy either geometrical or maximum force requirements for plane strain (after Bradley, 1991).

5.5.2 *Loading rate*

Reducing the loading rate also has the effect of reducing the yield stress of the epoxy and therefore increasing the measured toughness (Kinloch, 1985). The rate sensitivity of epoxies depends upon a number of parameters such as the test temperature and the cross-link density. At room temperature, epoxies with a higher degree of cross-linking, and therefore an inherently stiffer structure, do not exhibit distinct rate-dependent characteristics (Barbezat, 1990) but low cross-linked epoxies do (Barbezat, 1990).

Gledhill *et al.* (1979) undertook a series of long-term constant load (static fatigue) tests on aluminium tapered double cantilever beam (TDCB) specimens bonded using an amine-cured DGEBA epoxy. During the course of testing, none of the specimens failed and, indeed, no crack growth was observed. The test specimens were therefore unloaded after a prescribed time and tested to rupture. The resulting fracture toughness data indicated that G_{Ic} increased steadily with time under load being almost double its initial short-term value after 10^6 seconds (Figure 5.10). The authors noted that since the compressive yield stress of this material decreased linearly with increasing logarithmic time-of-test, the formation of a large plastic zone was easier at longer times resulting in increased crack tip blunting.

Hunston *et al.* (1984) recognized the inter-relation between rate and temperature and attempted to express it in a more quantitative manner. They determined the linear viscoelastic properties for a number of unmodified and rubber-toughened systems over a wide range of frequencies and temperatures. These viscoelastic properties were then treated by applying a time-temperature superposition procedure to the experimental data. The resulting master curve showed surprisingly little scatter with the

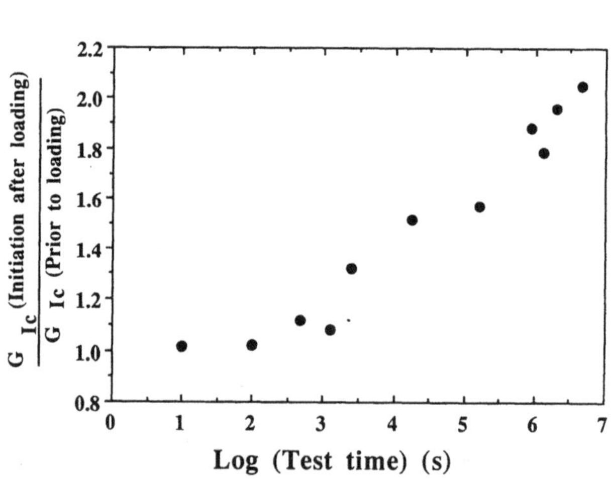

Figure 5.10 Static fatigue tests on a tapered double cantilever beam showing the increase in fracture toughness with loading time (after Gledhill *et al.*, 1979).

data for both the unmodified resins and the toughened systems falling on what appears to be a unique curve. The shift factors determined from these viscoelastic analyses were then applied to the fracture toughness data yielding a G_{Ic} master curve.

This type of data reduction technique is potentially very powerful since it permits the toughness of an epoxy to be predicted for any given rate/temperature combination.

5.6 Microstructural effects

Most of the fundamental physical and mechanical properties of an epoxy resin are determined by the nature and density of its cross-linked network. Although a significant amount of work has been undertaken in an attempt to relate the mechanical properties of these materials to their microstructure, a fundamental understanding of the effect of varying basic parameters is still lacking. Clearly, the molecular architecture of an epoxy will directly influence its ability to carry load and absorb energy. A detailed discussion regarding the influence of varying the network structure of an epoxy on its subsequent mechanical behaviour is beyond the scope of this chapter. For a more detailed appraisal the reader is directed to more extensive works (Morgan, 1985; Fischer, 1992).

In general terms, the degree of cross-linking influences the glass transition temperature of the polymer, its ability to undergo localized plastic deformation and therefore its toughness characteristics at a given temperature (Nielsen, 1969). Available experimental evidence suggests that mechanical properties such as tensile strength, modulus and failure strain do not exhibit any appreciable cross-link density dependence (Misra *et al.*, 1979). Epoxies that have high cross-link densities tend to have high T_gs but are frequently prone to brittle failure. A low cross-link density yields a less rigid network structure that is capable of undergoing greater amounts of plastic deformation which in turn may result in a higher fracture toughness. Bell (1970) as well as Misra *et al.* (1979) reported significant increases in impact strength with increasing molecular mass between cross-links (lower cross-link density). LeMay and Kelley (1986) established a direct relationship between the square root of the average molecular mass of the network strands and the fracture energy for crack arrest (Figure 5.11). Although not plotted, their results also indicated that K_{Ic} at initiation increased with increasing molecular mass of strands. Levita (1989) made similar observations following fracture mechanics tests on a DGEBA-DDS epoxy. Here, a 100% increase in fracture toughness was reported over a range of molecular weights from approximately 300 to 3800 g/mole. The experimental results also suggested a linear dependence over this range of molecular weights.

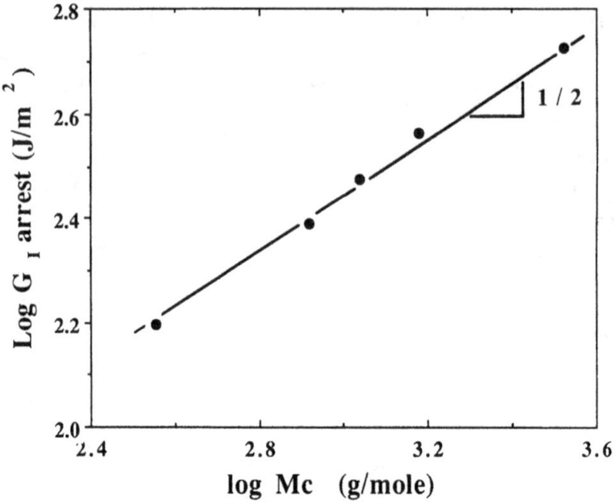

Figure 5.11 The dependence of the crack arrest fracture energy G_I on the average molecular weight between cross-links M_c for a series of stoichiometric DGEBA/DDS epoxies (after LeMay and Kelley, 1986).

In a detailed study, Glad (1986) showed that the extension ratio, λ, within the plane stress damage zone in thin epoxy films could be directly correlated to the maximum extension ratio of a single network strand, λ_{max}, by the relationship:

$$\lambda - 1 = 0.32(\lambda_{max} - 1)$$

This appears to be the first time that the deformability of an individual epoxy network strand has been related to the deformation process at the crack tip.

Fischer (1992) showed that the calculated crack tip opening displacement could be directly related to the effective molecular mass between cross-links. He also suggested that the width and length of the deformation zone at the crack tip are directly proportional to the molecular mass of the network strands.

Considerable debate has raged in the literature over many years regarding the existence or non-existence of a very fine nodular structure within epoxy resins (Racich and Koutsky, 1976; Morgan and O'Neal, 1977; Dušek *et al.*, 1978; Mijovic and Koutsky, 1979). It has been suggested that due to the fact that epoxies are normally made by the addition of a curing agent to a pre-polymer, a two phase structure could occur (Morgan and O'Neal, 1977). Mijovic and Koutsky (1979) argue that nodules correspond to high cross-linked regions ranging in size from 6 nm to 10 μm. Much of the evidence for the existence of this fine second phase structure is based on observations obtained from scanning electron micrographs and replication techniques. If such a fine structure does indeed exist in these thermosetting polymers, it is certainly conceivable that they would have a significant influence upon the mechanical response of the material itself. Indeed, Mijovic and Koutsky (1979) established a correlation between the morphology of a DETA-cured DGEBA epoxy and their ultimate mechanical properties. They suggested that the size of the nodules, resulting from the use of various curing-agent concentrations, can be quantitatively related to the amount of plastic flow occurring during crack initiation. Dušek *et al.* (1978) etched the fracture surfaces of an amine-cured DGEBA and revealed globular structures some 20 to 40 nm in size. Similar structures were observed, however, following etching of amorphous polystyrene and PMMA. Further, the small angle X-ray scattering curves for the cured epoxy did not differ from those of common amorphous polymers. The authors concluded that nodules do not exist in simple amine-cured epoxies of this sort. They did point out, however, that with more complicated systems a more pronounced inhomogeneity might be caused by thermodynamic incompatibility or by non-alternating mechanisms of the curing reaction. It is also possible that the presence of such a secondary phase may be due to incomplete mixing of the hardener and resin during manufacturing (Ellis, 1992).

5.7 Fractography of epoxy resins

Considerable information concerning the cause of fracture as well as the damage mechanisms occurring during the failure process can be obtained from a detailed examination of the fracture surfaces. Although a significant number of articles have addressed this problem, the wide range of terms and expressions currently employed to describe fracture surface characteristics still result in considerable confusion and a lack of a detailed understanding.

Even a cursory examination of almost any epoxy fracture surface highlights the presence of a number of distinct regions exhibiting different features and textures. One of the most effective ways of understanding and explaining the formation of such zones is to systematically vary test parameters such as temperature and rate whilst measuring the variation of crack velocity during failure. A number of techniques are available for measuring the velocity of a crack in a brittle polymer, possibly the best of these being the graphite gauge technique developed by Stalder (1985). Here a very thin layer of graphite is sprayed around the anticipated crack trajectory. Following this, two silver electrodes are painted on the specimen parallel to the crack path. Now, since a linear relationship exists between the relative crack length and the relative conductance, the determination of the variation of potential with time permits the crack velocity to be measured at any moment during fracture. This technique has been successfully applied to double-torsion (Stalder, 1985), single edge notch (Cantwell *et al.*, 1990), three point bend (Stalder, 1985) and compact tension test geometries (Cudré-Maroux, 1988).

In Figure 5.12 two single edge notch fracture surfaces are shown along with the associated crack velocity profiles. Figure 5.12a shows a smooth fracture surface with few apparent markings. The corresponding crack velocity profile shows that the crack accelerated to 230 m/s and then stabilized. The second fracture surface, Figure 5.12b, is much rougher in appearance due to extensive crack bifurcation. The velocity profile again shows a region of crack acceleration, however, in this case the crack reaches 430 m/s, the limiting velocity in this material. Crack branching and therefore a very rough three dimensional fracture surface occurs only when the crack

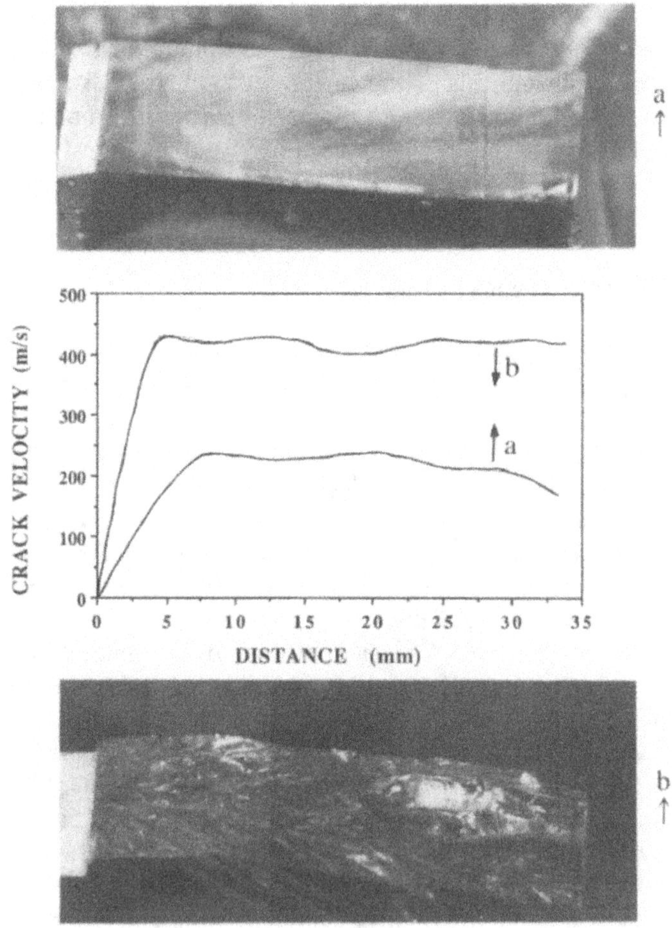

Figure 5.12 Fracture surfaces of single edge notch Araldite B specimens showing the variation of crack velocity during failure. Top: a smooth fracture surface; bottom: a rougher surface due to extensive crack bifurcation.

has reached the maximum velocity (determined by the sonic velocity of the material). Since any excess energy within the specimen cannot be dissipated by further accelerating the crack, the crack must bifurcate creating multiple fracture surfaces and greater energy dissipation.

Under certain conditions, other features may be prominent on the fracture surface of fracture mechanics-type specimens. One example is parabolic markings similar to those presented in Figure 5.13. Parabolic markings are associated with the initiation of secondary cracks in front of the primary crack in the region of fast unstable crack propagation. The precise shape and form of such markings depend upon the relative velocities of the primary and secondary crack fronts. If both fracture fronts move with the same velocity a hyberbola is formed whereas an ellipse is created if the primary crack propagates more rapidly than the secondary (Döll, 1989). In this case, the distance between the vertices of the markings is between 50 and 100 microns. Following impact tests on notched PMMA samples, Van Noort and Ellis (1984) observed a lower density of markings with the average distance between vertices being approximately 200 microns.

The above observations have been used to explain certain macroscopic features on the fracture surfaces of unmodified and particulate-filled epoxies (Cantwell *et al.*, 1990). Figure 5.14 shows the fracture surfaces of a hydantoin-based epoxy tested at a number of temperatures between 23 and 105°C (T_g of the resin = 125°C). A detailed examination of the failed surfaces highlighted four zones: a defect (not always in evidence), a smooth mirror-like region, a slightly less smooth area and a rough three-dimensional zone. The mirror-like zone was found to correspond to the region of slow crack growth where the crack advanced at a speed of the order of several millimetres per minute. Fine lines oriented parallel to the direction of crack propagation are sometimes observed in this zone. These lines or 'river-markings' occur as a result of crack propagation on two slightly different

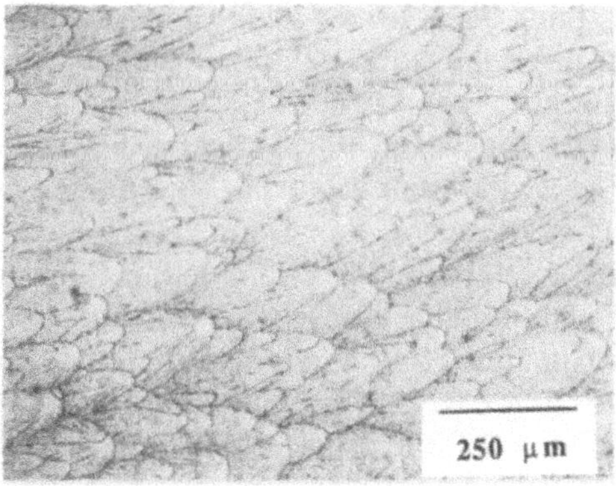

Figure 5.13 Parabolic markings on the fracture surface of a DGEBA epoxy (photo courtesy of Lu Fan).

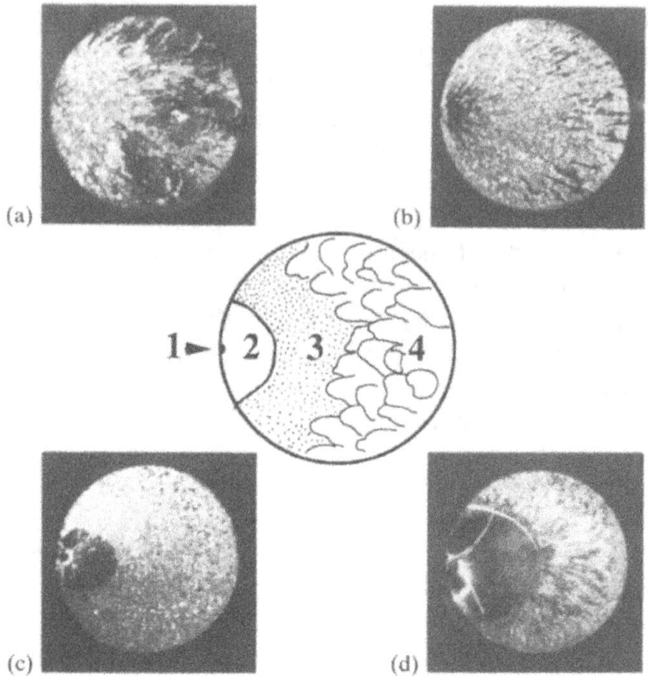

Figure 5.14 Tensile fracture surfaces of a hydantoin-based epoxy resin (type XB 2900) cured with a formulated anhydride hardener (HY 925) as a function of test temperature. Four regions have been identified: (1) a defect (not always in evidence) (2) a slow growth mirror-like region (3) a smooth region where the crack accelerates in an unstable manner and (4) a rough three dimensional area. (a) 23°C; (b) 50°C; (c) 85°C; (d) 105°C.

planes. The intersection of these two planes creates a welt frequently of the order of tens of microns in diameter. River markings are useful in as much as they indicate the direction of slow crack propagation. These lines can then be traced back enabling the defect or cause of failure to be identified.

Beyond the mirror region extends another flat zone exhibiting parabolic markings similar to those reported above. It was established that this region corresponded to the zone over which the crack was accelerating in an unstable manner. Once the crack achieved its limiting velocity (typically 400 to 500 m/s in an epoxy tested at room temperature) and was therefore no longer capable of accelerating, it bifurcated creating multiple fracture surfaces and a rough fracture zone. The velocity of sound within this material is approximately 1650 m/s suggesting that the limiting crack velocity is some 25–30% of this value. The rough zone was not observed in the specimens tested at 85 and 105°C suggesting that the crack had either stabilized at a velocity below the limiting value or else that the required acceleration zone was larger than the specimen diameter.

5.8 Toughening strategies for epoxy resins

The data presented in Figure 5.1 indicate that a large number of epoxies offering an equally wide range of toughnesses are presently available. Unfortunately, almost all of the epoxies for use in high temperature load-bearing applications are brittle offering values of G_{Ic} of 300 J/m^2 or less. In recent years, a considerable amount of work has been undertaken in an attempt to enhance the toughness of these materials. These studies have tended to concentrate on the addition of a second phase such as rubber particles, thermoplastic particles or mineral fillers such as silica flour.

5.8.1 Mineral filler-modified epoxies

Work by Moloney *et al.* (1984) and Spanoudakis and Young (1984) has shown that the modulus and toughness of an epoxy can be increased significantly through the introduction of small glass spheres. The effect of the incorporation of solid fillers on mechanical properties of epoxy resins is summarized in Table 5.1 (Moloney *et al.*, 1987). Clearly, the degree of enhancement in mechanical properties depends upon a number of parameters of which the volume fraction of filler as well as the surface treatment are amongst the most important. Typically, the addition of 40% by volume of glass beads results in a doubling of the Young's modulus and a 400% increase in K_{Ic} whilst incurring little or no loss in tensile strength (Moloney *et al.*, 1984; Spanoudakis and Young, 1984). Similar observations have been made following tests on silica-modified epoxy resins (Cantwell *et al.*, 1990). Some controversy still exists regarding the precise mechanisms responsible for the observed improvements in toughness in these materials. Several authors have suggested that the presence of the rigid particles serves to pin the advancing crack forcing it to bow out and hence increase the fracture energy (Spanoudakis and Young, 1984; Moloney *et al.*, 1983).

Table 5.1 Parameters determining the modulus, toughness and strength of particulate filled epoxies (from Moloney *et al.*, 1987)

Property	Effect on:		
	Modulus	K_c	Strength
Increase volume filler	Increase	Increase	~ Constant at matrix value
Particle size	Constant	Constant	Decrease
Increase aspect ratio	–	Increase	Increase
Increase strength and modulus filler	Increase	Increase	Increase
Improved adhesion	Constant	Constant	Increase
Tougher matrix	Small decrease	Increase	Decrease

Evans *et al.* (1985) suggested that substantial debonding of the second phase glass particles may occur around the crack tip creating microcracks leading to a reduction in the magnitude of the stresses in the process zone. A model based on an idealized constitutive law was developed which predicted the effect of filler volume fraction on toughness with some success. More recent work (Smith, 1989) has suggested that the presence of these solid particles results in a significant stress concentration in the surrounding matrix. At high loads, the local stress exceeds the yield stress of the matrix resulting in the formation of localized shear bands similar to those presented in Figure 5.15. The presence of such shear bands serves to blunt the propagating crack and thereby enhance the measured toughness. Under conditions of extreme crack tip ductility, such as at high temperatures and low strain rates, extensive particle-matrix debonding may occur. Under these circumstances, the presence of such a myriad of micro-cracks in a zone immediate to the primary crack reduces the inherent toughness of the material resulting in a significant drop in K_{Ic} (Cantwell *et al.*, 1990).

Low (1990) showed that residual stresses play a significant role in the failure process of filled epoxy systems. These residual stresses arise from the thermal expansion mismatch between the epoxy and the dispersed phase. He showed that compressive radial and tensile tangential stresses are generated at the interface when the filler has a lower thermal expansion coefficient that the matrix. It was indicated that the presence of such stresses tends to enhance interface debonding and encourages catastrophic unstable fracture resulting in a flat fracture surface.

5.8.2 *Thermoplastic-modified epoxies*

A number of workers have examined the feasibility of toughening epoxies using a thermoplastic second phase such as poly-ether-imide, poly-ether-

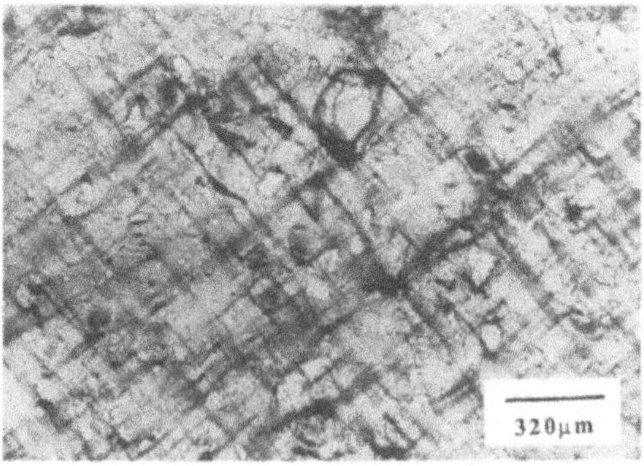

Figure 5.15 Shear bands in a 16% silica-filled epoxy resin tested in tension at 85°C (courtesy of Josh Smith).

sulphone, polysulphone and poly-phenylene-oxide (Hedrick *et al.*, 1985; Bucknall and Partridge, 1986; Pearson and Yee, 1991). The advantage of thermoplastic modifiers such as those reported above is that their incorporation into epoxy resins does not result in the reduction in modulus, reduced high temperature properties or poorer creep resistance associated with their rubber-modified counterparts. The early studies resulted in relatively small increases in toughness (Bucknall and Partridge, 1983; Hedrick *et al.*, 1985). More recent work by Pearson and Yee (1991) using a number of different thermoplastic modifiers to toughen a low T_g epoxy resulted in some cases in a toughness enhancement close to that achieved using CTBN (Figure 5.16). Thin sections removed from the necked region of a series of tensile specimens exhibited signs of extensive shear banding when viewed under polarized light. Surprisingly however, sections removed from the crack-tip regions of mechanics-type fracture specimens did not indicate the presence of this form of localized inelastic deformation. Instead, all of the samples showed signs of profuse microcracking, the exact nature of which depended upon the type of modifier used. The authors suggested that the presence of this damage zone at the crack tip served to shield the main crack from the applied stress intensity factor thus improving the measured toughness.

5.8.3 *Rubber-modified epoxies*

Extensive work has shown that the fracture toughness of epoxy resins can be significantly enhanced through the incorporation of a second rubbery phase. Although rubber has a low shear modulus, its bulk modulus is comparable to

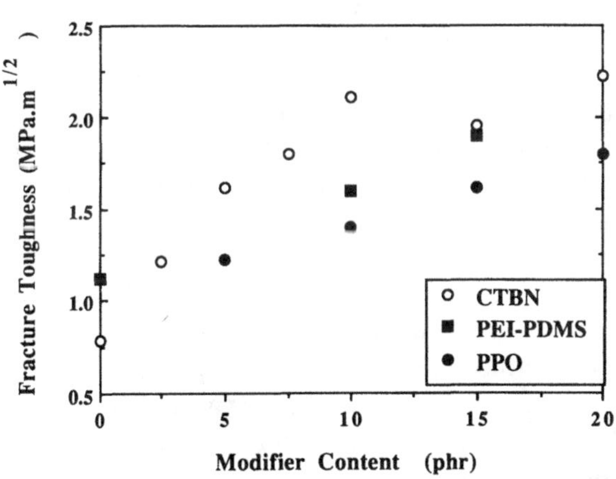

Figure 5.16 Toughness enhancement of an epoxy resin using thermoplastic particles and CTBN. The epoxy used for the CTBN-modified system was DER331 (a DGEBA from the Dow Chemical Co.) and that for the thermoplastic-modified materials was DER332 (also a DGEBA from the Dow Chemical Co.) (after Pearson and Yee, 1991).

that of an epoxy and rubber inclusions are therefore capable of bearing significant load within the material. Many workers have attempted to identify the precise mechanisms responsible for the significant toughness-enhancement associated with rubber modification. Kunz-Douglass *et al.* (1980) suggested that rubber particles enhanced toughness by bridging the crack and applying surface tractions which effectively reduce the stress intensity factor at the crack tip. Unfortunately, the rubber bridging model fails to predict the effect of varying test temperature. Since the tearing energy of rubber increases as the temperature is reduced below room temperature, the fracture toughness of rubber-modified epoxies should also increase with decreasing temperature. However, the available experimental evidence shows that this is not the case. Rubber bridging may indeed contribute to the energy-absorbing process but it is unlikely to be the principal mechanism.

Evans *et al.* (1986) attempted to quantify the energy dissipated in plastic dilation of the matrix following debonding of the rubber particle. They concluded that this mechanism may be important when it occurs local to a stress concentration or crack. Kinloch (1987) suggested that two important deformation processes occur in rubber-toughened epoxies. Firstly, the presence of the rubber particle creates a stress concentration which acts as a site for plastic shear deformation. It is not clear whether these shear bands initiate at the equator of the particle where the tensile stress concentration is a maximum or nearer the pole where the maximum shear stress concentration occurs. Since there are a large number of these stress-concentrating particles within the volume of the material, considerable energy dissipation occurs. Secondly, the tri-axial stress state existing at the tip of the crack causes the particles to fail and cavitate creating voids within the bulk of the material. Under continued loading, these voids grow and dissipate energy. Further, the cavitation process results in a reduction extent of stress-tri-axiality at the crack tip which in turn leads to a reduction in the yield stress and therefore greater yielding in areas of severe stress concentration. Chen and Jan (1991) used a number of epoxy end-capped CTBN rubbers to influence the cavitation process in a DGEBA epoxy. They showed that end-capping CTBN in this way resulted in large cavitated zones and a considerably higher fracture toughness.

Huang and Kinloch (1992) attempted to quantify the relative contributions of particle bridging, shear yielding and void growth as a function of temperature in a piperidine-cured DGEBA. Their analysis indicated that, at low temperature, shear yielding represents the main energy-absorbing mechanism whereas void growth and yielding are equally important at elevated temperatures (Figure 5.17). It is interesting to note that rubber bridging makes an appreciable contribution to the toughness of the polymer at sub-zero temperatures.

The degree of toughness enhancement appears to depend strongly upon the epoxy's ability to undergo plastic deformation and, therefore, upon its

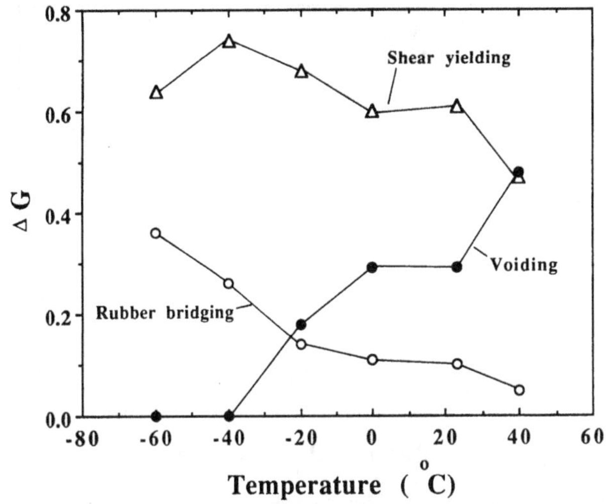

Figure 5.17 The relative contributions of shear yielding, rubber bridging and void growth as a function of temperature in a CTBN-modified piperidine-cured DGEBA (data taken from Huang and Kinloch, 1992).

microstructure. An early study by Meeks (1974) showed that the extent to which toughening can be induced in a rubber-modified epoxy depends upon the curing agent used. Anhydrides were shown to be the most useful in this respect with only modest toughening being observed with many of the other curing agents examined. By employing different cure schedules, Kinloch *et al.* (1987) varied the degree of cross-linking in a CTBN-modified DGEBA resin. Although the addition of rubber particles resulted in significant increases in G_{Ic} for both the low and high molecular weight resins, the relative increase was greatest in the latter. A more detailed study by Pearson and Yee (1989) clearly showed that the degree of toughness enhancement depends strongly upon the monomer molecular weight of the resin employed. For low values of molecular weight the increase in G_{Ic} was less than 100% whereas at higher values a thirty-five fold increase was reported. These observations were supported by scanning electron micrographs of the fracture surfaces which showed little plastic deformation in the high cross-link density material but extensive plastic flow and void enlargement in lower cross-linked epoxy. Levita (1989) also identified a strong resin molecular weight-dependency in CTBN-modified DGEBA-DDS (Figure 5.18), although his results suggested that K_{Ic} tends to level off at high molecular weights.

5.8.4 *Effect of particle size and volume fraction*

The aforementioned studies as well as that of Bradley *et al.* (1992), have highlighted the fact that high T_g epoxies such as those destined for use in the

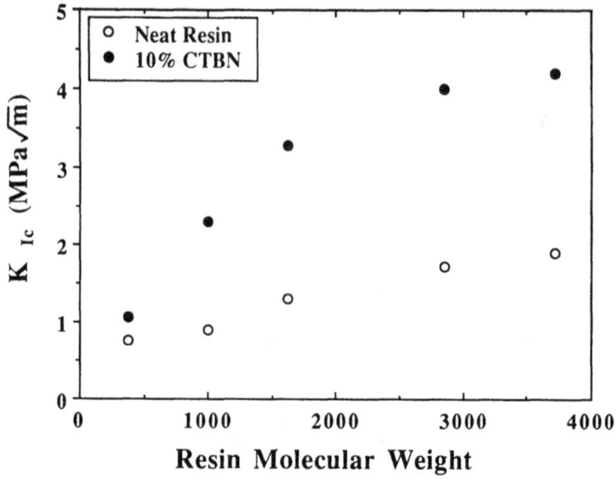

Figure 5.18 Fracture toughness vs resin molecular weight for neat and CTBN-modified DGEBA-DDS (after Levita, 1989).

aerospace industry often prove extremely difficult to toughen. The highly cross-linked nature of many of these epoxies limits their ability to undergo significant localized plastic deformation and therefore respond positively to rubber-toughening. Bradley *et al.* (1992) pointed out that the interparticle distance may well be greater than the size of the process zone at the crack tip. Consequently, the rubber inclusions might not interact with the crack process zone and under such circumstances their effect would be only marginal. The interparticle distance can be estimated from a simple two-dimensional analysis:

$$d_p = r_p\left[\left(\frac{2\pi}{3V_F}\right)^{1/2} - \frac{\pi}{2}\right]$$

where r_p is particle radius, and V_F the volume fraction of particles. This distance can therefore be decreased by increasing V_F for a given particle size or decreasing r_p for a given volume fraction. Unfortunately, increasing the volume fraction of rubber particles often results in an increase in particle diameter. Therefore, identifying and separating the effects of r_p and V_F are often difficult. In an early study, Sultan and McGarry (1973) suggested that particles around one micron in size encouraged crazing in epoxy resins whereas those considerably smaller than this value provoked shear yielding. It is generally accepted, however, that crazing does not occur in these cross-linked thermosets. Kunz-Douglass and co-workers (1980) reported that large rubber particles were not as efficient toughness-enhancers as smaller one micron inclusions. Hwang *et al.* (1989) claimed that sub-micron particles are less effective tougheners than those in the 1 to 5 micron range. The work of Bascom *et al.* (1981) suggested that a bimodal distribution of particles was more effective than a unimodal phase. More recently, Pearson (1990) showed that 0.2 micron particles were much more effective for toughening a DGEBA epoxy than either 2 or 50 micron particles. He suggested that large particles merely act as bridging particles which provide only a modest increase in toughness whereas small particles within the crack-tip process zone cavitate, relieving the plane strain constraint thereby facilitating yielding.

Kinloch and Hunston (1986) showed that the influence of volume fraction on G_{Ic} is complex and depends upon the particular test conditions. By applying a time-temperature superposition analysis similar to that reported above (section 5.4.2), they showed that at low temperatures varying the volume fraction of rubber particles in a piperidine-cured DGEBA epoxy has only a slight effect on toughness. At higher temperatures, however, doubling the volume fraction of particles had a similar effect on G_{Ic}. They concluded that the relationship between G_{Ic} and V_F depends upon the ability of the epoxy to respond to the presence of the rubber particles by undergoing shear-yielding.

5.8.5 *Hybrid systems*

A number of workers have examined the feasibility of adding both rigid and soft fillers to epoxy resins in order to achieve a significant enhancement in both modulus and fracture toughness (Young *et al.*, 1986; Smith, 1989). Smith (1989) showed that the K_{Ic} of a CTBN-silica hybrid epoxy was up to 50% higher than the standard silica-filled system. The elastic moduli of the hybrid and standard systems were, however, similar. Examination of the crack-tip region (Figure 5.19) indicated that damage took the form of shear yielding, debonding and cavitation, i.e. the deformation and failure mechanisms observed in the standard CTBN and silica-modified materials. Interestingly, rubber particle cavitation occurred preferentially at the poles

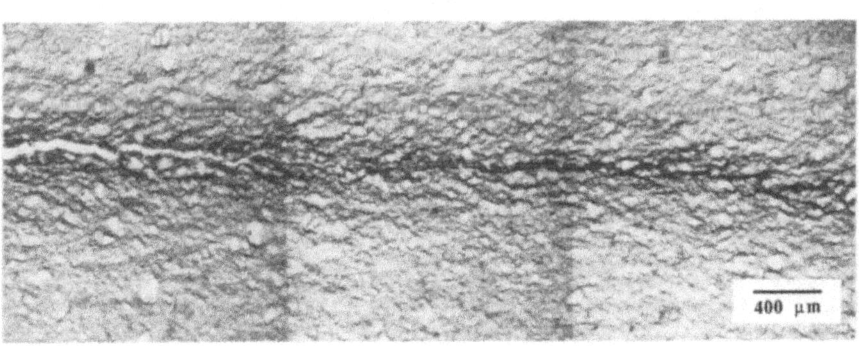

Figure 5.19 The crack tip region of a silica/CTBN-modified epoxy resin tested at 105°C (photo courtesy of Josh Smith).

of the silica particles. These observations support the numerical analysis of Guild and Young (1989) sho showed that the maximum stress concentration in a rigid filler-modified epoxy occurs at the poles of the inclusion.

Clearly, considerable progress has been made in the area of epoxy toughening and fracture suppression. More recently, a number of workers have been assessing the potential of core-shell modifiers consisting of a hard thermoplastic shell and a soft rubber core. The particles are generally sufficiently small and closely spaced to interact with the crack tip plastic zone, yielding appreciable toughness enhancement even in notoriously brittle high T_g systems.

5.9 Conclusions

Epoxy resins have proven themselves to be excellent engineering polymers finding extensive use in areas such as coatings, adhesives and matrices for aerospace composites. A large number of research studies have enabled the effects of varying both basic material parameters and operating conditions to be studied and characterized. The mechanisms by which these materials deform and fail are now well understood. However, more work is required before a full understanding of the influence of the network structure on many of the basic mechanical properties is achieved. The inherent brittle nature of these thermosetting polymers has been suppressed by incorporating a uniformly-dispersed second phase such as rubber particles, thermoplastic spheres or mineral fillers.

Acknowlegement

WJC is grateful to the Fonds National Suisse for financing a sabbatical at the Czechoslovakian Academy of Sciences, Prague, during which time part of the text was prepared.

References

Barbezat, M. (1990) *PhD Thesis,* Ecole Polytechnique Fédérale de Lausanne.
Bascom, W.D., Ting, R.Y., Moulton, R.J., Riew, C.K. and Siebert, A.R. (1981) *J. Materials Sci.* **16**, 2657–2664.
Bell, J.P. (1970) *J. Appl. Poly. Sci.* **14**, 1901–1906.
Bradley, W.L. (1989) In *Thermoplastic Composite Materials*, ed. Carlsson L.A. Elsevier, Applied Sci. Pub.
Bradley, W.L. (1991) *Proc. of the 8th Int. Conf. on Deformation, Yield and Fracture of Polymers*, PRI paper 36.
Bradley, W.L., Schultz, W., Corleto, C. and Komatsu, R. (1992) to be published in *Rubber-Toughened Plastics*, ed. Riew, C.K.
Bucknall, C.B. and Partridge, I.K. (1983) *Polymer* **24**, 639–644.
Bucknall, C.B. and Partridge, I.K. (1986) *Polymer Eng. and Sci.* **26**, 54–60.
Cantwell, W.J., Roulin-Moloney, A.C. and Kaiser, T. (1988a) *J. Materials Sci.* **23**, 1615–1631.

Cantwell, W.J., Roulin-Moloney, A.C. and Kausch, H.H. (1988b) *J. Materials Sci. Letters* **7**, 976–980.
Cantwell, W.J., Smith, J.W., Kausch, H.H. and Kaiser, T. (1990) *J. Materials Sci.* **25**, 633–648.
Chen, T.K. and Jan, Y.H. (1991) *J. Materials Sci.* **26**, 5848–5858.
Cherry, B.W. and Thomson, K.W. (1981) *J. Materials Sci.* **16**, 1913–1924.
Cudré-Maroux, N. (1988) *PhD Thesis,* Ecole Polytechnique Fédérale de Lausanne.
Döll, W. (1989) In *Fractography and Failure Mechanisms of Polymers and Composites*, Ch. 10, ed. Roulin-Moloney, A.C. Elsevier Applied Science.
Dušek, K., Plestil, J., Lednicky, F. and Lunak, S. (1978) *Polymer* **19**, 393–397.
Ellis, B. (1992) Private communication.
Evans, A.G., Williams, S. and Beaumont, P.W.R. (1985) *J. Materials Sci.* **20**, 3668–3674.
Evans, A.G., Ahmad, Z.B., Gilbert, D.G. and Beaumont, P.W.R. (1986) *Acta Metall.* **34**, 79–87.
Feuerhahn, M. (1986) *Kunstoffe* **76**, 872–873.
Fischer, M. (1992) In *Advances in Polymer Science 100*, Springer-Verlag, Berlin, pp. 313–355.
Glad, M.D. (1986) *PhD Thesis*, Cornell University.
Gledhill, R.A. and Kinloch, A.J. (1976) *Polymer* **17**, 727–731.
Gledhill, R.A., Kinloch, A.J., Yamini, S. and Young, R.J. (1978) *Polymer* **19**, 574–582.
Gledhill, R.A., Kinloch, A.J. and Shaw, S.J. (1979) *J. Materials Sci. Letters* **14**, 1769–1772.
Griffith, A.A. (1920) *Phil. Trans. Roy Soc.* **A221,** 163–198.
Guild, F.J. and Young, R.J. (1989) *J. Materials Sci.* **24**, 298–306.
Hedrick, J.L., Yilgör, Wilkes, G.L. and McGrath, J.E. (1985) *Polymer Bulletin* **13**, 201–208.
Hibbs, M. and Bradley, W.L. (1987) *Proc. Soc. Exp. Mechanics*, Fall Meeting, Georgia.
Huang, Y. and Kinloch, A.J. (1992) *J. Materials Sci. Letters* **11**, 484–489.
Hunston, D.L., Kinloch, A.J., Shaw, S.J. and Wang, S.S. (1984) in *Adhesive Joints* ed. Mittal, K.L., Plenum Press, pp. 789–807.
Hwang, J.F., Manson, J.A., Hertzberg, R.W., Miller, G.A. and Sperling, L.H. (1989) *Poly. Eng. Sci.* **29**, 1466–1476.
Kinloch, A.J. (1985) In *Advances in Polymer Science* **72**, ed. Dušek, K. Springer-Verlag, Berlin, pp. 46–67.
Kinloch, A.J. (1987) In *Rubber-Toughened Plastics* ed. Riew, C.K. Advances in Chemistry Series 222 American Chem. Soc. 1989 Ch. 3.
Kinloch, A.J. and Hunston, D.L. (1986) *Polymer* **5**, 1207–1209.
Kinloch, A.J. and Williams, J.G. (1980) *J. Materials Sci.,* **15**, 987–996.
Kinloch, A.J., Shaw, S.J. and Hunston, D.L. (1983) *Polymer* **24**, 1355–1363.
Kinloch, A.J., Gilbert, D.G. and Shaw, S.J. (1986) *J. Materials Sci.* **21**, 1051–1056.
Kinloch, A.J., Finch, C.A. and Hashemi, S. (1987) *Polymer Comm.* **28**, 322–325.
Kunz-Douglass, S., Beaumont, P.W.R. and Ashby, M.F. (1980) *J. Materials Sci.* **15**, 1109–1123.
LeMay, J.D. and Kelley, F.N. (1986) In Adv. in Polymer Science, **78** ed. Dušek, K. pp. 116–148.
Levita, G. (1989) In *Rubber-Toughened Plastics* ed. Riew, C.K. Advances in Chemistry Series 222 American Chem. Soc. 1989 Ch. 4.
Lilley, J. and Holloway, D.G. (1973) *Phil. Mag.* **8**, 215–220.
Low, I.M. (1990) *J. Materials Sci.* **25**, 2144–2148.
Marshall, G.P., Coutts, L.H. and Williams, J.G. (1974) *J. Materials Sci.* **9**, 1409–1419.
Meeks, A.C. (1974) *Polymer* **15**, 675–681.
Mijovic, J.S. and Koutsky, J.A. (1979) *J. Appl. Polym. Sci.* **23**, 1037–1042.
Misra, S.C., Manson, J.A. and Sperling, L.H. (1979) *ACS Symp. Series* **114**, 137–156.
Moloney, A.C., Kausch, H.H., Kaiser, T. and Beer, H.R. (1987) *J. Materials Sci.* **22**, 381–393.
Moloney, A.C., Kausch, H.H. and Stieger, H.R. (1983) *J. Materials Sci.* **18**, 208–216.
Moloney, A.C., Kausch, H.H. and Stieger, H.R. (1984) *J. Materials Sci.* **19**, 1125–1130.
Morgan, R.J. (1980) In *Developments in Reinforced Plastics* **1**, ed. Pritchard, G. App. Sci. Pub., Ch. 7.
Morgan, R.J. (1985) In *Adv. in Polym. Sci.* **72**, ed. Dušek, K. Springer-Verlag, Berlin, pp. 3–43.
Morgan, R.J. and O'Neal, J. (1977) *J. Materials Sci.* **12**, 1966–1980.
Morgan, R.J., O'Neal, J. and Miller, D.B. (1979) *J. Materials Sci.* **14**, 109–124.
Narisawa, I., Murayama, T. and Ogawa, H. (1982) *Polymer* **23**, 291–294.

Nielsen, L.E. (1969) *J. Macromol. Sci.-Revs Macromol. Chem.*, **C3(1)**, 69–103.
Pearson, R.A. (1990) *PhD Thesis*, Univ. of Michigan.
Pearson, R.A. and Yee, A.F. (1989) *J. Materials Sci.* **24**, 2571–2580.
Pearson, R.A. and Yee, A.F. (1991) *Proc. of the 8th Int. Conf. on Deformation, Yield and Fracture of Polymers*, PRI paper 40.
Phillips, D.C., Scott, J.M. and Jones, M. (1978) *J. of Materials Sci.* **13**, 311–322.
Racich, J. and Koutsky, J. (1976) *J. of Appl. Polym. Sci.* **20**, 2111–2129.
Rooke, D.P. and Cartwright, D.J. (1976) *Compendium of Stress Intensity Factors*, HMSO, London.
Scott, J.M., Wells, G.M. and Phillips, D.C. (1980) *J. Materials Sci.* **15**, 1436–1448.
Sih, G.C. (1973) *Handbook of Stress-Intensity Factors for Researchers and Engineers*, Lehigh Univ.
Smith, J.W. (1989) *PhD Thesis,* Ecole Polytechnique Fédérale de Lausanne.
Spanoudakis, J. and Young, R.J. (1984) *J. Materials Sci.* **19**, 473–486.
Stalder, B. (1985) *PhD Thesis*, Ecole Polytechnique Fédérale de Lausanne.
Sultan, J.N. and McGarry, F.J. (1973) *Poly. Eng. and Sci.* **13**, 29–34.
Tada, H., Paris, P. and Irwin, G.R. (1973) *The Stress Analysis of Cracks Handbook*, Del Research Corp., Hellertown.
Thomson, K.W. (1981) *J. Materials Sci. Letters* **16**, 1435–1436.
van den Boogaart, A. (1966) *Inst. of Physics Conf. Proc.* London, pp. 167–175.
Van Noort, R. and Ellis, B. (1984) *J. Materials Sci. Letters* **3**, 1031–1034.
Williams, J.G. (1973) *Stress Analysis of Polymers*, Longmans, London.
Williams, J.G. (1984) *Fracture Mechanics of Polymers*, Ellis Horwood Ltd, Chichester.
Yamini, S. and Young, R.J. (1977) *Polymer* **18**, 1075–1080.
Yamini, S. and Young, R.J. (1979) *J. Materials Sci.* **14**, 1609–1618.
Yamini, S. and Young, R.J. (1980) *J. of Materials Sci.* **15**, 1823–1831.
Young, R.J. (1980) In *Developments in Reinforced Plastics – 1*, ed. Pritchard, G. App. Sci. Pub. Ch. 9.
Young, R.J. and Beaumont, P.W.R. (1976) *J. Materials Sci. Letters* **11**, 778–779.
Young, R.J., Maxwell, D.L. and Kinloch, A.J. (1986) *J. Materials Sci.* **21**, 380–388.

6 Electrical properties of epoxy resins

G.P. JOHARI

6.1 Introduction

Epoxy resins are widely used for coating and encapsulating electrical circuit components and electronic devices, where they serve to isolate a device from adverse environmental effects of atmospheric gases, moisture, current leakage, solvents, micro-organisms, mechanical shock and vibrations. In a more recent application, known as micro-electronics packaging, they provide the required insulation between closely-packed, fragile electrical elements with good adhesion to both the substrate and the elements. It is this application that now requires that the changes in the electrical properties of epoxies during their curing and post-curing be understood at a fundamental level, and that any effects of ageing on these properties be studied in enough detail that procedures for eliminating these effects may be developed.

Although dielectric measurements for studying the progress of polymerization were made as early as 1934 by Kienle and Race (1934), who investigated polyesterification reactions, extensive literature on the subject has appeared only since 1968. An excellent review of these studies up to 1986 has been given by Senturia and Sheppard (1986). For the reader's convenience they have included, in this review, the details of measurement methods and a comprehensive bibliography with 118 references on the dielectric studies of various pure epoxies and their composites.

Since 1986, studies focused on dielectric monitoring of the progress of chemical reactions during the processing of various epoxies have been made by several groups (Day, 1986; Lane and Seferis, 1986; Bidstrup *et al.*, 1989; Gatro and Yandrasits, 1989; Lane *et al.*, 1989; Sheppard and Senturia, 1989; Mangion, 1990; Kranbuehl, 1991; Parthun and Johari, 1992a; Alig and Johari, 1992; Tombari and Johari, 1992). They have related the change in the electrical properties to the increase in the viscosity of the epoxy resin, thereby deducing that the increase in the glass transition temperature with time allows modelling of the isothermal curing process – a subject of considerable technical importance – by the Vogel–Fulcher–Tamman equation. A collection of papers related to these observations has been published in a special issue of *Polymer Engineering and Science* (vol. 29, no. 5, 1989).

A somewhat different but more general treatment is based upon the premise that in all addition reactions the growth in the size of a macromolecular addition product slows the rate of the very chemical reactions by which it grows, and that this continues until neither chemical reactions nor molecular diffusion occur at a perceptible rate. In this way, Johari and coworkers (Johari and Mangion, 1991; Parthun and Johari, 1992a) prefer to envisage the curing of an epoxy resin as a negative feedback process between chemical reactions and molecular diffusion, which ultimately and irreversibly vitrifies or solidifies a liquid. Thermodynamic evidence for this vitrification is now available from the careful heat capacity measurements by Cassettari *et al.* (1992); there is an irreversible, abrupt decrease in the heat capacity of an epoxy resin over a narrow range of time, which remarkably resembles the (reversible) decrease in heat capacity observed on supercooling a liquid and rubbery polymer through their glass transition temperature, T_g. The time at which this decrease occurs becomes less when the reaction occurs at a higher temperature.

This chapter reviews the developments, since 1986, in our knowledge of both the dielectric effects during the curing process of epoxy resins and the electrical properties of their cured state. Much recent progress has come from work on amine-cured epoxies, but many of the principles that have been discovered relate equally to other reactions of linear chain and network polymerizations, a subject of our continuing studies.

This chapter is organized such as to provide first (section 6.2) general features of changes in the physical properties during the curing of a thermoset polymer, in particular an epoxy resin. This is followed in section 6.3 by the description of a general formalism for the electrical properties of an epoxy resin as they change with the curing time, temperature and frequency used for the measurement. Section 6.4 provides the changes in the dielectric permittivity, and section 6.5 those in the ionic conduction during the process of sol–gel–glass conversion. The time and temperature evolution of the dielectric properties is given in section 6.6 which is followed by the chemical kinetics and dielectric behaviour in section 6.7. Sections 6.8 and 6.9 provide a description of the sub-T_g relaxations in partially-cured and fully-cured epoxy resins and the effects of post-cure and physical ageing on their electrical properties. A summary of the electrical applications of epoxy resins is given in section 6.10.

6.2 Physical changes during the epoxy curing

A variety of mechanical relaxation measurements on thin wire braids impregnated with epoxies, carried out by Gillham and coworkers (for a review see Aronhime and Gillham, 1986), and of viscoelasticity measurements on pure epoxies by Choy and Plazek (1986) have shown a general pattern of physical changes that occurs during the curing of an epoxy. As an epoxy group opens to form one OH group and one chemical bond with another molecule during the curing process, the viscosity increases in a manner such that the original liquid or sol state is converted to the gel state of an infinitely-connected network structure with mechanical modulus of less than 10^6 Pa. Here the viscosity approaches a formally infinite value, but localized molecular motion and diffusion of unreacted component persist in the network structure. As the curing progresses, the cross-links density increases with further reactions, the cross-linked network structure densifies and its mechanical modulus approaches that of a vitrified or glassy solid, i.e. about 10^{10} Pa. Gillham and coworkers (Enns and Gillham, 1983) have deduced the transformation–time–temperature diagrams of several epoxy resins by using such results obtained from their torsional-braid analysis.

Once an epoxy resin has vitrified, chemical reactions continue to occur, albeit at a much slower rate, which further raise its density, glass transition temperature, T_g, and mechanical modulus – an effect known as post-cure. In addition, a second type of spontaneous densification, which does not involve chemical reactions, occurs when the high-energy, kinetically metastable, amorphous structure tends, as in a chemical-stable glass, toward a lower-energy, thermodynamically metastable structure of the corresponding equilibrium state. Known as structural relaxation, it also increases both the solid's density and mechanical modulus. Thus the effects of isothermal ageing of a cured epoxy resin involve two processes; first a chemical one that raises the cross-link's density during the early period of ageing, and second a physical or structural one (relaxation) that does not alter the chemical structure, but has the same effects on the density and mechanical modulus as the chemical process, and which ultimately determines the physical behaviour of a cured epoxy resin. Although the two processes during isothermal ageing have qualitatively similar effects on the density and mechanical modulus, they have quite different effects on the electrical behaviour of epoxy resins (section 6.8).

6.3 Theoretical formalism for electrical properties

As an epoxy liquid and a proton-containing curing agent chemically react, the isothermal growth of a macromolecule causes the molecular diffusion coefficient to irreversibly decrease with the passage of time until the reactions that allow the growth come to a virtual halt. The concomitant changes in the molecular or structural state and chemical composition of the liquid irreversibly alter its electrical properties, we deduce, in at least seven ways, namely:

(i) A general decrease in the dc conductivity as the diffusion coefficient or impurity ions in the epoxy resins decreases with increase in its viscosity and any proton transfer along H–bonds in the epoxy mixture is virtually eliminated by the formation of cross-links.

(ii) An increase in the molecular diffusion or relaxation time as a result of which the dielectric permittivity measured at a fixed frequency monotonically decreases toward the value corresponding to the infrared region.

(iii) A change in the number of dipoles per unit volume, and therefore in $\Delta\varepsilon$, the contribution to permittivity from orientation polarization as a result of both the chemical reactions that alter the dipole moment and the volume contraction that raises the number density of dipoles.

(iv) A change in the dielectric relaxation function as the chemical structure of the liquid changes and its viscosity and density increase.

(v) A change in $\Delta\varepsilon_{ir}$, the contribution to permittivity due to infrared polarization, as the vibrational frequencies of the various modes in the structure change on cross-linking and densification.

(vi) A change in the optical refractive index or optical polarization as the network structure forms and densifies.

(vii) The splitting of a unimodal relaxation function into a bimodal relaxation function.

Thus in addition to the frequency and temperature dependence, the complex dielectric permittivity, ε^*, of an epoxy resin is also time-dependent and may be expressed by a general equation:

$$\varepsilon^*(i\omega, T, t_c) = \varepsilon'(\omega, T, t_c) - i\varepsilon''(\omega, T, t_c) \qquad (6.1)$$

where the dielectric permittivity, ε', and loss, ε'', contain contributions from both the dipolar orientation polarization and dc or ionic conduction, ω is the angular frequency of the applied electric field and T and t_c refer to the temperature and time since the beginning of the curing process of a thermoset polymer. (Terms in parentheses refer to independent variables in such studies.)

In ion-containing materials, the measured permittivity and loss include contributions from both the space charge effects and dc conduction which are due to diffusion of ions and the dipolar polarization, such that

$$\varepsilon'(\omega, T, t_c) = (Z_0 \sin(\alpha\pi/2)\, \omega^{-(1+\alpha)})G^2(\omega, T, t_c)/C_0^+ + \varepsilon'_{dip}(\omega, T, t_c) \qquad (6.2)$$

and,

$$\varepsilon''(\omega, T, t_c) = (\sigma_0(T, t)/\omega e_0) - (Z_0 \cos(\alpha\pi/2)\omega^{-\alpha})G^2(\omega, T, t_c)/C_0 + \varepsilon''_{dip}(\omega, T, t_c) \qquad (6.3)$$

where the first term on the right hand side of equation 6.2 and the second in equation 6.3 refer to the contribution from a 'constant phase element' for an interfacial impedance which is in series with the bulk properties of a material. In these first terms in equations 6.2 and 6.3 Z_0 is the characteristic impedance of the electrode-insulator interface, G is the measured conductivity which is equal to the reciprocal of resistivity, C_0 is the geometric capacitance of the sample, and $\alpha = 0.5$, as Johnson and Cole (1951), McDonald (1987) and Cole and Tombari (1991) have shown. In the last terms of equation 6.2 and 6.3 ε'_{dip} and ε''_{dip} are the contributions to ε' and ε'' from dipolar reorientation alone. σ_0 in equation 6.3 is the dc conductivity and e_0 the permittivity (= 8.8514 pF m^{-1}) of free space.

For epoxy resins with relatively low ionic concentration or dc conductivity, contributions from space-charge polarization are generally insignificant and can be neglected, so that ε' and ε'' may be written as,

$$\varepsilon'(\omega, T, t_c) = \varepsilon'_{dip}(\omega, T, t_c) \tag{6.4}$$

and

$$\varepsilon''(\omega, T, t_c) = \varepsilon''_{ion}(T, t_c) + \varepsilon''_{dip}(\omega, T, t_c) \tag{6.5}$$

where $\varepsilon''_{ion} = (\sigma_0/\omega e_0)$ is the contribution to ε'' from dc conductivity. Here it is clear that the permittivity of an epoxy resin during its curing process is determined only by the dipolar orientation polarization but the loss is determined by both the ionic or dc conduction and dipolar reorientation process.

We first consider the change in the dc or ionic conductivity contribution, ε''_{con}, during the curing of an epoxy resin. A number of the author's studies on diamine-cured epoxies (Mangion and Johari, 1990c, 1991a, b, c; Johari and Mangion, 1991; Parthun and Johari, 1992a, b) have shown that the decrease in the dc conductivity of an epoxy resin during its curing is described by a power law,

$$\sigma_0(T, t_c) = \sigma_0(T, t_c \to 0)\left[\frac{t_{gel}(T) - t_c(T)}{t_{gel}(T)}\right]^{x(T)} \tag{6.6}$$

where $\sigma_0(T, t_c \to 0)$ is the dc conductivity at the beginning of the curing reaction at a temperature T, t_{gel} is the time to reach the gelation point and x is the critical exponent of the scaling equation used as a generalized property function by Stauffer *et al.* (1982) and Djabourov (1988). (An alternative equation $\sigma_0(T, t_c) = A \exp(B/(t_0 - t))$, which implies a singularity at $t = t_0$, was used also by Johari and Mangion (1991), Mangion and Johari (1991a, b, c) and Parthun and Johari (1992a, b), but was found to be a less satisfactory description of the measured conductivity than the power law.) The significance of x in equation 6.6 lies in recognizing that for a given time to reach gelation, its value determines the rate at which σ_0 decreases with time, i.e. the higher the value of x, the more rapid is the decrease of σ_0 with the curing time. The description of the data by equation 6.6 allows extrapolation of the measured dc conductivity so that the time for the gelation of an epoxy resin during both its isothermal and ramp-curing can be estimated.

By combining equations 6.5 and 6.6, the dielectric loss during the curing of an epoxy is given by:

$$\varepsilon''(\omega, T, t_c) = \frac{\sigma_0(T, t_c \to 0)}{\omega e_0}\left[\frac{t_{gel}(T) - t_c}{t_{gel}(T)}\right]^{x(T)} + \varepsilon''_{dip}(\omega, T, t_c) \tag{6.7}$$

We now consider the dipolar contributions to ε' and ε''. According to the classical theory of dielectrics, the complex permittivity due to orientation polarization alone in a material is given by

$$\varepsilon^*(i\omega, T, t_c) = \varepsilon_\infty(T, t_c) + [\varepsilon_0(T, t_c) - \varepsilon_\infty(T, t_c)] \times \int_0^\infty \exp - i\omega t \left(- \frac{\partial \Phi(t)}{\partial t} \right) dt \tag{6.8}$$

where ε_0 and ε_∞ are the limiting low- and high-frequency permittivities respectively, t refers to the time for the decay of polarization for a chemical structure that remains unchanged during the measurement time, or when the curing reaction occurs too slowly in comparison with the time for measurement. In equation 6.8 Φ represents a relaxation function whose normalized form is given by

$$[\Phi(t)]_{t_c} = \exp[-(t/\tau)^\beta]_{t_c} \tag{6.9}$$

where β is an empirical parameter. For a Debye relaxation process, i.e. with a single relaxation time, β is equal to unity. For the majority of viscous liquids and solids, β is in the range zero to unity and here it is referred to as Kohlrausch–Williams–Watts (KWW) function (Williams and Watts, 1970), or stretched exponential parameter.

For a chemically and physically stable material, t and τ in equations 6.8 and 6.9 have a strict meaning. It is the time for the observation of the decay of polarization during which τ, the average relaxation time, remains invariant with time. For the curing process of a thermoset, τ is not invariant of time and therefore an approximation needs to be made. For frequencies of 1 kHz or more which are often used in the studies of epoxy resins, t is of the order of 1 ms or less and the effects from a change in both the chemical structure and physical state of the epoxy resin during this period of 1 ms can be justifiably assumed to be negligible at the usual curing temperatures when the curing time is about one hour or more. This approximation allows us to write equation 6.9 in a form used for a system with a time-invariant chemical structure, as follows:

$$N^*(i\omega\tau) = \int_0^\infty \exp - i\omega t \left(- \frac{\partial \Phi}{\partial t} \right) dt \tag{6.10}$$

or

$$N'(\omega\tau) - iN''(\omega\tau) = \mathscr{L}\left(- \frac{\partial \Phi}{\partial t} \right) \tag{6.11}$$

where the normalized quantity N^* is itself a function of the product of ω and τ of the structural state of the epoxy resin at any given instant since the beginning of the reaction, $\mathscr{L}$ represents a one-sided Laplace transform and ε' and ε'' are written in terms of N' and N'' by:

$$\varepsilon'(T, t_c) = \varepsilon_\infty + \frac{A}{T} N'(\omega\tau(T, t_c)) \tag{6.12}$$

and

$$\varepsilon''(T, t_c) = \frac{A}{T} N''(\omega\tau(T, t_c)) \tag{6.13}$$

where $(\varepsilon_0 - \varepsilon_\infty) = A/T$ with A as a constant according to the Curie Law.

For a few epoxy resins whose ε_0 could be measured during the curing process, Huraux and Sellaimia (1973), Sheppard and Senturia (1986), Mangion and Johari (1991a,b) and Tombari and Johari (1992) have found that the total decrease in ε_0 is by about 10% or less. This decrease therefore has much less effect on ε^* according to equation 6.8 in comparison with the effect from the increase in τ, which occurs by several orders of magnitude, typically from 10^{-9} s to 10^{+4} s. As a first approximation, therefore, consider that both ε_0 and ε_∞ remain constant with t_c and equations 6.8 and 6.10 may be combined to obtain

$$\varepsilon^*(\mathrm{i}\omega, T, t_c) = \varepsilon_\infty(T) + (\varepsilon_0(T) - \varepsilon_\infty(T))N^*[\mathrm{i}\omega\tau(T, t_c)] \tag{6.14}$$

It is easy to see that equation 6.14 is phenomenologically invariant of ω and τ as the choice of experimental variable. This means that any of the several shapes of the relaxation spectrum or Cole–Cole plots (Cole and Cole, 1941) can be obtained by varying either ω or τ. Thus during the curing of an epoxy resin, which has been envisaged as a negative feedback between molecular diffusion and chemical reactions (Johari and Mangion, 1991), the shape of the plots of ε' and ε'' against the reaction time resembles the dielectric dispersion and loss spectrum (i.e. plots of ε' and ε'' against the logarithmic frequency) of a chemically and physically stable substance. However, for epoxy resins where curing causes τ to irreversibly increase, the detailed shape of the complex plane plots of ε' and ε'' is mathematically described by a relaxation function, when measurements are made for a fixed frequency, for which the form of equation 6.9 is to be written as:

$$\Phi(t_c) = \exp\left[-(t/\tau(t_c))^{\gamma}\right], \tag{6.15}$$

where the $\tau(t_c)$ is now a pseudo-equilibrium average value of τ at instant t_c, which is determined by using the limiting short-time and long-time values of $\varepsilon^*(\mathrm{i}\omega, t_c)$ with τ as a variable of t_c. The new parameter, γ, has been named the curing parameter by Mangion and Johari (1990b, c).

Before considering the application of this formalism, one further aspect needs to be discussed here; that is that the dc conductivity of a time-variant system such as the curing of epoxy resin is not easily measurable because of the requirement that measurement be made as a function of decreasing frequency till a plateau value at a low enough frequency is reached and that during the period of measurement the chemical structure of the substance remain unaltered. This requirement cannot be fulfilled for the curing of epoxy resins. Therefore, it is necessary to devise an alternative procedure to determine whether a fixed value of frequency used for the measurement is low enough for $\varepsilon''(\omega, T, t_c)$ to be equal to $\sigma_0(T, t_c)/\omega e_0$ within the experimental error. We describe this procedure as follows.

The measured values of ε' and ε'' of an epoxy resin may be transformed into the complex electrical modulus, M^*, formalisms by the equations:

$$M^*(\mathrm{i}\omega, T, t_c) = (\varepsilon^*(\mathrm{i}\omega, T, t_c))^{-1} = M'(\omega, T, t_c) + \mathrm{i}M''(\omega, T, t_c) \quad (6.16)$$

where

$$M' = \frac{\varepsilon'}{\varepsilon'^2 + \varepsilon''^2} \quad \text{and} \quad M'' = \frac{\varepsilon''}{\varepsilon'^2 + \varepsilon''^2}. \quad (6.17)$$

M' and M'' are the real and imaginary components of the complex electrical modulus. When M^* is entirely due to ionic conduction, it is related to Maxwell or conductivity relaxation time, τ_σ, by the equation,

$$M^*(\mathrm{i}\omega, T, t_c) = M_0\left(\frac{\mathrm{i}\omega\tau_\sigma(T, t_c)}{1 + \mathrm{i}\omega\tau_\sigma(T, t_c)}\right) \quad (6.18)$$

where $\tau_\sigma = e_0\varepsilon_0/\sigma_0$, and $M_0 = \varepsilon_0^{-1}$. In equation 6.18, M^* is invariant of ω and τ_σ (or equivalently σ_0) as one's choice of variables and M' and M'' obey the expression

$$[M''(\omega, T, t_c)]^2 + [M'(\omega, T, t_c) - \tfrac{1}{2}M_0(T, t_c)]^2 = [\tfrac{1}{2}M_0(T, t_c)]^2 \quad (6.19)$$

Accordingly, a complex plane plot of M'' against M' would be a semicircle with a radius equal to $\tfrac{1}{2}M_0$ and centres on the M' axis, provided ω, T and t_c were such that no processes other than the ionic conduction contributed to $\varepsilon^*(\mathrm{i}\omega, T, t_c)$. If other contributions were present, the plot would deviate from the semicircular shape. Such a complex plane plot can therefore be used to determine the time during the curing process of an epoxy resin up to which its measured conductivity for a fixed frequency is equal to its dc conductivity.

6.4 Dielectric effects of sol–gel–glass conversion

Since the rate of chemical reactions during the curing process of an epoxy resin is sensitive to temperature, the change in the electrical properties with both the temperature and time become important variables in the curing process. Therefore, measurements are often carried out as a function of both temperature and time. As a typical example of the behaviour observed, measurement of the changes in electrical properties during the course of curing of a stoichiometric mixture of diglycidyl ether of bisphenol A (DGEBA) and hexamethylenediamine (HMDA) at six different temperatures are shown in Figure 6.1, where the real and imaginary components of dielectric permittivity and electrical modulus are plotted against the logarithmic curing time.

For short periods of curing, ε' slightly decreases first towards a plateau value and thereafter in a stepwise manner to ε' of about 4.4 as the curing time, t_c, approaches infinity. This step shifts towards longer t_c as the isothermal temperature for curing is decreased. The corresponding value of ε'' first decreases from a near plateau value to reach a local minimum, which is followed by a peak. As $t_c \to \infty$, ε'' decreases to reach a limiting low value of less than 0.02, which corresponds to that of a vitrified solid at a high temperature. As the isothermal curing temperature is decreased, the minimum becomes shallower, the peak becomes higher and both the minimum and the peak shift towards longer curing times.

The corresponding plots of M' in Figure 6.1 show an increase in M' with curing time, that occurs in two steps towards an ultimate value M_∞. Both steps shift to longer times with decrease in the curing temperature, and the first step becomes smaller in height than the second step, while the corresponding first plateau becomes broader on a logarithmic scale. The M'' plots in Figure 6.1 show two peaks whose positions shift to longer t_c and the width of the first peak increases.

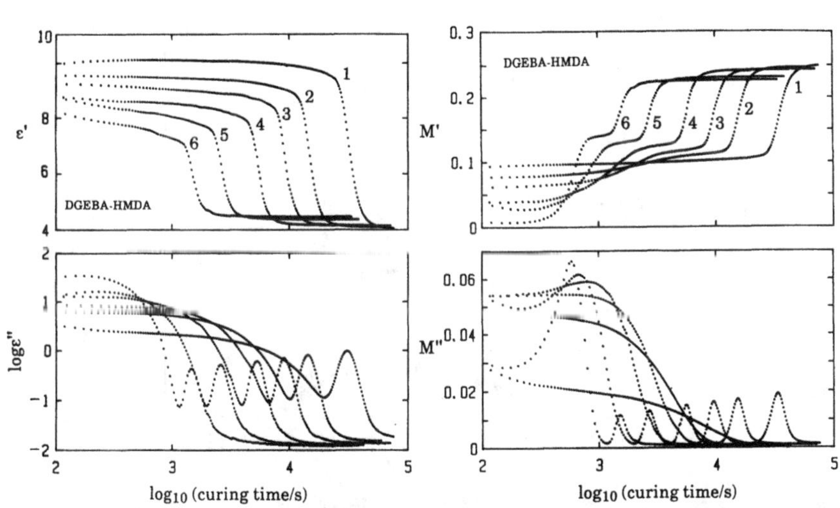

Figure 6.1 The real and imaginary components of dielectric permittivity and electrical modulus of the DGEBA cured with hexamethylenediamine measured for a fixed frequency of 1 kHz are plotted against the curing time. The isothermal temperature for curing are: (1) 284.3 K, (2) 296.5 K, (3) 304.2 K, (4) 312.3 K, (5) 324.3 K and (6) 336.5 K.

The theoretical formalism given in the preceding section here is more appropriately considered by examining the shape of the complex plane plots of ε'' and ε' and of M'' and M', which are shown in Figures 6.2 and 6.3. The shape of the ε'' against ε' plots resembles that of the Cole–Cole plots of chemically and physically stable materials with a relatively large ionic conductivity. Nevertheless, it is advisable to maintain a distinction between these plots, in which changes in the ε^* and M^* occur when the dielectric relaxation time irreversibly (and spontaneously) increases and each data corresponds to a certain t_c, and Cole–Cole plots, where the changes in the ε^* and M^* are produced by varying the measurement frequency. The width of the plot, i.e. $(\varepsilon'(t_c \rightarrow 0) - \varepsilon'(t_c \rightarrow \infty))$, increases with decrease in the curing temperature, as does the contribution to ε'' from ionic conductivity, which, although large in the beginning of the curing process, becomes too small to alter the shape of the complex plane plot as the curing proceeds.

The corresponding complex plane plots of M' and M'' measured for a fixed frequency of 1 kHz in Figure 6.3 have the shape of a semicircle, with

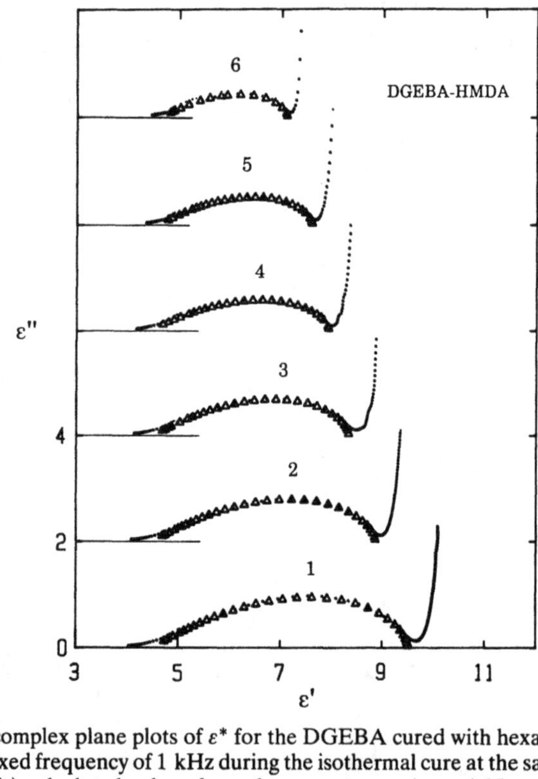

Figure 6.2 The complex plane plots of ε^* for the DGEBA cured with hexamethylenediamine measured for a fixed frequency of 1 kHz during the isothermal cure at the same temperature as in Figure 6.1. (△) calculated values from the parameters, $\Delta\varepsilon = 4.85$ and $\gamma = 0.32$ for (1); 4.35 and 0.32 for (2); 3.90 and 0.32 for (3); 3.48 and 0.31 for (4), 3.00 and 0.30 for (5) and 2.45 and 0.30 for (6).

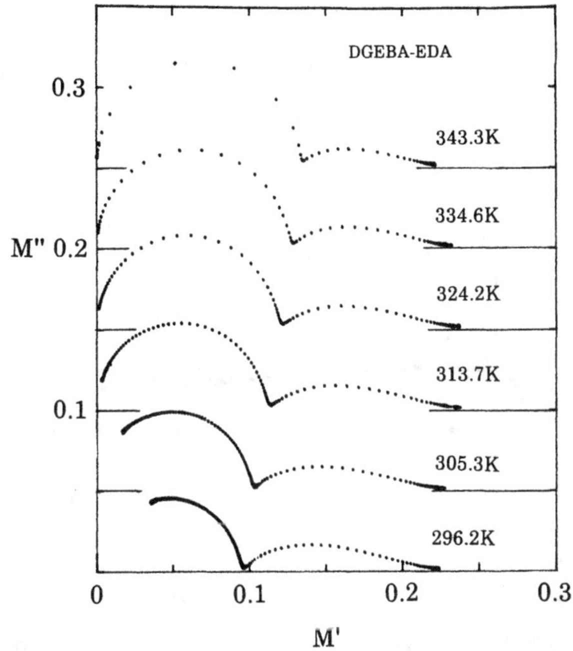

Figure 6.3 The complex plane plots of M^* for the DGEBA cured with ethylenediamine measured for a fixed frequency of 1 kHz during its isothermal curing at the temperature indicated. The semicircle is due to conductivity relaxation and the skewed arc due to dipolar relaxation.

centre on the axis, which is followed by an apparently skewed arc. However, as $t_c \rightarrow \infty$, progressively more deviations from the shape of the skewed arc appear. A change in the temperature for curing also affects the shape of these plots in that both the diameter of the semicircle and the width of the skewed arc generally increase as the temperature for isothermal curing of the epoxy resin is increased.

The shape of the complex plane plots of ε^* and M^* indicate that at the initial stages of an epoxy resin curing, it is the dc conductivity rather than the dipolar reorientation which dominates the dielectric behaviour. As the curing time increases, deviations from a semicircular shape of the M'' against M' plot occur and a new shape of a skewed arc emerges, which is a reflection of a dipolar relaxation process. Therefore, only part of the measured conductivity data is mainly due to ionic conduction and this part lies at times of cure shorter than the time t_{min}, at which a minimum appears in Figure 6.3. The conductivity, $\sigma(t_c)$, is equal to the dc conductivity when $t < t_{min}$ but exceeds the true $\sigma_0(t_c)$ as t_c approaches t_{min}, and this limitation should be considered in the analysis of the kinetic effects during the epoxy resin curing.

Typical plots for the measured conductivity against the curing time for one epoxy resin, DGEBA cured with 4,4′-diaminodiphenylmethane are shown

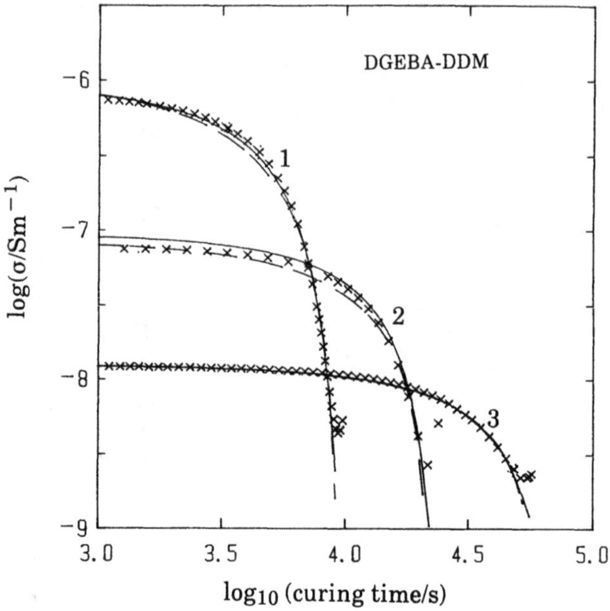

Figure 6.4 The logarithmic plots of the measured conductivity against the curing time for the DGEBA cured with 4,4′-diaminodiphenylmethane at temperatures indicated. The dashed line was calculated from equation 6.6 or power law, and the full line from equation 6.20, or singularity equation (after Mangion and Johari, 1991).

in Figure 6.4. The curves calculated from equation 6.6 for values of $\sigma_0(\rightarrow 0)$, x and t_{gel} have been shown (by the dashed line) to indicate the adequacy of equation 6.6 as a description of curing kinetics.

Another description, which is an alternative to the power law of equation 6.6 has also been used by Mangion and Johari (1991a, b, c) and Parthun and Johari (1992a). Its plausibility lies in recognizing that the increase in viscosity during the early stages of the curing process determines the ionic diffusion according to the Stokes–Einstein equation. This is particularly so in view of Johari and Pathmanathan's (1990) argument that the power law of equation 6.6, when applied to temperature-dependence of relaxation time, is equivalent to the well-known Vogel–Fulcher–Tamman equation over a narrow range of measurements. This alternative equation implies an approach of $\sigma_0(t)$ towards a singularity during the curing process according to

$$\sigma_0(T, t_c) = A(T) \exp \{-B(T)/[t_0(T) - t_c)]\} \tag{6.20}$$

where t_0 is the point of singularity or the time taken to reach a value $\sigma_0(T, t_c) = 0$ and A and B are temperature-dependent empirical parameters which determine the rate at which conductivity approaches the singularity at t_0. The data calculated from equation 6.20 are shown as a solid line in Figure

6.4. Mangion and Johari (1990b, c, 1991a, b, c) and Parthun and Johari (1992a, b) have found that the value of t_0 obtained from equation 6.20 is close to the time of cure when the ε'' peak for 1 kHz frequency measurement appears, and which in turn is closer to the vitrification time than to the gelation time. In addition, Parthun and Johari (1992a, b) have found that for DGEBA cured with a variety of diamines, the value of t_{gel} of equation 6.6 is virtually independent of the frequency used for the measurement, whereas that of t_0 of equation 6.20 systematically changes. Figure 6.5 for the DGEBA cured with ethylenediamine shows the variation of $\sigma(t_c)$ measured at

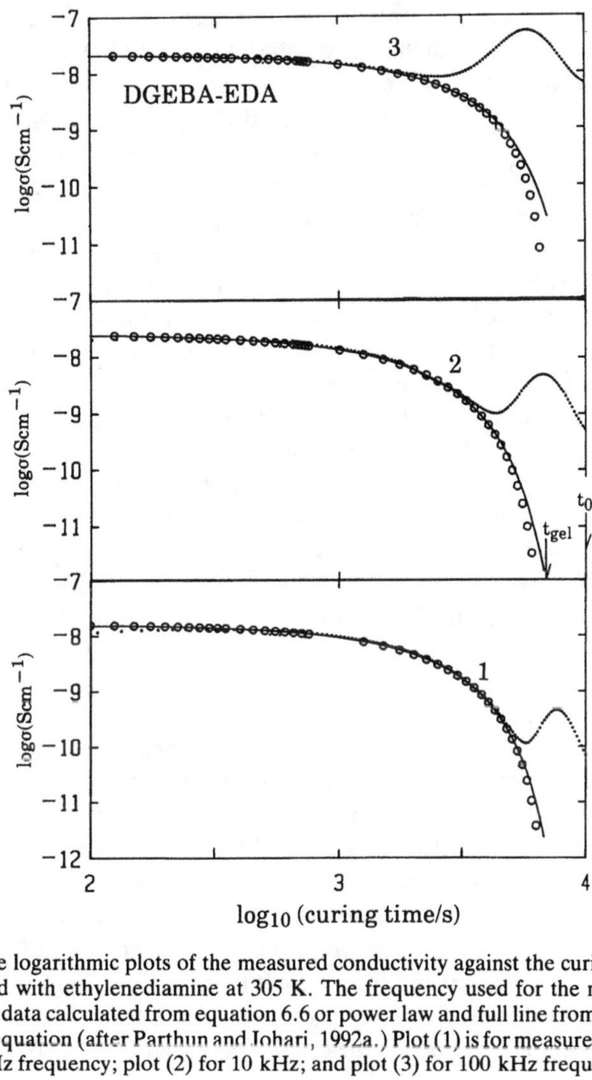

Figure 6.5 The logarithmic plots of the measured conductivity against the curing time for the DGEBA cured with ethylenediamine at 305 K. The frequency used for the measurement is indicated. (O) data calculated from equation 6.6 or power law and full line from equation 6.20, or singularity equation (after Parthun and Johari, 1992a.) Plot (1) is for measurements made for 1 kHz frequency; plot (2) for 10 kHz; and plot (3) for 100 kHz frequency.

different frequencies against the curing time of the epoxy resin, where the calculated t_{gel} and t_0 are marked by arrows.

The changes in ε^* of an epoxy resin during its curing, but after the dc conductivity's contribution to ε'' has become negligible, are of some interest here. These changes as seen in Figure 6.2 show that the fixed frequency values of ε' and ε'' of an epoxy resin during its curing resemble the ε' and ε'' plots of chemically and physically stable dipolar liquids and solids measured at a fixed temperature but with varying frequency. This, as mentioned in section 6.3, is a reflection of the fact that relaxation functions are invariant of ω or τ as the choice of independent variable. Mangion and Johari (1990b, c) have used equations 6.14 and 6.15 to calculate γ, the curing parameter, by using a modification of procedure given by Moynihan *et al.* (1973) and Bendler and coworkers (Montroll and Bendler, 1984; Dishon *et al.*, 1985; Weiss *et al.*, 1985), and Parthun and Johari (1992a,b) have calculated it by a procedure given by Muzeau *et al.* (1991). The data calculated by Parthun and Johari have been shown along with the experimental values in Figure 6.2. The agreement seen here demonstrates the adequacy of the formalism, given in section 6.3, for the curing behaviour of epoxy resins.

Given that the measured dielectric data are satisfactorily described by equation 6.15, it now becomes possible to calculate the relaxation time at any instant during the curing of an epoxy resin. This is so because in the plots of Figures 6.1 and 6.2, each data point corresponds to a unique value of $\omega\tau(t_c)$ in equation 6.14, which in turn corresponds to a unique value of N^*. Thus both the time-dependence of ε^*, and of $\tau(t_c)$ can be calculated when $\Delta\varepsilon$ and γ are known. Johari and Mangion (1991) were first to show the adequacy of such calculations, which was followed by the extensive work by Parthun and Johari (1992a, b) who used a variety of curing agents to demonstrate the variation of $\tau(t_c)$ with the curing time of an epoxy resin. As a typical example of this calculation taken from the work of the latter group, the calculated values of dielectric relaxation time, and the permittivity and loss of the DGEBA cured with propylenediamine have been plotted against the curing time in Figure 6.6.

The range of relaxation time as determined by the above procedure is of course limited by the insensitivity of $N^*(\omega\tau)$ to $\omega\tau$ for a given value of γ in equation 6.15 but $\omega\tau$ can also be varied by independently varying ω which then allows the determination of a wider range of τ over a broader time scale during the curing process. This is a remarkable feature in that it allows the determination of virtually the entire change in τ from picoseconds to kseconds as an epoxy resin reaches its ultimately cured state. By using frequencies of 50 Hz, 1 kHz and 100 kHz, Parthun and Johari (1992a, b) have been able to obtain the curing time dependence of τ of a number of epoxy resins, a typical example of which is shown in Figure 6.7. The data obtained from different frequencies of measurement overlap over a narrow region of curing time, but all lie on the same curve which is now much

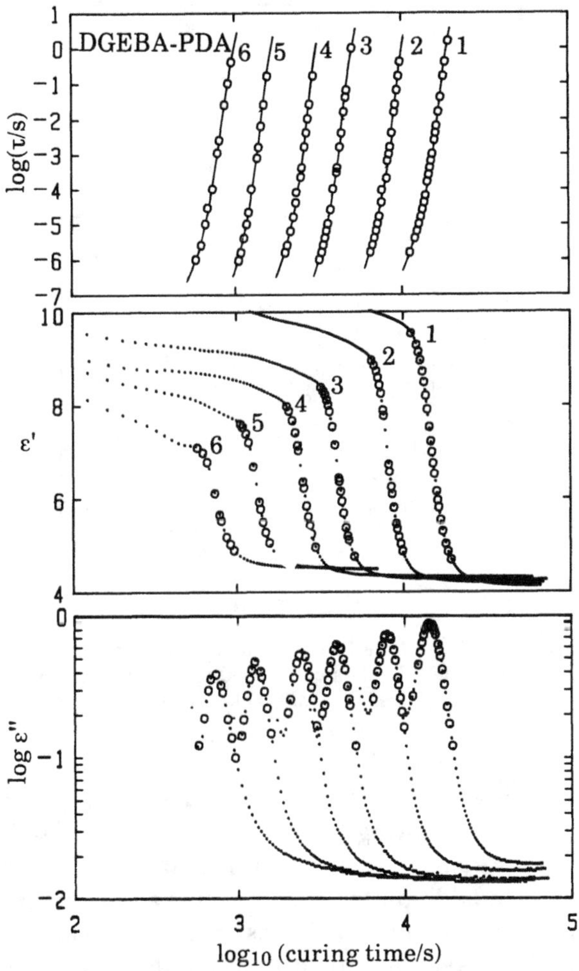

Figure 6.6 The relaxation time, the permittivity and loss of the DGEBA cured with propylenediamine measured for a fixed frequency of 1 kHz is plotted against the curing time during their isothermal curing at (1) 294.9 K, (2) 303.8 K, (3) 315.5 K, (4) 323.7 K, (5) 335.9 K and (6) 346.0 K. (○) calculated, and (●) measured, data.

broader, covering a range of 10^{-9} to 10 seconds. The single plot of the relaxation time against the curing time seems to be a satisfactory demonstration of the correctness of the formalism given in section 6.3.

6.5 Ionic conductivity and sol–gel conversion

In the production of diglycidyl ether of bisphenol A as in other types of epoxies made from epichlorhydrin and bisphenol A, HCl is produced as a

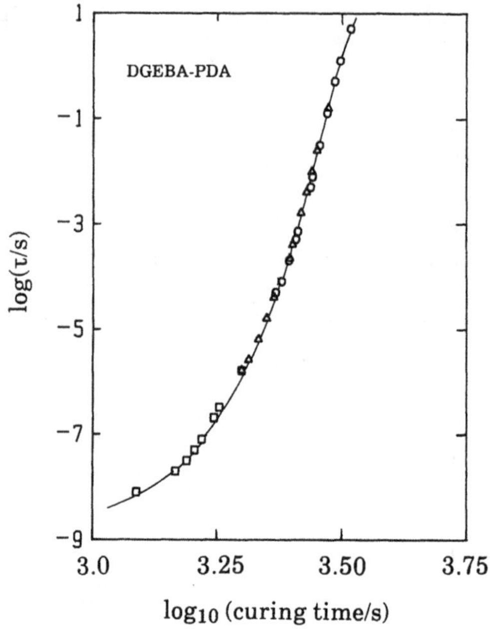

Figure 6.7 The relaxation time of the DGEBA cured with propylenediamine at 324.0 K is plotted against the curing time. The data correspond to measurements made at 50 Hz (○), 1 kHz (△) and 100 kHz (□).

byproduct which is subsequently neutralized with an alkali. Even after considerable effort in removing NaCl, certain amounts of sodium, chloride and hydrogen ions remain as impurities in the commercially available uncured epoxies. In addition the curing agents also contain impurity ions such as ammonium and hydroxyl ions in amines and anhydrides and hydrogen and hydroxyl ions in alcohols. Altogether these impurities substantially contribute to the dc conductivity of an epoxy resin according to the equation,

$$\sigma = \sum_{i} n_i z_i \mu_i \tag{6.21}$$

where n_i is the concentration of ions with ionic charge z_i and mobility μ_i. μ_i is often related to the inverse of viscosity according to Stokes' law and assuming that n_i remains unchanged during the curing process, σ is regarded as inversely proportional to the viscosity and its decrease to the increase in the viscosity of the epoxy resin. This subject has been ably reviewed by Senturia and Sheppard (1986) who have correctly pointed out that the variation of σ with viscosity is an oversimplification for an epoxy resin, for here the viscosity approaches a formally infinite value at gelation, while sufficient ionic and molecular degrees of freedom persist in the gelled state. After a reanalysis of the work on this subject, they deduced that the decrease in σ with curing time is related to the difference $(T - T_g)$, which itself rapidly decreases as curing progresses. Thus after the maximum rate of conversion which was taken to be at about 30–40%, the conductivity becomes more sensitive to changes in T_g with the curing time. Its magnitude, they concluded, provides no indication of the approach of gelation of epoxy resin.

An alternative conclusion has come from the detailed studies of Mangion and Johari (1990b, 1991a, b, c) and Parthun and Johari (1992a, b) who have shown that the dc conductivity decreases according to a scaling law of equation 6.6, and its value approaches zero at the gelation time of the epoxy resin. However, this approach towards a zero value, they point out, is interrupted by the contribution from ac conductivity which rises, with the progress of cure, towards a peak value in a fixed frequency measurement. They have argued that for the anticipated appearance of a singularity for σ, the time required to reach gelation should be independent of the frequency used for the dielectric measurement, provided the change in the frequency does not alter the relative contributions from other additional relaxation processes. This is, of course, expected since the dc conductivity due to ionic diffusion is invariant of the frequency for measurement, or ε'' inversely scales with the frequency of measurement. They have therefore measured the dielectric properties for a frequency of 50 Hz, 1 kHz, 10 kHz and 100 kHz as the curing of DGEBA with ethylenediamine, propylenediamine and hexamethylenediamine progressed and have found that the gelation time, t_{gel}, obtained from equation 6.6 is independent of the frequency used for the measurement.

A note of caution is necessary here, namely, that this new approach to determining the time for gelation should not be regarded as undermining the value of the now widely quoted (by Senturia and Sheppard, 1986) previous efforts which concluded that no electrical 'events' accompany gelation, for none does. The determination of gelation time by electrical measurement described here relies on accurate determination of the dc conductivity and the analysis of its variation with the curing time, which could not be done until recently. It is also necessary to recognize the fact that the (event of) approach towards formally a zero conductivity at gelation is prevented by the significant contribution from ac or dipolar conductivity of the gelled state. There is, of course, a need for confirmation of the value for gelation time obtained from equation 6.6 by measurements using a suitable mechanical relaxation procedure.

6.6 Time and temperature evolution of the dielectric properties

The fact that a single frequency measurement during the isothermal curing can provide information on both the gelation time and the relaxation time is valuable in determining how the various relaxation processes evolve on curing an epoxy resin. In virtually all physical measurements at a certain instant during the curing of an epoxy resin, the data obtained correspond to relaxation characteristics of its structure at that instant of cure but for the frequency used in the measurement. At any instant of cure this (fixed) structure is expected to relax with at least a bimodal distribution of relaxation times, known as α- and β-relaxations (Johari, 1973, 1976), when T_g of the thermoset is near or below its curing temperature, but with only one distribution of relaxation times when T_g is much greater than the curing temperature, i.e. under conditions when the α- and β-relaxation processes are merged. Thus as the curing progresses or t_c increases, the structural state of an epoxy resin changes from that of a liquid when the curing temperature is much greater than T_g to a rigid solid when it is equal to or lesser than T_g. For a fixed frequency measurement, as the curing progresses, changing properties belonging to each of the continuously different structural states would be observed as these states traverse with time from the ones with a very short relaxation time of a fluid, say 10^{-9} s, to the ones with a very long relaxation time, say 10^4, of its vitrified state. This is illustrated in Figure 6.8 where the anticipated increase in the relaxation time for the α- and β-relaxation processes with decreasing temperature are illustrated for measurement made at different instants t_0, t_1, t_2, etc. of cure. For simplicity, the plots of β-relaxation processes are drawn to have the same slopes and to merge at a frequency of measurements of about 10^7 Hz. The pre-exponential factor is necessarily kept the same for all plots. In Figure 6.8 the measured ε' and ε'' at 1 kHz frequency for a given value of γ would initially correspond to a structure and relaxation at a point T_c^{-1} on curve t_0. With the passage of time, and in a continuous manner, this would be followed with those properties which correspond to the structure and relaxation times for points at T_c^{-1} on curves t_1, t_2, ..., etc., and ultimately on curve t_{vit}, the vitrification time. As t_{vit} is reached, all contributions from the α-relaxation process may not yet reach their minimum value if the frequency of measurement corresponds to the β-relaxation rate of the network structure. In section 6.4 we surmised that the deviations of ε'' and ε' from a stretched exponential decay function in the plots of Figures 6.2 and 6.3 as $t_c \rightarrow \infty$ are due to the contribution from β-relaxation, and earlier studies (Mangion and Johari, 1990c) have shown that the strength of β-relaxation initially increases with time during the post-curing at the expense of the height of another sub-T_g, γ-relaxation, whose strength decreases. Simultaneously, the α-relaxation process shifts to lower frequencies and thus its contribution to ε'' decreases. In measurements made at a fixed instant after t_{vit}, where post-cure occurred sufficiently slowly to allow the measurement of the spectrum, as shown in earlier studies (Mangion and Johari, 1990a), a peak in the frequency spectrum of ε'' is found and a shift in the position of the β-relaxation peak towards high frequencies with increase in the

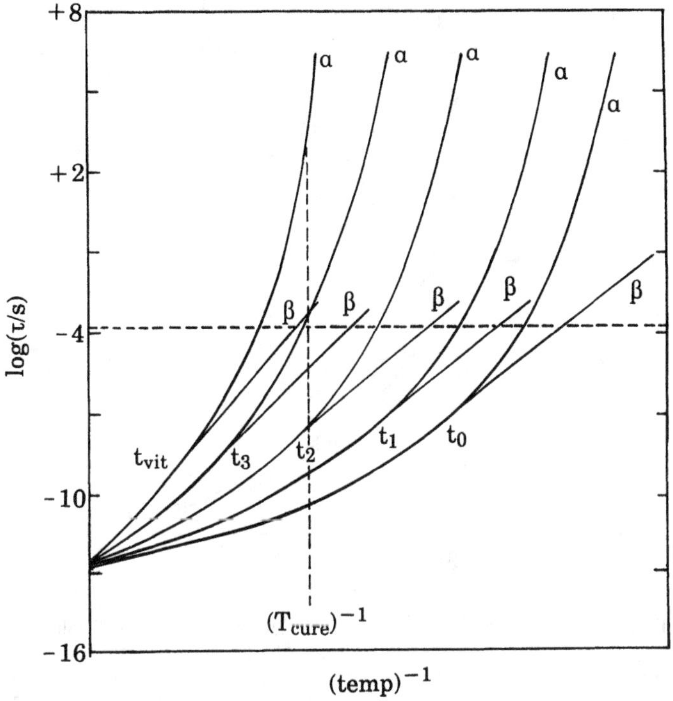

Figure 6.8 An illustration of the change in relaxation time for the α- and β-relaxation processes of the network structure at a given instant during the curing of an epoxy resin. For simplicity, the plots for the β-process are drawn to have the same slope and to merge with the α-process at a frequency of 10^7 Hz. The pre-exponent for all plots has been kept the same.

temperature. Alternatively, if isothermal curing was done at the same temperature but different frequencies were used for measurement of the thermoset's dielectric properties, it could be deduced from Figure 6.8 that at $t_c > t_{vit}$, the measured ε'' is frequency-dependent, showing a peak whose position would shift to a higher frequency with increase in the temperature of the isothermal cure.

Both from the theoretical and experimental, or technical, points of view, it is instructive to determine how $\Delta\varepsilon(= \varepsilon'(t \to 0) - \varepsilon'(t \to \infty))$, γ, t_{gel}, and $t_{peak}(\varepsilon'')$ change with change in the temperature of the isothermal cure. Plots of these quantities against the curing temperature in Figure 6.9 clearly show that their magnitude monotonically decreases with the increase in the curing temperature. The observed decrease is qualitatively consistent with the expectation that an increase in temperature decreases the number of dipoles per unit volume through the Kirkwood–Fröhlich equation (Fröhlich, 1949) and therefore $\varepsilon'(t \to 0)$ decreases. An increase in temperature also decreases the optical refractive index and the infrared polarizability, and therefore, $\varepsilon'(t \to \infty)$ decreases, but this decrease is much less than that in

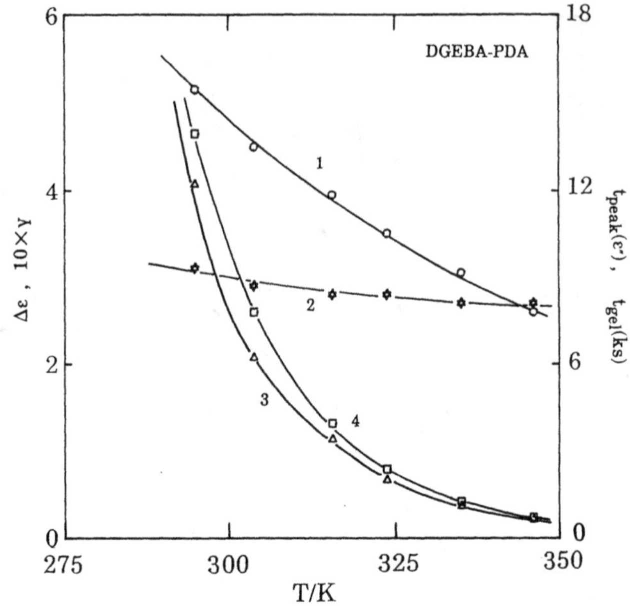

Figure 6.9 The temperature-dependence of the dielectric parameters obtained from 1 kHz measurement during the isothermal curing of DGEBA with propylenediamine. $\Delta\varepsilon = (\varepsilon'(t \to 0) - \varepsilon'(t \to \infty))$, γ = curing parameter, t_{gel} = gelation time and t_{peak} (ε'') = time when the relaxation time of the epoxy resin is equal to 159 ms. (1) $\Delta\varepsilon$; (2) γ; (3) t_{gel}; (4) t_{peak} (ε'').

$\varepsilon'(t \to 0)$. The decrease in t_{gel}, and $t_{peak}(\varepsilon'')$ with increasing temperature is obviously a reflection of faster chemical reaction rates with increase in the curing temperature, that convert the sol to a gel and gel to a vitreous state. The decrease in γ with increase in the curing temperature is unexpected in that for chemically and physically stable substances the parameter for the stretched exponential relaxation function, β, approaches unity as the temperature is increased. In this respect, the curing parameter, γ, differs from the K–W–W parameter, β, and this difference justifies both its distinction from β, and its significance as a kinetic parameter for a kinetic process. In technical terms, the decreasing value for γ implies that $\varepsilon''(t_c)$ becomes more spread out during the curing period and that ε'' of the cured sample is greater for a higher curing temperature than for a lower curing temperature, if contributions from the β-relaxation process remained constant. Note that the use of γ in equation 6.15 does not imply that its Fourier transform into the frequency domain is meaningful.

6.7 Chemical kinetics and dielectric behaviour

Let us now consider whether a useful relation between the chemical kinetics measured in terms of the extent of reaction and the electrical properties of an epoxy resin can be found. Attempts for seeking such a relation require that each of the seven effects noted in section 6.3 be considered, as both the dynamic and static properties from near zero frequency to optical frequencies are affected by the chemical reactions. Although in principle the magnitude of these effects can be determined both theoretically and experimentally, the effort required for doing so is enormous. So, as a compromise, a procedure that includes only the most prominent effects can provide a useful relation between the chemical kinetics and dielectric properties of an epoxy resin, and these effects are changes in the conductivity, σ, and the average dipolar relaxation time, τ, of an epoxy resin.

A number of studies up to 1986, which have been ably reviewed by Senturia and Sheppard (1986), have related the resistivity, or the inverse of conductivity, to the extent of epoxide conversion, α. By assuming that α is inversely proportional to log (σ), they have provided a model for the curing kinetics expressed by the equation,

$$(\partial \log \sigma/\partial t)_T = -K(1-\alpha)^m \tag{6.22}$$

where σ is the reciprocal of resistivity measured for a fixed frequency, K is the rate constant and m is the empirical reaction order parameter whose value varies substantially depending upon the epoxy resin, the experimental method and the temperature range.

In a fundamentally different approach Mangion and Johari (1991a, b) have concluded that the dipolar relaxation time, rather than the conductivity, is related to the reaction kinetics. Their studies have shown that the dipolar relaxation time increases with the progress of reaction and that the logarithmic plot of τ against t_c is sigmoidal in its shape up to $\tau = 10^3$–10^4 s, beyond which measurement of τ is prevented by the limit on the time itself which is required for such measurements. Parthun and Johari (1992a) have analytically examined the dependence of τ on the curing time and have shown that the shape of the plot of the extent of epoxy conversion against logarithmic curing time is related to that of the corresponding plot of the average dielectric relaxation time, instead of the resistivity of an epoxy resin measured for a fixed frequency. This relation is given by an empirical equation:

$$\tau(T, t_c) = \tau(T, 0) \exp [S\alpha^n(T, t_c)], \tag{6.23}$$

$$\alpha = [(\ln \tau(T, t_c) - \ln \tau(T, 0))/S]^{1/n} \tag{6.24}$$

where $S = \ln(\tau(T, \infty)/\tau(T, 0))$ is a constant for a given epoxy and n is an empirical parameter. The factor S normalizes the plot of ln $(\tau(T, t))$ against t_c. This normalization allows, for a direct comparison of the plot of ln $\tau(T, t_c)$, and the plot of α, against log (t_c), by allowing the values of S^{-1} ln $(\tau(T, t_c)/\tau(T, 0))$ to vary between zero and unity. The parameter n

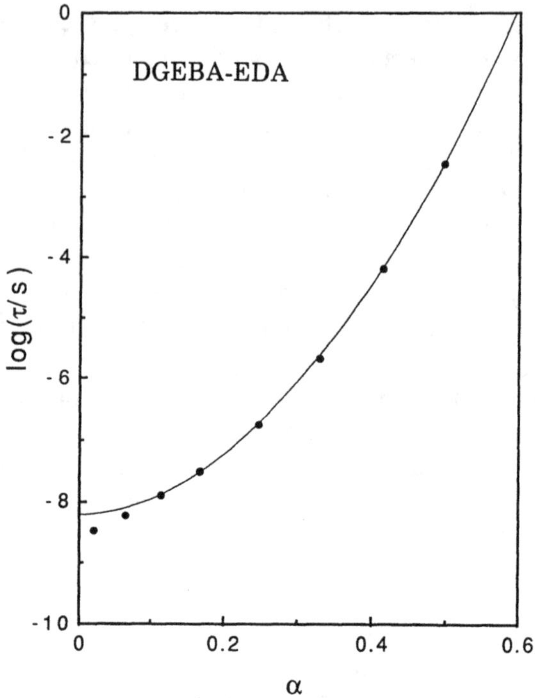

Figure 6.10 The relaxation time calculated from the dielectric behaviour of DGEBA cured with ethylenediamine at 296.2 K (•) compared against the prediction from equation 6.22. Full line is calculated from equation 6.23 with $\tau(T,0) = 6$ ns, $\tau(T,\infty) = 52$ s and $n = 1.95$.

alters the shape of the sigmoidal plot so that $\tau(T, t_{gel})$ can be made to correspond with $\alpha(T, t_{gel}))$. When $n = 1$, the extent of cure is directly proportional to ln $(\tau(T, t_c))$.

The measurement in the MHz frequency range by Tombari and Johari (1992) has confirmed that for the curing of epoxy resin DGEBA with ethylenediamine at 296.2 K, equation 6.23 or 6.24 provides a satisfactory description of the calorimetrically measured extent of reaction and the dipolar relaxation time with $\tau(0) = 6.0$ ns, $\tau(\infty) = 52$ s and $n = 1.95$. A typical plot showing such an agreement is given in Figure 6.10. There is a need for further measurements of dielectric and calorimetric behaviours of the epoxy for demonstrating the usefulness of this relation.

6.8 Curing and the high-frequency relaxation process

Careful studies of the dielectric properties in the frequency range 0.01–20 MHz by Tombari and Johari (1992) have led to observations of the dielectric loss spectrum of chemically and structurally arrested states, at different time periods, for DGEBA-ethylenediamine epoxy resin during its curing at 296.2 K. They demonstrate that these spectra are similar to the corresponding plots generally obtained for structurally stable liquids and solids and yield the same value of τ as do the fixed frequency measurements made by Mangion and Johari (1991a). One such spectrum in its normalized form which is based on their studies is shown in Figure 6.11. In addition to the similarity with the spectra of structurally stable liquids and solids, Figure 6.11 shows a remarkable feature, namely, the presence of a second relaxation process which comes into evidence after a curing period of 1.25 h, and a decrease in the strength of relaxation as the curing progresses. Thus what appears to be a single relaxation peak at the beginning of the curing splits into two peaks which separate progressively more from each other as the curing progresses and as the viscosity of the epoxy resin increases. These relaxations are referred to as α- for the low-frequency and β- for the high-frequency processes.

At first glance, it would appear that, in Figure 6.11, β-relaxation is the new process in that it seems to be the one that emerges and separates from the α-process. However, a spectrum normalized with respect to the β-relaxation

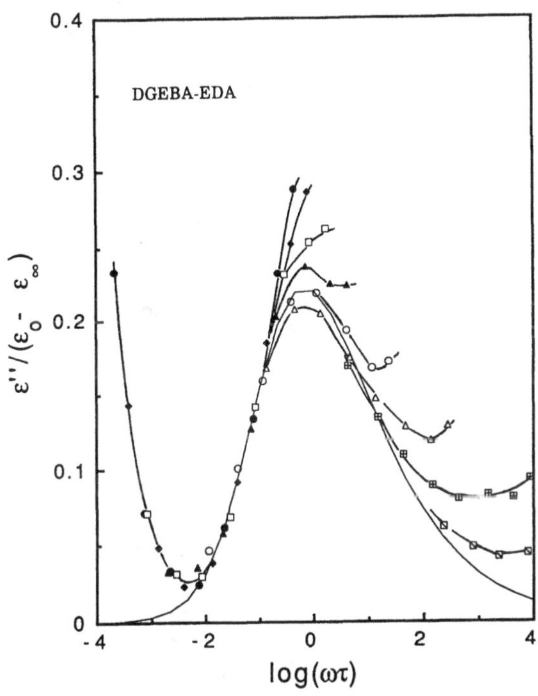

Figure 6.11 The normalized plots of dielectric loss against frequency for the DGEBA-ethylenediamine mixture at several fixed instants during its curing at 296.2 K. The instant of measurement for each plot in ks is (●) 0.9; (◆) 2.7; (□) 4.5; (▲) 6.3; (○) 8.7; (△) 11.1; (⊞) 13.5; (⧅) 15.9.

peak, which is not shown here, would indicate that it is instead the α-process which emerges and progressively separates from the β-process as it shifts towards a lower frequency. For clarity, these features are illustrated in Figure 6.12, where the variation of the relaxation rates for the α- and β-processes is shown against the reaction time at a fixed temperature. The plot illustrates a pattern of the development of relaxation processes and their separation when an epoxy resin undergoes isothermal curing. As the reaction proceeds, the relaxation process with a unimodal distribution of times acquires a bimodal distribution of the α- and β-relaxation processes, each showing a peak for a fixed frequency of measurement. This feature of splitting of the relaxation process, we note, bears a remarkable resemblance to that observed on supercooling a liquid whose α-relaxation process also progressively separates, as *reversible* molecular clustering raises the liquid's viscosity, from the β-relaxation process in a time-temperature plane (Johari, 1976).

Recent studies have shown an additional remarkable feature of the sub-T_g relaxation in epoxy resins, namely the development of the slower or high temperature β-relaxation process at the expense of a faster or low temperature, γ-process. This is shown in Figure 6.13 where plots of the ε'' against the temperature after different curing times are drawn. In this figure

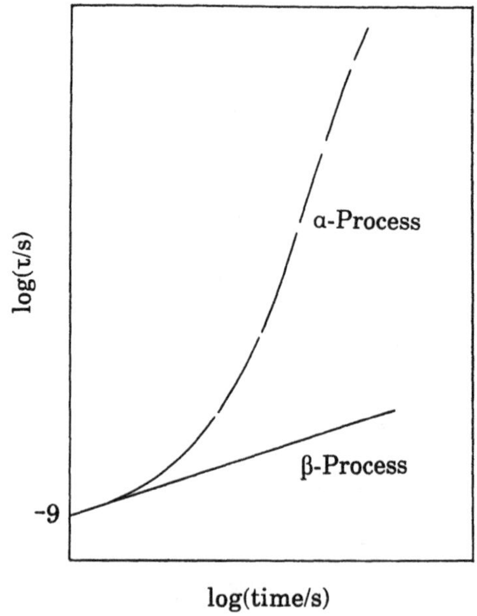

Figure 6.12 An illustration of the change in relaxation time with increase in the reaction time of an epoxy resin, as the α- and β-processes evolve at a fixed temperature.

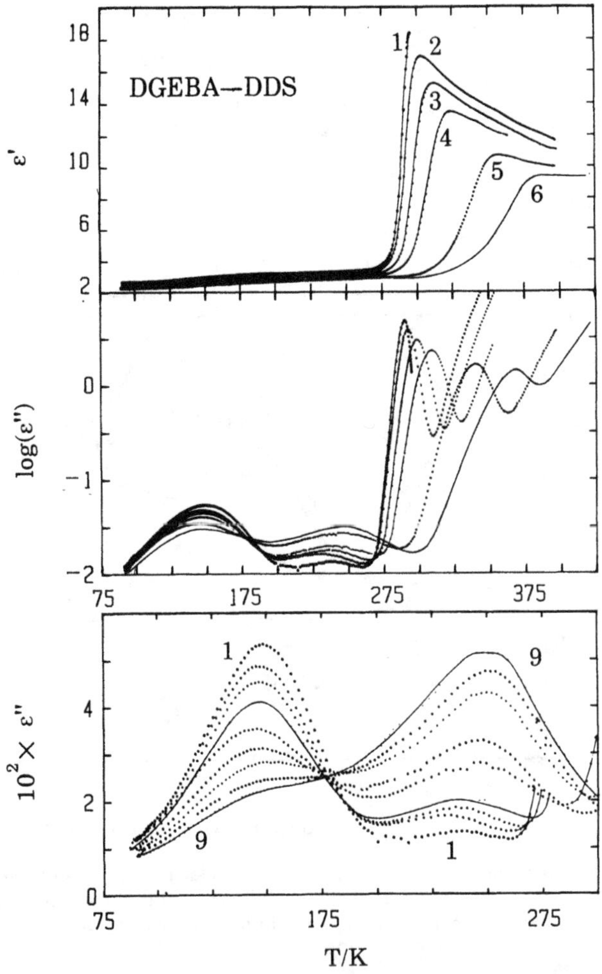

Figure 6.13 The dielectric loss of the sub-T_g relaxation regions (i.e. γ and β) and the α-relations regions of the DGEBA cured with 4,4′-diaminodiphenyl sulphone. The bottom figure is for the enlargement of the corresponding regions of the γ- and β-relaxation processes. Plots 1 to 6 in both figures refer to an accumulated equivalent ageing time of 0.2, 1.7, 3.5, 5.6, 9.0 and 15 ks, respectively, and plots 7 to 9 to 24, 45 and 84 ks all at a temperature of 398 K.

each curve corresponds to an accumulated equivalent curing time at 398 K, a concept which takes into account both the curing which occurs at certain heating and cooling rates and that which occurs on isothermal hold at a certain temperature (Mangion and Johari, 1991a, b). The decrease in the height of the α- and γ-relaxation peaks and the increase in that of the β-relaxation peak for the DGEBA cured with 4,4′-diaminodiphenyl sulphone (Figure 6.14) clearly show the evolution of the high-temperature sub-T_g relaxation processes as an epoxy resin progressively cures.

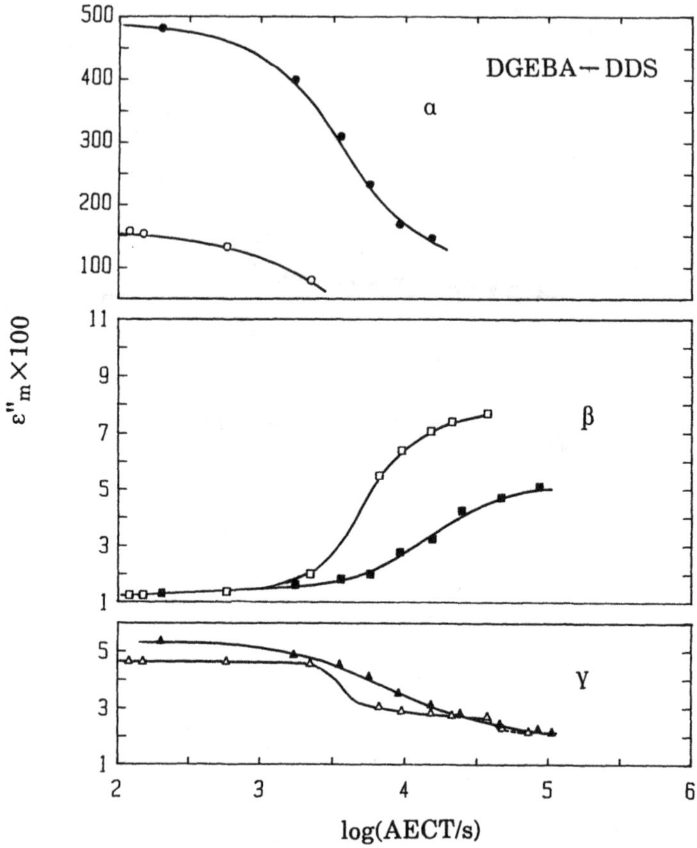

Figure 6.14 The peak value for ε'' for the α-, β-, and γ-relaxation processes during the curing of the DGEBA with 4,4′-diaminodiphenyl sulphone (solid notations) and with 4,4′-diaminodiphenyl methane (open notations) is plotted against the accumulated equivalent curing time (log (AECT)) at 398 K.

6.9 Ageing effects on electrical properties

The vitrification of an epoxy resin generally occurs before the chemical reactions that raise its viscosity reach completion, i.e. that unreacted components persist in the cured epoxy resin. Therefore, when an incompletely cross-linked state of a macroscopically rigid, approximately fully cured epoxy resin is kept at a temperature below its T_g, localized rotational and translational diffusions of its reacting molecules or chain segments tend to form (ultimately) a stoichiometric product (Choy and Plazek, 1986; Plazek and Frund, 1990). This chemical process increases the number density of cross-links and thereby the T_g of the cured epoxy resin. For cases where an increase in the number of cross-links increases the number of dipoles which can undergo hindered rotation, the ageing of the amorphous network structure is expected to increase, as a result of chemical reactions, the dielectric relaxation strength of a sub-T_g relaxation process. Since this increase competes against the decrease in the dielectric relaxation strength that generally occurs on spontaneous densification or structural relaxation towards a lower-energy, thermodynamically metastable structure of an amorphous solid (Tool and Eichlin, 1931; Johari, 1985; Plazek and Berry, 1986; Scherrer, 1986), the net change in the electrical properties of a cured epoxy resin is expected to depend on the relative magnitude of the effects of the chemical reaction and the structural relaxation.

Sidebottom and Johari (1990) and Mangion and Johari (1990c, 1991b) have provided careful measurements of this effect for the DGEBA cured with 4,4′-diaminodiphenyl methane or DDM. Figure 6.15 shows typical results of a study in which the cured epoxy was aged for a total period of about 500 h at 403 K, about 25 K below its calorimetric T_g, and its dielectric properties were measured at predetermined intervals. The cured epoxy resin showed only one sub-T_g relaxation peak at 268 K, and the height of this peak first increased with increase in the ageing time, attained a peak value between the period of 48 and 168 h and thereafter decreased. An analysis of this peak in terms of the stretched exponential relaxation function was done by Sidebottom and Johari (1990) and the dielectric strength $\Delta\varepsilon$ of the sub-T_g relaxation process was determined. This is also plotted against the ageing time in Figure 6.15.

In the chemical reactions of curing of epoxy resin with a diamine, the epoxy group opens to form one OH group and one covalent bond between

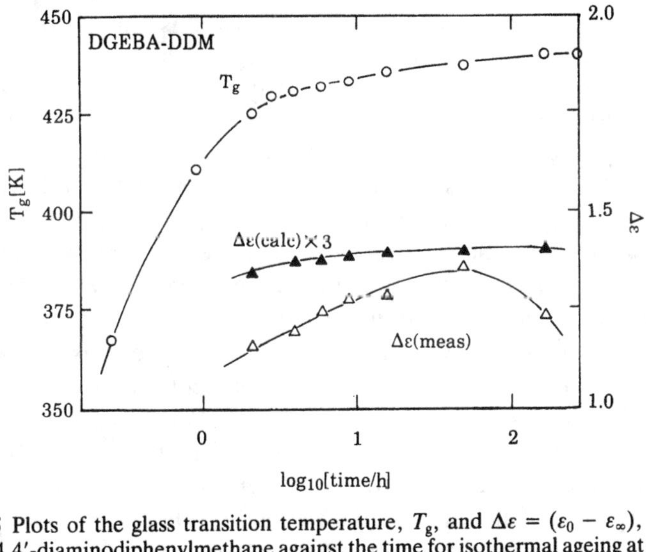

Figure 6.15 Plots of the glass transition temperature, T_g, and $\Delta\varepsilon = (\varepsilon_0 - \varepsilon_\infty)$, of DGEBA cured with 4,4′-diaminodiphenylmethane against the time for isothermal ageing at 403 K. Data for T_g (○) are taken from Plazek and Frund (1990). (△) the $\Delta\varepsilon$ values as measured, (▲) the $\Delta\varepsilon$ values as calculated. The scale for $\Delta\varepsilon$ is on the right.

the end carbon atom of the epoxy molecule and the terminal atom of the curing agent. The volume swept out by the hindered rotation of the OH group about the C–O bond is much smaller than the volume swept out by rotation of the epoxide-resin group about the newly formed C–N covalent bond. Therefore, the increase in $\Delta\varepsilon$ at the early stage of ageing in Figure 6.15 seems at first sight to be mainly due to an increase in the number of OH groups formed as a result of chemical reactions, and in fact, the mechanical relaxation peak corresponding to the β-relaxation process in epoxy resins observed by Ochi *et al.* (1987) has been attributed to the motion of hydroxyl groups. However, a variety of studies on nonstoichiometric compositions of cured epoxy resins by Mangion *et al.* (1992b) have shown that the height of the dielectric β-relaxation peak remains the same both for stoichiometrically starved (2 moles DGEBA–0.75 moles DDM) and saturated (2 moles DGEBA–1.25 moles DDM) epoxy resin, although the number of –OH groups formed on curing in the former case is much less than in the latter case. This suggests that the dielectric β-peak is not due to the motions of only the –OH group in the network structure; other groups also move and contribute to the β-process.

Sidebottom and Johari (1990) have calculated the increase in the $\Delta\varepsilon$ of the sub-T_g relaxation which is based on the premise that the number of OH groups formed is equal to the number of cross-links in the network structure and that the latter can be satisfactorily determined from the observed increase in the glass transition temperature using the configurational entropy theory by DiMarzio (1964). They found that the calculated $\Delta\varepsilon$ is one third the value evaluated from the experiment as shown in Figure 6.15. Therefore, the formation of an OH group as a result of chemical reactions in the amorphous solid state of a cured epoxy resin is only partly responsible for the increase in $\Delta\varepsilon$. Other local dipolar segments also undergo rotational diffusion. These dipolar segments are likely to be in the vicinity of the OH group, but their exact chemical constitution cannot be unambiguously identified by dielectric measurements alone. It is probable that the formation of OH groups creates localized regions of relatively loose molecular packing, where molecular or segmental diffusion persists.

One of the remarkable features of the change in the electrical properties of cured epoxy resins is that when the T_g of the network structure approaches a constant value, with increase in the ageing time, as is seen in Figure 6.15, the $\Delta\varepsilon$ of the sub-T_g relaxation process begins to decrease. This effect is qualitatively similar to that observed for molecular liquids (Johari, 1973, 1976) and linear chain amorphous polymers (Pathmanathan *et al.*, 1989; Muzeau and Johari, 1990). It now seems sufficiently well-established that the nature of changes in the electrical properties of cured epoxy resins over a very long period of ageing is similar to that for other amorphous polymers, although over a short period of ageing the nature of these changes in the two cases contrast with each other.

6.10 Electrical applications of epoxy resins

Epoxy resins are now increasingly used in the various areas of electronics, a use that is partly due to the versatility of the procedures by which they can be adapted in a controlled manner for encapsulating, potting, thin-film coating and embedding or packing electronic circuits or light emitting diodes, and partly due to their suitable electrical properties. One of the earlier monographs on the subject is *Plastic Coatings for Electronics* (Licari, 1970). Since then a number of edited books (May, 1988; Lai, 1989; Seraphin *et al.*, 1989) containing articles on the subject of electrical applications of epoxy resins have become available. Amongst these, perhaps the most comprehensive account, with 190 references, is given by Lee (1988) in chapter 9 of the second edition of *Epoxy Resins, Chemistry and Technology* (ed. May, 1988). This may be consulted for details of the characteristics of epoxy resins for electronic applications and procedures for their use.

Amongst thc primary reasons for the use of epoxy resins in electronics is their dielectric permittivity which should be 3–6 at ambient temperatures and low frequencies, their dissipation factor, tan δ, of 10^{-3} to 10^{-2} for 60 to 1000 Hz frequency, their dielectric strength of about 120–180 kV/mm and the volume resistivity of 10^{19}–10^{12} ohm/m, with relatively low sensitivity to temperature. In addition, an epoxy resin should possess a number of other physical properties in order to make it suitable for electrical and electronic applications, such as good adhesion to metals and silicon, low permeability

Table 6.1 Typical epoxy resin compositions for 100 parts by weight of DGEBA epoxy resin

Application	Inorganic filler (by part)	Curing and other reagents (by part)
Low temperature potting epoxy	Glass beads (18) (0.125 mm dia)	Butyl glycidyl ether (2) Diethylamino propylamine (2)
Flame-resistant epoxy	–	Chlorendic anhydride (90)
Miniaturized circuits epoxy	–	Allyl glycidyl ether (10) Polysulphide polymer (10) 2,4,6-Tris-(dimethylamino-methyl)phenol (10)
Low exotherm potting epoxy	Silica (100)	Phthalic anhydride (30)
Flexible potting epoxy	–	Polysulphide polymer (50) Triethylene tetramine (10)
Electrical encapsulating-compound	Silica (5) clay (75)	Diethylene triamine (12)
Machinable compound	Powdered aluminium (200)	Allyl glycidyl ether (10) Diethylene triamine (11)
High heat distortion (casting)	Silica (100)	*m*-Phenylene diamine (14)
For low exotherm (casting)	Powdered aluminium (80) Pumice (70)	Polysulphide polymer (25) Dimethylaminomethyl phenol (10)

or absorption of water (typically 0.1% to 0.25%), greater stability against chemicals and solvents to prevent both a decrease in the surface resistivity and corrosion. There are other requirements that epoxy resins fulfil in order to be used in electronic packaging, namely, their low viscosity prior to curing at a sufficiently low temperature, low stress resistance that develops between the electronic components and the epoxy resin, low heat of reaction and high thermal conductivity to present possible damage to the device under protection, and minimum effects of ageing. All of these take on a particular importance in the use of epoxies for encapsulating electronic implants such as a heart pacemaker where DGEBA epoxy cured with diethylenetriamine is used. Table 6.1 provides a partial list of typical epoxy resin compositions according to their applications as given by Lee (1988). Other compositions can be obtained on request from the various producers of epoxy resins.

Acknowledgements

I am grateful to Drs M.B.M. Mangion and E. Tombari and Mr M.G. Parthun for allowing me here to include figures which have resulted from their research collaboration with me.

References

Alig, I. and Johari, G.P. (1992) *Macromolecules* **25**.
Aronhime, M.T. and Gillham, J.K. (1986) *Adv. Polymer Sci.* **78**, 83–115.
Bidstrup, S.A., Sheppard, N.F. Jr. and Senturia, S.D. (1989) *Polym. Eng. Sci.* **29**, 325–328.
Cassettari, M., Salvetti, G., Tombari, E., Veronesi, S. and Johari, G.P. (1992) *J. Polym. Sci. B. Polym. Phys.* (in press); (1992) *IRNCOVO Cimento* **14D**, 763–766.
Choy, I.-C. and Plazek, D.J. (1986) *J. Polym. Sci. B. Polym. Phys.* **24**, 1303–1320.
Cole, R.H. and Tombari, E. (1991) *J. Non-Cryst. Solids* **131–133**, 969–972.
Cole, K.S. and Cole, R.H. (1941) *J. Chem. Phys.* **9**, 341–351.
Day, D.R. (1986) *Polym. Eng. Sci.* **26**, 362–366.
Dimarzio, E.A. (1964) *J. Res. Natl. Bur. Std.* **65A**, 611–617.
Dishon, M., Weiss, G.H. and Bendler, J.T. (1985) *J. Res. Natl. Bur. Stand.* **90**, 27–39.
Djabourov, M. (1988) *Contemp. Phys.* **29**, 273–297.
Enns, J.B. and Gillham, J.K. (1983) *J. Appl. Polym. Sci.* **28**, 2567–2591.
Fröhlich, H. (1949) *Theory of Dielectrics* Oxford University Press, Oxford.
Gatro, J. and Yandrasits, M. (1989) *Polym. Eng. Sci.* **29**, 278–284.
Huraux, C. and Sellaimia, A. (1973) *C.R. Acad. Sci. Paris* **277** (Serie B), 691–695.
Johari, G.P. (1973) *J. Chem. Phys.* **58**, 1766–1770.
Johari, G.P. (1976) *Ann. N.Y. Acad. Sci.* **279**, 117–140.
Johari, G.P. (1985) *Jou. de Physique* **C8**, **46**, 567–572.
Johari, G.P. and Mangion, M.B.M. (1991) *J. Noncryst. Solids* **131–133**, 921–929.
Johari, G.P. and Pathmanathan, K. (1990) *Phil. Mag. B.* **62**, 225–228.
Johnson, J.F. and Cole, R.H. (1951) *J. Amer. Chem. Soc.* **73**, 4536–4540.
Kienle, R.H. and Race, H.H. (1934) *Trans. Electrochem. Soc.* **65**, 87–107.
Kranbuehl, D.E. (1991) *J. Noncryst. Solids Part I*, **131**, 134, 930–934.
Kranbuehl, D.E., Delos, S., Hoff, M., Haverty, P., Freeman, W., Hoffman, R. and Godfrey, J. (1989) *Polym. Eng. and Sci.* **29**, 285–289.
Lai, J.H. (ed.) (1989) *Polymers for Electronic Applications*, CRC Press, Boca Raton.

Lane, J.W., Khattak, R.K. and Dusi, M.R. (1989) *Polym. Eng. Sci.* **29**, 339–346.
Lane, J.W. and Seferis, J.C. (1986) *Appl. Polymer Sci.* **31**, 1155–1167.
Lee, S.M. (1988) In *Epoxy Resins: Chemistry and Technology*, ed. May, C.A. CRC Press, Boca Raton.
Licari, J.J. (1970) *Plastic Coatings for Electronics* McGraw-Hill, New York.
McDonald, J.R. (ed.) (1987) *Impedance Spectroscopy* Wiley, New York.
Mangion, M.B.M. (1990) Ph.D. thesis, McMaster University, Hamilton, Ont. Candada.
Mangion, M.B.M. and Johari, G.P. (1990a) *J. Polym. Sci. B., Polym. Phys.* **28**, 71–83.
Mangion, M.B.M. and Johari, G.P. (1990b) *J. Polym. Sci. B. Polym. Phys.* **28**, 1621–1639.
Mangion, M.B.M. and Johari, G.P. (1990c) *Macromolecules* **23**, 3687–3695.
Mangion, M.B.M. and Johari, G.P. (1991a) *Polymer* **32**, 2747–2754.
Mangion, M.B.M. and Johari, G.P. (1991b) *J. Polym. Sci. B. Polym. Phys.* **29**, 437–449.
Mangion, M.B.M. and Johari, G.P. (1991c) *J. Polym. Sci. B. Polym. Phys.* **29**, 1117–1125.
Mangion, M.B.M. and Johari, G.P. (1991d) *J. Polym. Sci. B. Polym. Phys.* **29**, 1127–1135.
Mangion, M.B.M., Wang, M. and Johari, G.P. (1992a) *J. Polym. Sci. B. Polym. Phys.* **30**, 433–444.
Mangion, M.B.M., Wang, M. and Johari, G.P. (1992b) *J. Polym. Sci. B. Polym. Phys.* **30**, 445–454.
May, C.A. (ed.) (1988) *Epoxy Resins: Chemistry and Technology* Marcel Dekker, New York.
Montroll, E.W. and Bendler, J.T. (1984) *J. Stat. Phys.* **34**, 129–162.
Moynihan, C.T., Boesch, L.P. and Laberge, N.L. (1973) *Phys. Chem. Glasses* **14**, 122–125.
Muzeau, E. and Johari, G.P. (1990) *Chem. Phys.* **149**, 173–184.
Muzeau, E., Perez, J. and Johari, G.P. (1991) *Macromolecules* **24**, 4713–4723.
Ochi, M., Yoshizumi, M. and Shimbo, M. (1987) *J. Polym. Sci., B. Polym. Phys.* **25**, 1817–1827.
Parthun, M.G. and Johari, G.P. (1992a) *J. Polym. Sci. B. Polym. Phys.* **30**, 655–667.
Parthun, M.G. and Johari, G.P. (1992b) *Macromolecules* **25**, 3149–3155.
Parthun, M.G. and Johari, G.P. (1992c) *Macromolecules* **25**, 3254–3264.
Pathmanathan, K., Cavaille, J.Y. and Johari, G.P. (1989) *J. Polym. Sci. B. Polym. Phys.* **27**, 1519–1528.
Plazek, D.J. and Berry, G.C. (1986) In *Glass Science and Technology*, Vol. 3, ed. Kreidl, N.J. and Uhlmann, D.R. Academic Press, London.
Plazek, D.J. and Frund, J. Jr. (1990) *J. Polym. Sci. B. Polym. Phys.* **28**, 431–448.
Scherrer, G.W. (1986) *Relaxation in Glass and Composites* Wiley, New York.
Senturia, S.D. and Sheppard, N.F. Jr. (1986) *Adv. Polym. Sci.* **80**, 1–47.
Seraphim, D.P., Lasky, R. and Li, C-Y. (ed.) (1989) *Principles of Electronic Packaging* McGraw-Hill, New York.
Sheppard, N.F. Jr. and Senturia, S.D. (1989) *J. Polym. Sci. B. Polym. Phys.* **27**, 753–762.
Sidebottom, D. and Johari, G.P. (1990) *Chem. Phys.* **147**, 205–211.
Soualimia, A., Huraux, C. and Dispax, B. (1982) *Macromol. Chem.* **183**, 1803–1810.
Stauffer, D., Coniglio, A. and Adam, M. (1982) *Adv. Polym. Sci.* **44**, 103–158.
Tombari, E. and Johari, G.P. (1992) *J. Chem. Phys.* (in press).
Tool, A.Q. and Eichlin, C.G. (1931) *J. Amer. Ceram. Soc.* **14**, 276–308.
Weiss, G.H., Bendler, J.T. and Dishon, M. (1985) *J. Chem. Phys.* **83**, 1424–1427.
Williams, G. and Watts, D.C. (1970) *Trans. Faraday Soc.* **66**, 80–85.

7 Epoxy resin adhesives*

S.J. SHAW

7.1 Introduction

The use of structural adhesives in the manufacture of load-bearing components has grown extensively in recent years. This can be attributed to a number of desirable qualities which adhesive bonding allows in comparison with more traditional joining techniques such as riveting and welding. These include:

(i) Allowance of a relatively uniform stress distribution, resulting in improved fatigue performance.
(ii) The ability to join dissimilar substrate materials which, due to their dielectric nature, minimises the possibility of electrolytic corrosion between dissimilar metals.
(iii) Allows the joining of thin-gauge metals to each other, in particular honeycomb assemblies, resulting in the availability of lightweight structures exhibiting high strength to weight ratios.
(iv) Allows both increased design flexibility and the ability to fabricate complex shapes.
(v) The possibility of reduced production costs in comparison to welding and riveting.

Although there is a range of chemically different structural adhesives (Wake, 1982), those based upon epoxy resins have, over recent years, earned a reputation for combining both high load-bearing characteristics together with ease of processing, and can be regarded as workhorse products. A number of highly favourable characteristics can be regarded as responsible for this popularity.

(i) They exhibit excellent adhesion to most metallic alloys and various other substrate types.
(ii) They are capable of operation at temperatures up to approximately 150°C for both short- and long-term applications.
(iii) They are highly versatile in the sense that a wide range of processing, cure and property characteristics can be achieved.
(iv) They cure by reaction mechanisms which do not result in the generation of volatile by-products, e.g. water. Thus processing is

relatively easy, without the necessity for high applied pressures during the bonding operation.

(v) They exhibit good wetting properties when applied to well prepared surfaces and exhibit relatively low shrinkage during cure.

A study of the literature concerning epoxies reveals the vast number of formulation variables which can be employed for many applications including adhesives (Lee and Neville, 1967; Potter, 1970; May, 1988). Many different types of epoxide resins exist ranging from the essentially workhorse epoxy, the diglycidylether of bisphenol A (DGEBA) to the more complex systems based upon tri- and tetrafunctional resins (see chapter 1).

Perhaps rather more daunting, a large number of curing agents can be employed to convert the epoxy prepolymer to a crosslinked network, with a large number of these being capable of use in adhesive formulations (chapter 2). In addition, adhesive formulations can contain and benefit from further additives including fillers, toughening agents, coupling agents etc. (chapter 4), as well as being available in liquid, paste, film or carried film form, thus demonstrating the wide variation in epoxy adhesive types available for the potential user.

In this chapter, an attempt will be made to discuss the factors pertinent to the successful use of epoxy-based adhesives. Within this context, consideration will be given firstly to the theories of adhesion, i.e. why materials stick together including the important aspect of substrate wetting. Substrate surface pretreatments, generally considered vital for adhesive bonding with all adhesive types, including epoxies, will be briefly discussed together with an account of the test methods available for assessing important mechanical properties. Formulation aspects pertinent to epoxy adhesive applications will also be briefly reviewed together with accounts of the properties and, of particular importance, durability characteristics. Finally some examples of where epoxy-based adhesives have been successfully employed will be discussed.

7.2 Theories of adhesion and wetting phenomena

Kinloch (1980) has defined an adhesive as being a material which, when applied to substrate surfaces, can join them together and resist separation.

To provide a fairly wide and balanced account of the various facets of epoxy resin adhesive technology it is of interest to outline briefly the various factors and theories which underpin this definition. This will involve a discussion of two major factors, namely theories of adhesion and wetting. The former will hopefully provide the reader with an insight into the major question: why do materials stick to one another, whilst the latter will address the important subject of wetting which has been described by Sharpe and Schonhorn (1964) as the single most important factor likely to influence the strength of an adhesive joint.

7.2.1 *Theories of adhesion*

Four main theories have been proposed to account for the phenomenon of adhesion: (a) mechanical interlocking, (b) electrostatic, (c) diffusion and (d) the adsorption theory.

7.2.1.1 Mechanical interlocking. The mechanical interlocking theory, as the title indicates, quite simply proposes that adhesion between an adhesive and substrate is primarily dependent upon mechanical keying of the adhesive into substrate surface irregularities. The fact that flat surfaces can, in the presence of an adhesive, exhibit good adhesion clearly denies any universal applicability. Many examples exist where a mechanical interlocking component clearly provides a major contribution to adhesion (Wake, 1982). For example, those familiar with carpentry will be well aware of the long established view that roughening of timber prior to 'glueing' provides a much stronger joint. Later in this chapter (section 7.3.3.2), the influence of various surface treatments on the adhesive bonding of aluminium alloys will be discussed. Here surface topography will be shown to be of importance, thus suggesting a significant contribution to adhesion by an interlocking component. Thus, although mechanical factors are generally accepted, in certain cases at least, as providing an improvement in adhesion and thus joint strength, it is important to recognise that it takes no account of the dominant factors which must clearly exist, on a molecular level, at an interface between an adhesive and substrate.

7.2.1.2 Electrostatic theory. The electrostatic theory of adhesion, expounded initially by Deryagiun and co-workers (1955, 1957, 1969) is based on the proposed existence of an electrical double layer across an interface. Although it is possible to demonstrate the existence of electrical charges within an adhesive joint, the most common example of this being the observation of electrical discharge across the gap when a flexible adhesive tape is rapidly removed from a substrate, it is now generally recognised that such phenomena do not provide a significant contribution to adhesion. Indeed Kinloch (1980) has suggested that this type of observation probably arises from, rather than results in, high levels of adhesion.

7.2.1.3 Diffusion theory. The diffusion theory of adhesion was initially proposed by Voyutskii (1963) in the then Soviet Union. The main idea here is that adhesion between two phases occurs as a direct result of intermixing of the two contact substrates at the molecular level. Obviously for this to occur, molecules existing within the interfacial zone must (a) have, or under certain conditions be capable of exhibiting, high molecular mobility and (b) exhibit a sufficiently high degree of compatibility. The latter requirement is unlikely to be fulfilled for the majority of bonding applications where an extraneous adhesive is employed to bond thermoplastic substrates since the required levels of compatibility, particularly at the high molecular weights present in most thermoplastic compositions, would be insufficient for bond formation via a molecular diffusion process. Having said this, a diffusion mechanism is generally considered responsible for the self adhesion, or autohesion, exhibited by many unvulcanised elastomers, with natural rubber being a well known example. In addition the solvent welding of thermoplastics, where the substrates are treated with a suitable solvent followed by contact and pressure/heat application, is generally believed to be due to a diffusion mechanism (Titow, 1978).

Although Voyutskii has not claimed a diffusion-related mechanism for the adhesion phenomena exhibited in epoxy–metal bonding, other Soviet researchers have suggested the need for the 'diffusion' of adhesive into the rough porous oxide layers found on many metallic alloys such as aluminium. Although such a 'diffusion' process is clearly necessary so as to maximise the required degree of overall area contact between adhesive and substrate, diffusion must be regarded as an intermingling of the two components of an interphase on the molecular level. Consequently, diffusion as described above, cannot be regarded as being of any significance in the bonding of structural materials such as metals and composites with epoxies or indeed any other type of structural adhesive.

7.2.1.4 Adsorption theory. Of all theories proposed to account for the phenomenon of adhesion, the so-called adsorption theory is the one which has achieved more general acceptability within the adhesion fraternity and which can be said to offer the greatest relevance to adhesive bonding with epoxies (Hunstberger, 1967; Kinloch, 1980; Wake, 1982).

The basic tenet of the adsorption theory is that, provided sufficiently intimate intermolecular contact is achieved between two materials, they will adhere because of surface force interactions between the atoms in the two contacting materials. The intermolecular forces responsible for this behaviour include van der Waals' forces which can include the universal dispersion forces that result from randomly fluctuating dipoles, to the stronger intermolecular forces associated with permanent dipoles and hydrogen bonds. Indeed it has been suggested that the excellent adhesive properties exhibited by epoxies is due in large part to pendent hydroxyl groups which exist within the molecular structures of most epoxies and which are considered capable of interacting, via hydrogen bond formation, with hydroxyl groups frequently found in abundance on most metal oxide surfaces under normal conditions (Potter, 1970).

Mention must also be made here of the process of chemisorption whereby adhesive application to a substrate leads to the possible generation of primary covalent bonds. Although adhesion arising from so-called secondary force interactions can alone result in very reasonable joint strength characteristics (and indeed can usually be regarded as the dominant factor in adhesion) the additional presence of covalent bonds across the interface is often regarded as providing enhancements in load-bearing capacity. In addition, as discussed later in this chapter (section 7.7.4.2), primary covalent interfacial bonding is generally believed, in certain circumstances, to offer substantial advantages in terms of environmental stability.

Finally, it is of interest to reconsider briefly the previously employed phrase 'intimate intermolecular contact' since it provides probably the key to successful adhesion and indeed in a sense provides the very reasons why adhesives, such as epoxies, are frequently required for the joining of materials. All intermolecular forces responsible for adhesion are very short range, exerting influences over Å distances only. All practical surfaces, no matter how hard attempts have been made to smooth and polish, are exceedingly rough. Thus, as eloquently described by Allen (1979), when two practical surfaces are brought together, this can be regarded as akin to inverting the Alps over the Himalayas. In other words, only very limited contact is achieved to an extent that only a very small proportion of the two surfaces are in true intermolecular contact. This, of course, is where an adhesive, having the ability to flow and achieve contact with the majority of the substrate surface comes into play. Under these circumstances, intermolecular force interactions exist throughout the majority of the contact area thus providing the dramatic improvements in load-bearing capacity.

7.2.2 *Wetting*

Possibly one of the single most important factors likely to influence the strength of an adhesive joint is the ability of the adhesive to wet and spread spontaneously on the substrate surface. This most important characteristic can be quantified in terms of the contact angle which a liquid (adhesive) forms when placed in contact with a solid surface as shown in Figure 7.1. Considered initially by Young (1805) he related contact angle to the surface energies of the two contacting materials, as indicated in equation 7.1 which has been given his name.

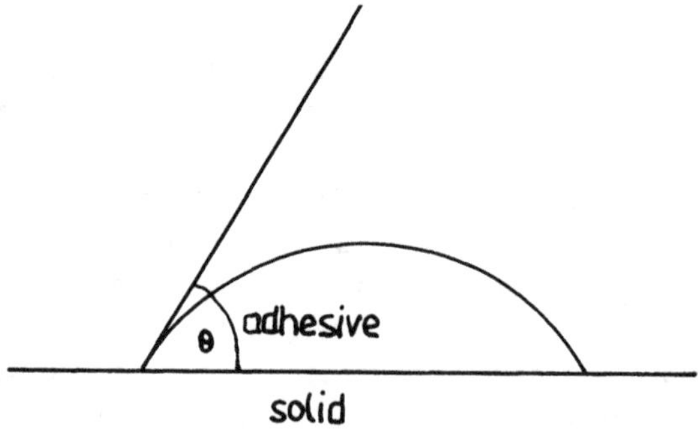

Figure 7.1 Contact angle θ, of a liquid (adhesive) in contact with a solid surface.

$$\gamma_{SV} = \gamma_{SL} + \gamma_{LV} \cos \theta \tag{7.1}$$

where γ_{SV} = surface free energy of the substrate in contact with the vapour of the liquid, γ_{SL} = interfacial free energy and γ_{LV} = liquid (adhesive) surface free energy.

Since a contact angle of 0 will clearly be a desirable objective, equation 7.1 can be employed to demonstrate a simple wetting criterion, namely that for the case where an adhesive will spontaneously spread on a substrate, then

$$\gamma_{SV} \geqslant \gamma_{SL} + \gamma_{LV} \tag{7.2}$$

If the interfacial free energy, γ_{SL}, is ignored, which may or may not be of importance, the criterion demonstrated in equation 7.2 can be simplified still further to state that spreading of adhesive on substrate will occur when:

$$\gamma_{SV} \geqslant \gamma_{LV} \tag{7.3}$$

i.e. when the surface free energy of the substrate is greater than that of the intended adhesive. Although highly simplified, the above offers, at the very least, an indication of the surface energy factors which are likely to be of importance for the successful utilisation of a structural adhesive. Indeed, the need for surface treatments with most substrate materials prior to adhesive bonding, be they metals, composites or plastics, can be attributed, partly at least, to the criterion indicated in equation 7.3, as will be discussed later in this chapter (section 7.3).

Although the surface tension of a liquid, γ_{LV}, can be measured using well established techniques, γ_{SV}, the surface free energy of a solid, is more difficult to determine. Although attempts have been made to assess γ_{SV} values for solids close to their melting points, such values can be misleading in adhesion-related studies since the wetting of a liquid adhesive on a solid is usually a room temperature process, or at least conducted at temperatures far removed from the melting point of the substrate. In recent years, however, contact angle values exhibited by a range of organic liquids on various substrate surfaces have been used to provide surface free energy values (Kaelble, 1971; Sherriff, 1976).

As far as adhesion and wetting are concerned, solid substrate surfaces can be grouped into those exhibiting low surface free energies which would include fluorinated and hydrocarbon polymers having values of approximately 20 mJ . m^{-2}. Of greater magnitude would be polymers exhibiting some considerable polar character followed by substrates such as inorganic materials and metals exhibiting values in excess of 500 mJ . m^{-2}. Thus bearing in mind equation 7.3, adhesives such as those based on epoxies, which exhibit surface free energies of approximately 45 mJ . m^{-2}, would not be expected to wet and adhere strongly to relatively low energy surfaces such as polyethylene. However, the reverse would be expected with high energy solids such as metals and it is generally found that epoxy adhesives show better performance with metals than with organic polymers.

Indeed with the latter, surface treatments designed to impart polar surface characteristics as a means of substantially increasing surface free energy are generally regarded as mandatory. Interestingly, the criterion demonstrated by equation 7.3 does in fact suggest that the application of molten polyethylene to a solid cross-linked epoxy should result in the thorough wetting of the latter by the former since polyethylene is now acting as the adhesive and the epoxy as the substrate. In other words the surface energetics are now reversed and experimental studies have in fact confirmed the validity of equation 7.3 in this case (Kinloch, 1980).

Although metals and metal oxides, in theory, are regarded as occupying the high surface energy end of the spectrum, in practice reality can be remote from this theoretical utopia. The very factors which promote wetting and good adhesion in the presence of an adhesive, also promote the adsorption of atmospheric contaminants such as hydrocarbons and moisture which in turn can convert the substrate surface from one of high surface energy to one of low surface energy. The importance of this phenomenon has been demonstrated by the work of Gledhill *et al.* (1977) who studied the effect of relative humidity on the wettability of mild steel surfaces. They found, as shown in Figure 7.2, that the presence of atmospheric moisture, even at low relative humidities, was sufficient to bring about a dramatic reduction in surface free energy from what would be expected for a totally dry, uncontaminated surface (>500 mJ . m^{-2}). In addition, increasing relative humidity from 7 to 88% was shown to result in a gradually decreasing value of γ_{SV}, suggesting that the ability of the surface to allow wetting by an epoxy

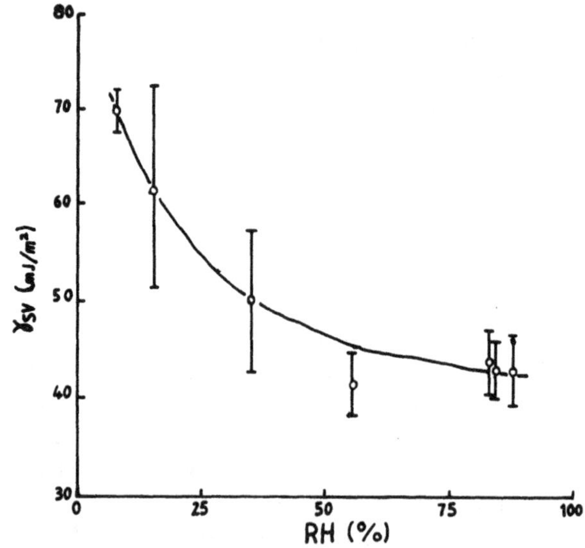

Figure 7.2 Influence of relative humidity on the surface free energy, γ_{sv}, of a grit-blasted mild steel surface.

adhesive would depend upon the relative humidity prevalent at the time of bonding. Joint strength values obtained from mild steel substrates bonded using an epoxy adhesive under two relative humidity environments demonstrated this to be the case, with joints bonded in the lower humidity environment exhibiting substantially greater strength values.

In addition to the thermodynamic factors discussed above, wetting as a kinetic process is also of importance, since although surface energetic considerations may indicate the potential establishment of intimate molecular contact between an adhesive and substrate, the kinetics of wetting may be the crucial parameter. For example, a rapid curing epoxy, or one having a particularly high viscosity, may not be able to fulfil the wetting potential implied solely by thermodynamic considerations.

Topographical factors can also influence wetting. Wenzel (1963) has shown that surface roughness can exert a strong influence on the contact angle an adhesive exhibits on that surface. He demonstrated this argument mathematically thus:

$$\cos \theta' = r \cos \theta \tag{7.4}$$

where r is defined as a roughness factor given by the ratio of the actual area to the projected area of the surface, and θ' and θ are the contact angle values for rough and smooth surfaces respectively. Thus, with a smooth surface, exhibiting θ less than 90°, roughening would result in a reduction in θ' with an apparent improvement in wetting. However, if θ were greater than 90° then an increase in surface roughness would have the reverse effect, with θ' appearing to increase still further.

7.3 Substrates and surface pretreatments

As mentioned previously, adhesion is a phenomenon which exists within a few molecular layers of the contacting materials. It is therefore necessary to produce substrate surfaces which will be conducive to efficient intermolecular interactions so as to promote both adequate initial joint strength and longevity in the intended service environment.

The degree to which surface preparation will be necessary will depend upon several factors including in particular the level of strength desired, longevity requirements (dependent upon service environment), level of contamination and the nature of the substrate. Numerous surface treatments exist for a wide range of substrates differing in complexity, efficiency and the nature of their effects on substrate surfaces. In this account an attempt will be made to describe briefly the main approaches to surface treatment which have been employed. This will be followed by more detailed accounts of methods investigated and employed with some of the more typical structural materials.

Surface treatments can be divided into three broad areas, these being associated with solvent cleaning, mechanical abrasion and chemical treatment.

7.3.1 Solvent cleaning

Solvent cleaning is concerned solely with the removal of organic contaminants (greases, oils, etc.) from the surface. These contaminants will reduce surface free energy and thus render the surface prone to poor adhesive wetting. Solvent treatment can be approached in several ways ranging from solvent wipe to ultrasonic vapour degreasing. The former, which simply involves wiping the substrate surface with a clean cloth or paper, although simple, suffers from the fact that the organic contaminants are essentially more evenly dispersed on the surface rather than removed. Although solvent immersion can be regarded as an improvement, solvent purity becomes vitally important since treated surfaces will only be as clean as the solvent. Vapour degreasing is the preferred variant. With this technique the disadvantages of the previously described methods are essentially removed since in this case substrate surfaces are continuously bathed in distilled uncontaminated solvent. Thus complete removal of organic matter is virtually assured. A more sophisticated form of solvent cleansing, ultrasonic vapour degreasing, utilises ultrasonic sound waves, provided by transducers incorporated into the degreasing tank, as a means of promoting rapid agitation of solvent which provides for efficient removal of organic matter.

Although solvent treatment is capable of removing organic contamination, it has no influence whatsoever on the oxide layer present on most metallic alloy surfaces. For many situations, removal of the perhaps relatively weak inherent surface oxide layer is regarded as important and in these circumstances more rigorous treatments such as mechanical abrasion and/or chemical treatment are necessary.

7.3.2 Mechanical abrasion

The mechanical abrasion approach to surface treatment can be regarded as incorporating processes by physical/mechanical means without having any significant impact on the chemistry of the surface. Paramount in this category are treatments involving dry abrasion, i.e. sanding of the substrate surface with abrasive paper, and grit blasting. Both techniques require solvent degreasing prior to abrasive treatment in order to prevent transfer of organic contaminants to the abrasive medium.

7.3.3 *Chemical pretreatment*

The third and perhaps the most important category, at least in terms of typical aerospace materials such as the aluminium and titanium alloys, concerns chemical pretreatments. With these alloys the prime objective is to (a) remove the existing, perhaps inherently weak and unstable oxide layer and (b) generate a new oxide layer conducive to both good adhesion and stability, particularly to hot/wet environments. As will be described, major advances have been achieved in recent years in developing efficient surface treatments for both aluminium and titanium alloys.

Although epoxy-based adhesives have shown themselves capable of bonding a wide variety of substrates, they have received more widespread use in bonding traditional structural materials such as various metallic alloys and, more recently, fibre-reinforced composite materials. Consequently in this necessarily limited review, discussion of surface pretreatments will be restricted to these more common materials.

7.3.3.1 Steel alloys. For the purposes of this discussion it is convenient to divide steel alloys into those vulnerable to rusting on the one hand and stainless steels on the other. With regards the former, extensive experience over many years has shown the simple solvent degrease/grit blast/degrease procedure to be most suitable for the majority of non-stainless steels.

Stainless steel alloys all contain, in addition to the main metallic component, relatively large amounts of chromium the presence of which in surface regions promotes the desired level of passivation. A number of studies have been conducted aimed at developing surface pretreatments for adhesive bonding. Work conducted by Allen and Alsalim (1977) has shown that any strong reducing acid will etch stainless steel surfaces with, in particular, sulphuric acid (10%) and hydrofluoric acid (4%) producing the most beneficial results in terms of improvements in joint strength beyond that which could be expected with a simple solvent degrease pretreatment. Using this approach the deposition of a carbon deposit following the etching process necessitated the need for a post-etch desmutting procedure. Nitric acid and chromic acid were found suitable for this purpose. In addition work by Smith and Haak (1979) showed that additional treatments involving chromic acid etching and anodising in nitric acid provided substrate surfaces beneficial to adhesion and adhesive bonding.

7.3.3.2 Aluminium alloys. As a result of their strength, low density and corrosion resistance, aluminium alloys have achieved a virtual monopoly in the aircraft industry. For many years aluminium, whether pure aluminium cladding, an alloy, or an unclad alloy, was pretreated for adhesive bonding using a chromic/sulphuric acid etching process. This treatment has, however, lost favour in adhesive-bonding applications because of problems that were encountered concerning poor durability of adhesive joints in warm/moist environments, this being particularly relevant with the relatively new low temperature curing, toughened epoxy adhesives. In comparison the more traditional phenolic-based adhesives have continued to exhibit excellent durability characteristics with aluminium subjected to the chromic/sulphuric acid etch (Schliekelmann, 1979; Albericci, 1983).

In an attempt to provide superior surface pretreatments, Boeing began a programme in the late 1960s aimed at examining the efficacy of an anodising process employing phosphoric acid. Greatly superior performance was achieved using this technique resulting in its commercial exploitation in 1974 (McMillan, 1979). It is now widely employed in aerospace applications in the United States. Contrary to this practice, chromic acid anodising has been the preferred pretreatment in Europe where it has been used successfully for many years.

With both anodising techniques, surface treatment would involve, in broad terms, a preliminary degrease in, for example 1,1,1-trichloroethane followed by an alkaline etch, the latter designed primarily to remove inherent oxide. This would be followed by etching in chromic/sulphuric acid followed in turn by the appropriate anodising procedure. Interdispersed between these main procedures would be various rinsing processes followed finally by drying and primer application.

7.3.3.3 Titanium alloys. Although aluminium alloys are currently the most widely used alloys in aircraft construction, the need for a higher temperature capability in, for example, supersonic aircraft applications, has resulted in titanium achieving a small measure of utilisation. As with aluminium, the desirability of adhesive bonding has resulted in numerous studies being conducted on titanium with a view to the development of efficient surface treatments. Most notable amongst these have been those based upon chromic acid/fluoride anodising (Moji and Marceau, 1976), an alkaline/hydrogen peroxide etch (Cotter and Mahoon, 1982) and a sodium hydroxide anodising process (Poole, 1985).

7.3.3.4 Fibre-reinforced composites. Composites based upon glass, Kevlar and particularly carbon-reinforced polymeric composites offer substantial advantages over traditional structural aerospace materials in terms of reduced weight together with potential improvements in specific strength and modulus. As a result of inherent problems associated with the joining of composite materials using conventional fastening techniques, adhesive bonding has been regarded as potentially a highly desirable alternative to mechanical joining.

In surface treatments the prime objective is, in part, similar to that previously described for metallic alloys, i.e. to produce a clean surface free of major extraneous contaminants. In addition, however, the polymeric component of fibrous-reinforced composites is, due to is organic character, prone to moisture uptake particularly when exposed to highly humid and warm environments. Studies have shown that absorbed moisture can result in the generation of voids in adhesive bondlines resulting in significant reductions in both initial strength and fatigue resistance (Sage and Tin, 1982; Parker, 1983). Thus drying at elevated temperatures prior to bonding so as to yield a water content of less than 0.4% is generally considered advisable. Apart from this, surface treatments are relatively simple in comparison to those employed with metallic substrates with a carefully controlled grit blast or even hand abrasion generally being considered suitable for the attainment of both a high initial joint strength and adequate long-term durability.

7.3.4 Primers

In addition to the surface pretreatments described above, the use of substrate surface primers is frequently recommended for the adhesive bonding of many metallic alloys. Primers can be employed to fulfil a number of requirements. These can include:

(i) *Protection.* Freshly prepared metallic surfaces are prone to contamination by atmospheric contaminants including water, hydrocarbons, etc. The virtually immediate coating of the pretreated surface with a primer can protect that surface from atmospheric contaminants by basically exchanging a surface of very high free energy, and thus prone to contamination, for one of more moderate surface free energy but highly compatible with the adhesive. After priming, the surfaces are less prone to contamination resulting in a substantial relaxation in the necessary time interval between surface pretreatment and bonding. Such primers are frequently solvent-based systems with solids similar to the intended adhesive.

(ii) *Wetting enhancement.* Adhesive systems used for metal bonding are often composed of films carried on cloth supports such as polyester or nylon. Because of their precise formulation, film adhesives often suffer from having only a short period of time at the appropriate bonding temperature before cure inhibits flow in a system which already has an inherently high viscosity. The application of a primer solution to a metallic substrate prior to adhesive bonding can therefore overcome the potential wetting problems which would clearly be associated with the use of such adhesives.

(iii) *Corrosion inhibition.* Corrosion-inhibiting primers, generally comprising an epoxy together with strontium chromate in a solvent base, are generally recommended for use with aluminium alloys following anodising pretreatments (Noland, 1976). This combined approach to surface pretreatment has been shown to provide the highest degree of moisture protection to bonded aluminium alloys, particularly when bonded using relatively low cure-temperature epoxy adhesives.

(iv) *Durability enhancement.* Under certain circumstances, water can cause the displacement of adhesive from a substrate producing a debond resulting in turn in a substantial reduction in load-bearing capacity. This effect has been shown to occur with steel joints bonded using simple epoxy adhesives (Gledhill and Kinloch, 1974). Studies using in particular organosilane compounds as the basis of primers, have shown that the application of organosilane solutions to substrate surfaces prior to adhesive bonding can result in substantial improvements in adhesive-substrate interfacial stability in the presence of moisture (Gledhill *et al.*, 1990). Durability related aspects such as this will be described in greater detail later in this chapter (section 7.7).

7.4 Methods of test

The mechanical testing of structural adhesive joints can be divided into two broad classifications namely (i) small coupon scale tests bonded and tested to well-defined specifications and (ii) evaluation at a structural level with either a complete structure or a sub-assembly being assessed. The former approach provides a means of assessing various adhesive joint qualities associated with material components, surface treatments and bonding procedures, allowing the generation of large amounts of data at relatively low cost. On the other hand structural level testing provides much more limited information at greater cost and is generally employed as a means of verifying design principles originally based, to a degree, on preliminary coupon type tests. Since tests at the structural level would generally be conducted on designs specific to a particular application, and therefore unique to that application, this section will be devoted to the small, coupon type test techniques which can be applied to epoxy adhesives in addition to other generic types.

Before briefly describing some of the main test methods which have been employed, it is of interest to mention the three main loading modes which adhesive joints can encounter in service and which test methods are designed to simulate. These are shown in Figure 7.3.

Although this approach is generally used in a fracture mechanics context (see chapter 5), it is useful to employ it here. The opening mode I situation, usually obtained in joints subjected to cleavage or peel, can be regarded as the most harmful loading mode which an adhesive joint can encounter. Conversely, mode II, an in-plane shear mode, represents the most desirable

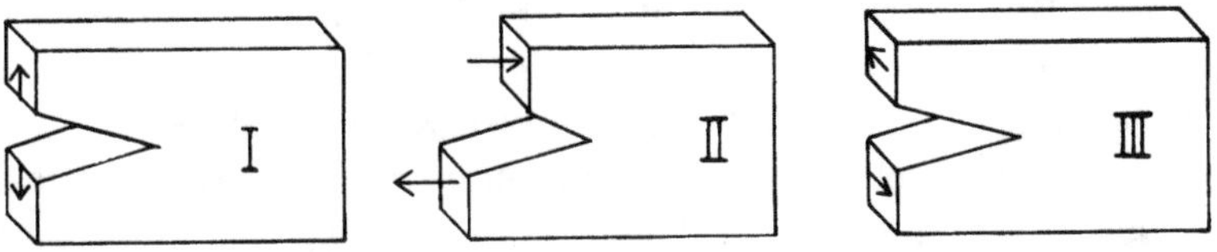

Figure 7.3 Modes of loading. I, tensile–opening; II, in-plane shear; III, tearing, antiplane shear.

means of load transfer with most adhesives, adhesive joints being optimised to maximise shear properties. Mode III, which represents a tearing mode of failure has been little studied and will not be considered further.

7.4.1 Conventional test techniques

Numerous adhesive joint configurations are available for evaluating the performance of adhesives, surface treatments and bonding processes, etc. In this necessarily limited account only the more common and most widely utilised of these configurations will be discussed. More extensive coverage can be found in the literature (Anderson *et al.*, 1977; Arnold, 1981; McMillan, 1981; Wake, 1982; Adams and Wake, 1986).

7.4.1.1 Lap-shear specimen. The lap-shear specimen, in its various forms (Figure 7.4) is probably the most commonly employed method for evaluating adhesive joint strength. Although the name implies a predomi-

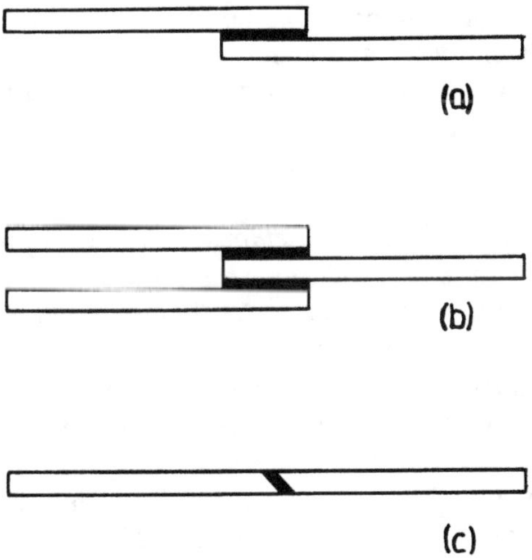

Figure 7.4 Typical tensile lap-shear joint geometries (a) single lap-shear joint; (b) double lap-shear joint; (c) scarfjoint.

nantly mode II shear stress application, this is a major oversimplification. As a result of its geometry, complications arising from, in particular, adhered bending can result in the generation of substantial cleavage stresses at the specimen ends in addition to a non-uniform shear stress distribution. Thus, to employ fracture mechanics terminology, the lap-shear configuration can be regarded as exhibiting combined mode I/II behaviour with significant peel and shear stress contributions. In spite of this complication, the lap-shear joint owes its popularity to ease of both preparation and test in addition to the fact that, in certain respects, its joint geometry simulates the combined mode I/II loading situation experienced by bonded joints in service. It thus allows an insight into the adhesives which are likely to perform well when subjected to this combined stressing situation.

7.4.1.2 Peel. Peel testing of adhesive joints is probably second only to lap-shear in popularity, this arising from both ease of manufacture and test. In addition, although exhibiting certain disadvantages, peel tests yield qualitative information regarding the ability of an adhesive joint to withstand the most damaging applied stress which an adhesive joint may be subjected to in service.

Three main types of joint configuration have been employed. These are generally referred to as the T-peel, Bell peel and the climbing drum method (Figure 7.5). The T-peel is probably the most widely used and is prepared by bonding together two adherents of similar thickness which, after cure, are pulled apart in the manner indicated. The Bell peel and climbing drum techniques involve the peeling of an adherend at constant radius around a steel roller. Both methods involve considerable energy absorption in the adherends so that the peel values obtained from these two techniques can be, and generally are, considerably higher than those obtained, for example, from the T-peel where deformation in the adherends is considerably less.

Due to the difficulties inherent in assessing both applied and failure stresses, results from peel tests are generally reported as linear values, e.g. $N.mm^{-1}$.

7.4.1.3 Other conventional test methods. A wide range of other joint geometries has been employed in assessing adhesive joint behaviour. These include butt-joints, napkin ring specimens and various modifications thereof. One further specimen worthy of note which has achieved some measure of prominence in recent years is the so-called thick adherend specimen (Figure 7.6). As indicated, the specimen is machined from two 25.4 mm thick aluminium plates 50.8 mm wide by 352.8 mm long, previously bonded together. A 25.4 mm overlap is machined into the specimen in the manner shown. This type of specimen has been employed to measure the shear properties of adhesive bonds, this being largely possible due to the thickness of the adherends restricting adherend bending. This in turn

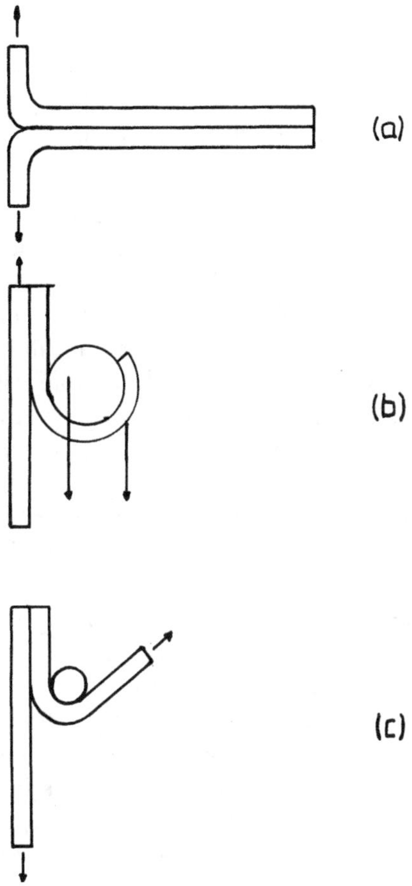

Figure 7.5 Typical peel joint configurations. (a) T-peel; (b) climbing drum; (c) Bell peel.

inhibits the generation of peel loads of the type found in other joint configurations such as the standard lap-shear specimen.

7.4.2 Fracture mechanics approach

Since structural adhesives such as those based on epoxies frequently fail by progressive crack propagation, failure criteria based upon both the initiation and propagation of flaws inherent in the joint can be considered relevant. Consequently fracture mechanics, where theory enables a mathematical evaluation of the conditions in which flaws will propagate under stress, has recently been considered as a means of assessing the load-bearing characteristics of adhesive joints.

The study of fracture mechanics can be conveniently divided into two inter-relatable parts. The first, arising from the initial work of Griffith

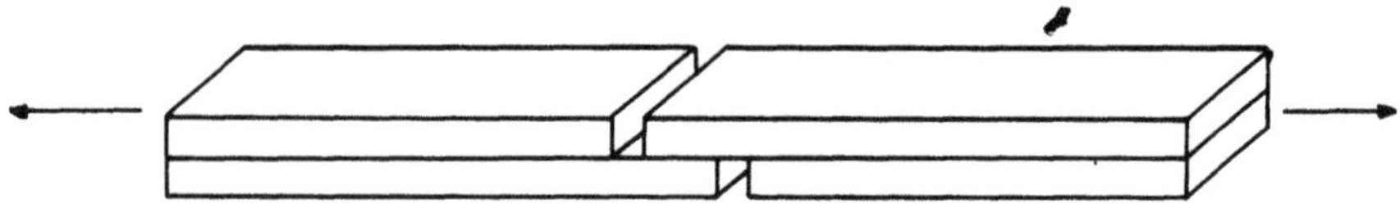

Figure 7.6 Thick adherend specimen.

(1920) quite simply proposes that fracture occurs when sufficient energy (denoted by the term *G*) is released from the stress field by growth of the crack to satisfy the requirements of the new fracture surfaces. The second, proposed initially by Irwin (1964), is based on the premise that the stress field at the crack tip can be defined by a parameter called the stress intensity factor, *K*, and states that fracture occurs when the value of *K* exceeds a critical value. Although both approaches have been employed in the study of adhesive joint fracture, the energy balance approach has received most attention.

In recent years several specimen geometries have been designed with a view to measuring the fracture resistance of structural adhesives (Kinloch and Shaw, 1981a); most studies being devoted to epoxy-based adhesive systems. Some of these designs are indicated in Figure 7.7. Probably the most popular specimen employed to assess mode I fracture energy, G_{Ic}, has been the contoured double cantilever beam specimen. Designed and initially employed by Mostoroy and Ripling (1966), this specimen is contoured in such a way as to provide a constant compliance. Since at a given applied load the value of fracture energy remains independent of crack length, *G* can be readily determined without knowledge of crack location; a feature which can be extremely useful where difficulty is encountered in determining the precise location of the crack tip. A number of variations of the above geometry have been designed and employed.

In addition to mode I, specimens have been developed for measuring mode II (in-plane shear) and combined modes I/II loads. With regards the last, the scarf-joint (Figure 7.8) has been designed to impart a combination of both mode I and mode II load conditions on the adhesive layer; the value of θ dictating the precise load combination (Trantina, 1972). The so-called independently-loaded mixed-mode specimen, also shown in Figure 7.8, was designed to allow independent application of cleavage and shear loads so as to provide both modes I and II or a controlled combination of the two (Bascom *et al.*, 1977).

Mode III loading has only received minor attention with regards adhesive joint fracture studies.

7.4.3 Environmental testing

One of the most hostile environments that an adhesively bonded structure can experience in service is one consisting of a high relative humidity

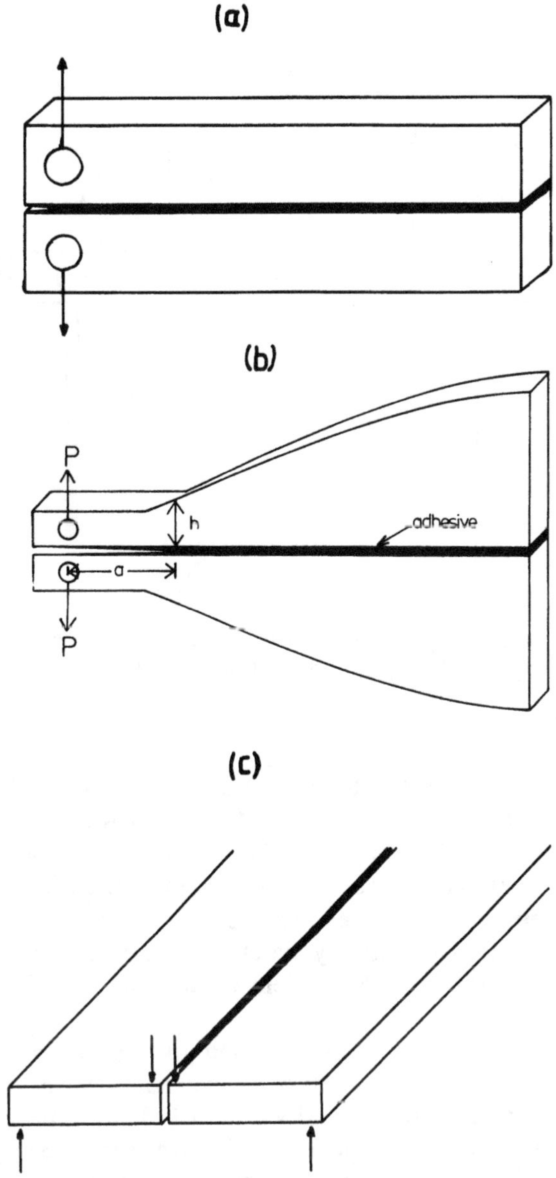

Figure 7.7 Typical mode I fracture mechanics adhesive joint specimens. (a) Parallel double cantilever beam; (b) contoured double cantilever beam; (c) double-torsion.

together with a high ambient temperature. Consequently investigations aimed at assessing the ability of bonded joints to withstand such harsh environments are reasonably common. Two approaches are generally employed. First, to submit bonded joints to exposure trials in 'real' environments, including in particular potentially damaging tropical

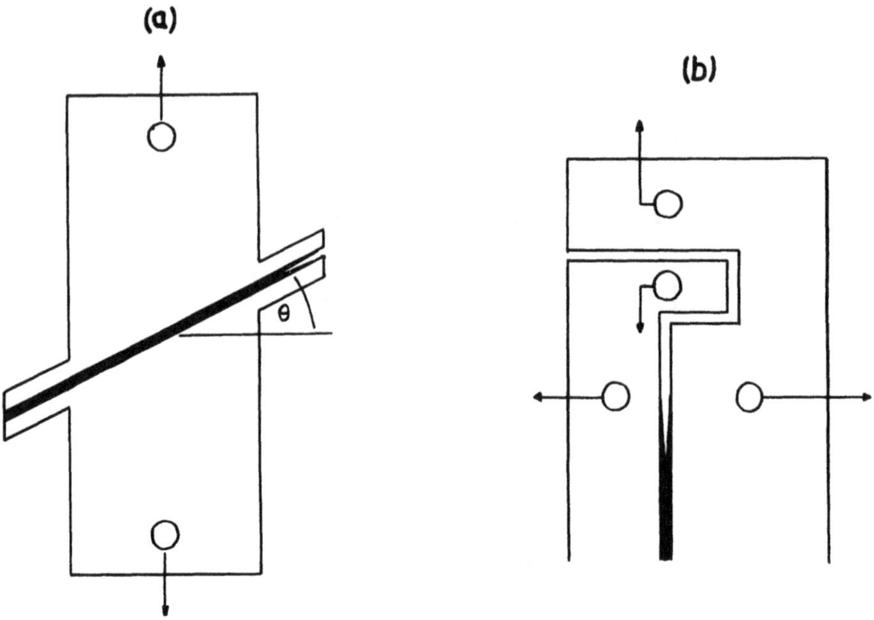

Figure 7.8 Typical mode II fracture mechanics adhesive joint specimens. (a) Scarfjoint; (b) independently loaded mixed-mode joint.

exposure. Second, to undertake accelerated tests with adhesive joints subjected to condition extremes beyond those that would be expected in actual service. Although not capable of providing an accurate measure of joint longevity, accelerated exposure trials allow, at the very least, insights into the bond components (adhesive, interface, oxide layer, etc.) which could be vulnerable to potentially harsh service conditions, as well as providing a fairly simple and convenient means for investigating ways of improvement.

Most of the joint configurations discussed above have been used, with varying degrees of success, in accelerated exposure trials. One particular specimen not described thus far and worthy of mention here is a double cantilever beam variant which has become known as the Boeing wedge test (Marceau *et al.*, 1977). This specimen is qualitative in nature and has as its main advantage both simplicity in preparation and use together with the speed with which information can be provided, particularly concerning environmental resistance. As indicated in Figure 7.9, the specimen is simply stressed by inserting a wedge into the bondline, thus requiring no external load application and enabling joints to be subjected to controlled environments. Durability assessments are made by relating crack extension measurements to time and the locus of failure can be observed. The wedge test is now commonly employed as a comparative test technique for

Figure 7.9 Boeing wedge test adhesive joint specimen.

assessing factors such as surface preparation, adhesive type, etc. (Marceau and Scardino, 1975; McMillan, 1979).

7.5 Epoxy adhesive formulation

Structural adhesives based on epoxies can be formulated to fulfil a wide range of applications together with cure and performance requirements. Indeed the number of formulatory variables can be regarded as immense providing the formulator with considerable scope, as well as difficulty, for achieving a specific set of requirements. Not only will characteristics be dependent upon the two main formulatory variables, i.e. resin type and curative but will also depend upon the selection of a wide range of available modifiers such as fillers, diluents, flexibilisers, toughening agents, etc. However, although a large number of epoxy-based adhesives are now available commercially, it will be likely that the majority of these will be based upon a surprisingly small number of the many available ingredients. In this necessarily limited account, an attempt will be made to outline the main formulatory materials available with examples given of the most common types employed together with the benefits they bestow. A more detailed account of the various types of additives and modifiers available for use with epoxy resins generally can be found in chapter 4.

Epoxies used in adhesive formulations can range from very simple, low molecular weight, primarily aliphatic resins to complex multi-functional predominantly aromatic types. However, by far the most common type employed is the diglycidyl ether of bisphenol A (DGEBA), having the following structure:

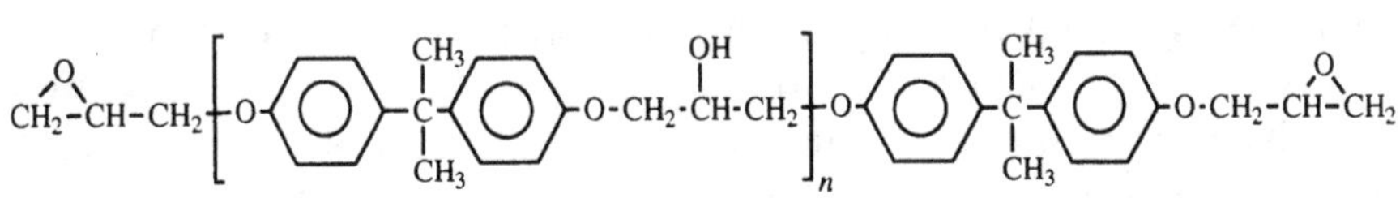

This material can, in fact, be regarded as the standard workhorse resin of the epoxy range and will be found in the majority of epoxy-based structural adhesive formulations. Epoxies of this type are available in varying molecular weights and can range from moderately viscous liquids (low values of n) to solids at room temperature (high n). Liquid and paste adhesive grades are generally composed of relatively low molecular weight epoxies. Higher molecular weight variants are frequently employed in film adhesives where a degree of structural integrity is required under use conditions.

In addition to DGEBAs, other resin variants are available, each capable of exploiting specific service requirements. These include interesting examples of epoxies being alloyed with other resin types including nylon-epoxies, polysulphide-epoxies and epoxy-phenolics. The former are based on a blend of epoxy with a soluble grade of nylon and have been shown to produce structural adhesives exhibiting toughness and a remarkable resistance to cleavage forces thus resulting in excellent peel and lap-shear strengths. Unfortunately the propensity to absorb large concentrations of moisture, thus rendering them susceptible to environmental degradation in particularly harsh hot/moist atmospheres, has somewhat diminished the euphoria which initially greeted their introduction. In addition blended resins such as these, together with polysulphide-modified variants are limited to a high temperature ceiling of approximately 70–80°C. However, epoxy-phenolic adhesives, produced by reacting a high molecular weight epoxy with a phenolic resin, are capable of operation at considerably higher temperatures, particularly for short periods of time.

Choice of curing agent can be equally as important as choice of resin and the adhesive formulator would need to consider several factors in choosing the appropriate curative regime for a specific formulation. These factors would include application conditions (particularly service temperature) together with cure and other processing requirements. Although literally hundreds of potential curing agents have been identified, only a relatively small number have been employed in adhesive formulations. By far the most common are the amines which can be categorised, chemically speaking, into the aliphatic primary and secondary polyamines, tertiary amines, polyamides, heterocyclic amines and aromatic amines (chapter 2).

Aliphatic amines of various types form the basis of most commercial room temperature-cure epoxy adhesives. Not surprisingly these are generally two-component systems. Although simple aliphatic amines such as diethylenetriamine (DETA) and triethylenetetramine (TETA) have been and still are to a limited extent employed in room temperature cure formulations, amines such as diethylene glycolbispropylamine, together with those based upon polyamide structures are nowadays more typical. The latter type offers the advantage of imparting improved flexibility to an otherwise rigid and possibly somewhat brittle cured resin. In addition, because of their reduced functionality in relation to aliphatic amines such as TETA, they can be used to good effect in equal proportions with epoxies, thus minimising potential mixing errors generally associated with curatives necessarily employed in low concentrations. This, together with lower volatility and toxicity in comparison with most other amines has made their use popular, particularly as the basis of the twin-tube epoxy adhesive kits now so prevalent in the domestic market.

Although two-part adhesive systems were originally the norm, one-component epoxy adhesives are now readily available, exhibiting long shelf-life characteristics, particularly when subjected to refrigeration or frozen storage. This clear advantage is achieved by the use of latent curing agents, with one particular amide, dicyandiamide (DICY), being the most common curative employed in one-component epoxy adhesive systems. DICY is generally employed in concentrations of approximately 5 phr (parts per hundred of resin by weight) and results in the rapid polymerisation and cross-linking of an epoxy on heating the formulation in excess of 150°C. In order to bring cure temperature down to more manageable levels, DICY-cured formulations generally require the use of a catalyst. Film adhesives, by necessity also employ latent curing agents so as to allow the required storage stability together with the requirement of being able to undergo film processing procedures without undergoing any significant cure.

Other curing agents which have been, and to a limited degree are still employed, include anhydrides and aromatic amines. These curing agents can be employed in hot-cure adhesive formulations and are capable of providing cured adhesives exhibiting higher temperature capabilities than the other curing agents discussed.

After resin and curing agent, fillers are the most common additive in epoxy structural adhesives where their incorporation can result in some not inconsiderable improvements in behaviour. Notable advantages include their ability to reduce shrinkage, thermal expansion coefficient, exothermic characteristics and, of course, cost. Their ability to reduce shrinkage can in fact be regarded as impressive. Some unfilled epoxy formulations are capable of undergoing about 5% shrinkage on cure which can exert a substantial influence on joint performance. The addition of filler in the appropriate concentrations can reduce shrinkage and result in quite a dramatic improvement in properties. In addition, a reduction in coefficient of thermal expansion can also reduce significantly the thermal expansion mismatch which would exist between a substrate and adhesive, thus minimising the generation of internal stresses during the cure process.

Filler incorporation has also been used to good effect to influence the flow characteristics of epoxy adhesives. Under certain circumstances thixotropy, a characteristic of particular practical importance in many paste adhesive formulations can be obtained by a suitable choice of filler. Such formulations offer the advantage of a fairly limited degree of flow when subjected to a moderate applied shear; the adhesive reverting to virtually zero flow when shear is removed.

In addition to these fairly broad effects, somewhat esoteric fillers can be employed to enhance or provide specific properties. As an example, the use of silver-filled epoxies has been employed in electrically conductive bonding applications. Although reasonably effective in this application, this can clearly be regarded as an example of filler addition not resulting in reduced formulation cost.

Fillers commonly employed in epoxy adhesive formulations include silica, aluminium and alumina powders.

Other possible additives include diluents, flexibilisers and toughening agents. Since these are discussed in some depth in chapter 4, they will not be mentioned further here. However, brief discussion of toughening agents is worthy at this point since they can, to a degree, be regarded as the one formulatory ingredient which has put the word 'structural' into the term 'structural epoxy adhesive'.

Epoxies, in common with many other thermoset polymers are prone to brittle behaviour which can result in relatively poor adhesive joint performance, particularly where significant peel forces exist. Toughening agents, most notably those based upon liquid reactive rubbers such as carboxyl terminated butadiene acrylonitrile rubbers (Ting, 1988) together with the use of thermoplastic-based modifiers such as polyethersulphone (Partridge and Bucknall, 1983) have been employed to excellent effect to dramatically increase toughness and improve resistance to peel forces; this generally being achieved with minimal effect on other important properties and parameters such as modulus and glass transition temperature, T_g.

Finally, a brief word regarding epoxy adhesive films. Adhesives of this type are generally available in both supported and unsupported forms; the former containing scrim carrier, generally based on nylon or polyester. Film adhesives of this type, as well as being relatively easy to handle, also offer more controlled regulation of bond thickness in comparison to their liquid/paste counterparts. They are generally employed in relatively large area bonding applications.

7.6 Properties of adhesive joints

The properties of an adhesive joint will depend upon several factors including, for example, the mechanical and physical characteristics of the bulk adhesive material, the degree of adhesion between the adhesive and substrate and the design of the joint. Factors likely to influence adhesion such as surface pretreatment have already been discussed and will therefore not be considered further. The highly important subject of adhesive joint design can also be considered beyond the scope of this chapter. This section will therefore be devoted to a consideration of the bulk properties of epoxy adhesive materials which in turn will be followed by some examples of typical adhesive joint properties which could be expected from a range of epoxy-based structural adhesives.

7.6.1 *Bulk properties of epoxy adhesive*

The properties likely to be of importance in determining the strength of a polymer, such as an epoxy, can be demonstrated by the use of the Griffith equation

$$\sigma_f = \left(\frac{2EG_c}{\pi a}\right)^{1/2} \tag{7.5}$$

where σ_f is the stress at fracture, E is modulus, G_c fracture energy and a regarded as a critical flaw size. As indicated, the Griffith equation shows clearly the direct positive relationship between stress at fracture, or the 'strength' of the material and the two other key properties of modulus and fracture energy. Although flaw size a can be regarded as a source of complication, the Griffith equation provides a clear indication as to the bulk mechanical properties which are likely to influence the load-bearing capacity of an adhesive material. Clearly high values of both modulus and fracture energy, all other factors being equal, are likely to enhance the load-bearing characteristics of a cured epoxy polymer.

The fracture energy, G_{Ic}, of all epoxy formulations will depend critically upon the precise formulation details, e.g. the type of resin and curing agent employed, together with other factors such as cure conditions. An examination of the literature provides a clear indication of the range of fracture energies obtainable by simply varying choice of resin and/or curing agent (Young, 1980; Kinloch and Young, 1983). Values varying between approximately 50 to in excess of 600 $J.m^{-2}$ have been obtained. Cure conditions, in particular cure temperature, have also been shown to promote large variations in fracture energy. For example work conducted by Shaw and Tod (1989) showed that by varying the temperature of cure between 120 and 160°C over cure times ranging from 2 to 6 hours fracture energy G_{Ic} varied between 180 and 360 $J.m^{-2}$ (Figure 7.10); the higher G_{Ic} values occurring at the higher cure temperatures. Likewise, modulus has also been shown to be dependent on the nature of both the resin and curing agent, although, provided gross undercure is avoided, variations in modulus from one formulation type to another are generally minor. Variations in cure temperature are also shown to provide significant but relatively minor changes in modulus.

The introduction in the late 1960s (McGarry and Willner, 1968) of rubber-modification as a means of enhancing toughness has done more than anything else to promote the view of epoxies as viable structural materials. Numerous studies (Kinloch and Shaw, 1981a; Kinloch *et al.* 1983) have

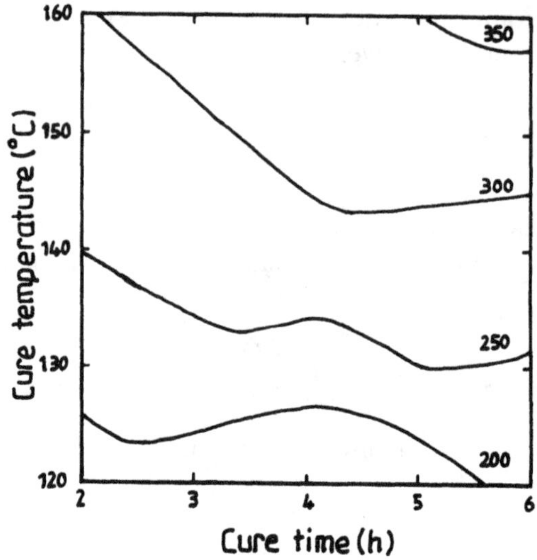

Figure 7.10 Effect of cure conditions (cure temperature and time) on the fracture energy ($J.m^{-2}$) of piperidine-cured DGEBA epoxy.

shown that the incorporation of reactive liquid rubbers such as carboxyl-terminated butadiene acrylonitrile rubbers (CTBN), can substantially elevate toughness whilst having a relatively minor influence on glass transition temperature, T_g, and properties (particularly modulus) associated with it. An example of the magnitude of toughness enhancement is shown in Figure 7.11.

As indicated, the incorporation of approximately 15 phr of CTBN can result in an approximately forty-fold improvement in G_{Ic}, thus clearly transforming an essentially brittle polymer into one exhibiting toughness characteristics similar, if not identical, to various thermoplastic polymers. Increases in CTBN concentration beyond approximately 20 phr have been shown to result in a decline in G_{Ic} and, particularly modulus, this being due to a process known as phase inversion which can result in flexibilisation of the system (Bascom *et al.*, 1975). For this reason, structural epoxy adhesive formulations of this type generally contain CTBN in concentrations of 10–15 phr.

Variation in cure conditions, as with unmodified epoxy formulations, has also been shown to have a strong influence on the toughness of rubber-modified epoxies. In particular the work of Shaw and Tod (1989) has shown, as indicated in Figure 7.12, that by simply varying cure temperature between 120 and 160°C over cure times ranging from 2 to 6 hours, G_{Ic} changes between 1.8 and 12 $kJ . m^{-2}$ are possible; as with unmodified systems, the higher cure temperatures producing by far the highest G_{Ic} values. Although users of epoxy adhesives would clearly prefer to operate at much reduced

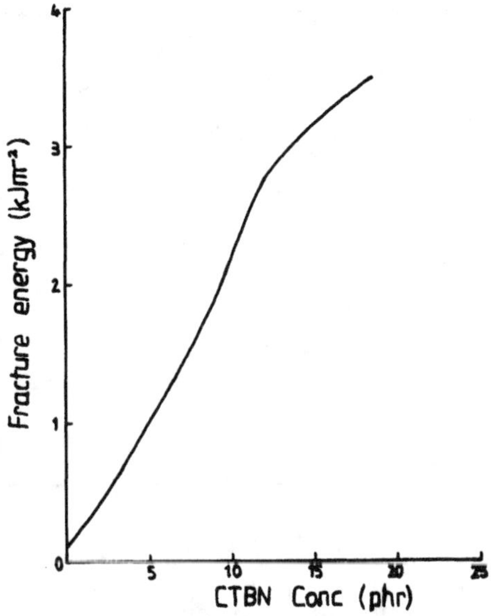

Figure 7.11 Influence of elastomer incorporation (CTBN) on the fracture energy of a piperidine cured DGEBA epoxy.

temperatures, preferably room temperature, studies such as these have clearly demonstrated the remarkable influence that cure temperature can have on the toughness of rubber-modified epoxy adhesives and, in turn, on adhesive joint strength parameters which would be expected to be dependent on toughness, e.g. peel and lap-shear strength. Indeed, experiments conducted by the author have shown that lap-shear strength values, obtained from joints produced from mild steel adherends using the epoxy adhesive formulation mentioned above, increased significantly with an increased temperature of cure.

Other epoxy resin characteristics likely to be of importance to the efficient and safe utilisation of adhesives based on them would include T_g and water absorption behaviour. To a first approximation, T_g will dictate the high temperature limit of an adhesive formulation and various factors such as the type of resin and curing agent employed in the formulation and the curing regime employed will dictate the level of T_g obtained. Generally, formulations employing multi-functional resins and curing agents, together with hot-cure as opposed to room temperature cure systems would result in relatively high T_g adhesives. For epoxy formulations generally, T_g values up to approximately 250–300°C have been obtained (Morgan, 1980). However, for the more common toughened adhesive systems based upon DGEBA resins cured with DICY and employing CTBN-based toughness modifiers, T_g values in the region of 100°C are more typical.

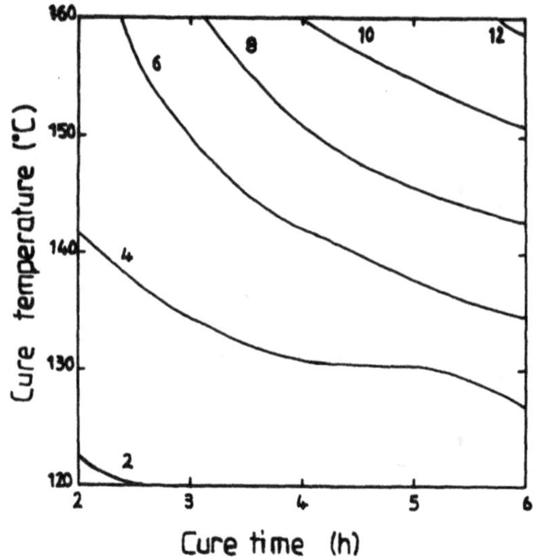

Figure 7.12 Effect of cure conditions (cure temperature and time) on a piperidine-cured CTBN-modified DGEBA epoxy.

Finally, it is important to recognise that a moisture laden atmosphere can exert a severely debilitating effect on adhesive joints, as is discussed in section 7.7. Epoxy adhesive materials, being predominantly organic in nature, will be prone to water absorption, both the amount absorbed and the rate of absorption being dependent upon formulation variables such as the epoxy resin and curing agent types employed, together with environmental variables such as temperature and relative humidity as well as cure conditions. A wide range of equilibrium water concentration values and diffusion coefficients have been quoted in the literature for an equally wide range of formulations and absorption conditions (Wright, 1981; Ellis and Rashid, 1984). Equilibrium concentrations of from 0.25% to in excess of 8% have been quoted. Although the absorption of water has, in a highly limited sense, been shown to be beneficial in terms of both improved toughness and static fatigue resistance (Gledhill *et al.* 1979), water absorption is generally considered harmful. As far as the bulk epoxy polymer is concerned, water can exert strong debilitating influences on mechanical properties such as modulus, as well as reducing T_g. A reduction in T_g of 20°C per 1% of water absorbed is a rule of thumb frequently applied for assessing the likely influence of moisture on the T_g of an adhesive and, in particular, properties associated with it (Wright, 1981). Since, as mentioned above, T_g can be regarded as a measure of a polymer's high temperature capability, the unexpected absorption of water into an adhesive layer, particularly one designed for use at elevated temperatures, could have particularly severe implications. The remarkably tough and peel-resistant epoxy-nylon adhesives which, as explained previously, entered into being under a wave of near euphoria, only to be brought down to earth with a bump following realisation as to their poor moisture resistance, serve to emphasise the importance of, at the very least, attempting to engineer into an adhesive system, an element of hydrophobicity.

7.6.2 Adhesive joint mechanical properties

In this section a brief account will be given of the adhesive joint mechanical properties which could be expected of a range of the more common types of epoxy-based adhesives. Although the examples given cannot be regarded as exhaustive, it is hoped that they will provide the reader with a feel for the strength properties available with the various epoxy adhesive types.

Figure 7.13 shows some typical lap-shear strength data for three hot-cure epoxy adhesive variants, in this specific case as a function of temperature. Several features are worthy of mention. First, the lap-shear strength attainable with epoxy adhesives toughened with elastomeric modifiers such

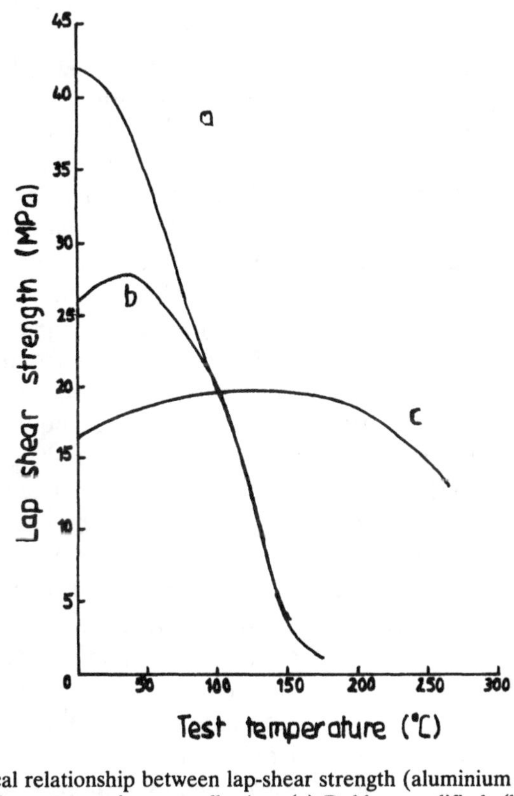

Figure 7.13 Typical relationship between lap-shear strength (aluminium substrates) and test temperature for three types of epoxy adhesive. (a) Rubber-modified; (b) one-part, general purpose; (c) epoxy phenolic.

as the CTBNs can be regarded as impressive, with lap-shear values in excess of 40 MPa at and below room temperature. Note how these values are in excess of these which could be expected of typical general-purpose unmodified epoxies, particularly at moderate test temperatures; this is a direct effect of the improvements of toughness and hence peel resistance brought about by rubber modification. The ability of epoxy-phenolics to retain significant strength at temperatures up to and in excess of 250°C is apparent. Although it is clear that significantly higher strengths can be obtained at lower temperatures for the 'general-purpose' and 'toughened' systems, in both cases temperatures in excess of 75–100°C cause a precipitous strength decline. Since the epoxy-phenolic is relatively unaffected by such temperatures, it clearly offers significant strength advantages at temperatures in excess of about 100°C. Note in particular how the epoxy-phenolic retains in excess of 50% of its room temperature strength at about 250–300°C (Wake, 1982). With all three systems the short-term high temperature capability will be related to T_g; epoxy-phenolics clearly exhibiting substantially higher T_g values than the other two systems. Water absorption, as mentioned previously, will generally result in a reduction in T_g, the magnitude of the decline being largely dependent upon the quantity of water absorbed. Thus, water absorption will generally lead to a deterioration in high temperature load-bearing capacity. It should also be stated that high temperature capability in this context refers simply to the ability of adhesive joints to resist short-term high temperature excursions. For longer-term applications the epoxy-phenolics are generally limited to a high temperature ceiling of about 250°C, since oxidative degradation results in a severe decline in mechanical properties at higher temperatures. For the more common DGEBA-based adhesives, long-term use at temperatures in excess of about 175°C is generally considered inadvisable.

Room temperature cure epoxy adhesives usually provide properties somewhat inferior to those indicated in Figure 7.13, particularly regarding the ability to operate at elevated temperatures. Film adhesives often provide superior strength characteristics in comparison to their liquid/paste counterparts since they generally contain higher molecular weight components which result in improvements in toughness and peel resistance.

In addition to the use of fracture mechanics as a means of studying the fracture characteristics of bulk epoxy systems, it has also been employed extensively to study the toughness and crack propagation characteristics of adhesive joints, as outlined in section 7.4. Much of this work has been concerned with assessing the influence of joint geometry on the toughness and crack growth behaviour of both unmodified and rubber-modified epoxy adhesives together with attempting to relate bulk to adhesive joint fracture. Joint geometry within this context refers primarily to the effects of bond thickness, i.e. thickness of the adhesive layer, and mode of loading.

The effect of bond thickness on the toughness and crack growth characteristics of unmodified epoxy resin adhesive joints has been considered by various workers. Mostovoy and Ripling (1971) for example, employed aluminium tapered double cantilever beam specimens bonded with amine- and anhydride-cured epoxies to study bond thickness effects. They found that, in the majority of cases, adhesive G_{Ic} was virtually independent of bond thickness from approximately 0.05 to 0.5 mm, i.e. the common range of thickness traditionally employed in practice. In addition the values of G_{Ic} obtained were usually similar in magnitude to those found for their 'bulk' counterparts. The same has not been found to be true for rubber-modified epoxies where considerably more complex behaviour has been observed in a number of studies (Bascom *et al.*, 1975; Kinloch and Shaw, 1981). This complexity is demonstrated in Figure 7.14 which shows the influence of bond thickness on the adhesive fracture energy of a rubber-modified epoxy adhesive formulation and its unmodified counterpart. The relative simplicity exhibited by the latter is clear with, as previously mentioned, no significant effect of bond thickness on toughness. With the rubber-modified system the dramatic improvement in toughness is immediately apparent. However, perhaps of greater interest is the pronounced influence that bond thickness has on fracture energy. As indicated, this parameter is shown to pass through a maximum value, which we can call G_{Icm} at a certain bond thickness, t_m. At thickness beyond t_m, G_{Ic} undergoes a reduction until a value is reached which remains essentially constant with increased thickness. This constant G_{Ic} has been shown to be similar, if not identical to the G_{Ic} values obtained from bulk adhesive specimens. Reasons for this bond thickness effect have been discussed in

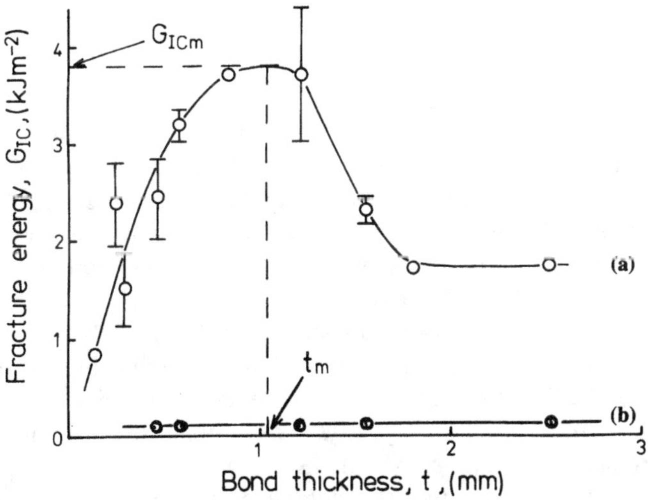

Figure 7.14 Influence of bond thickness on the adhesive fracture energy of a piperidine cured DGEBA epoxy. (a) Rubber-modified system containing 15 phr CTBN; (b) unmodified.

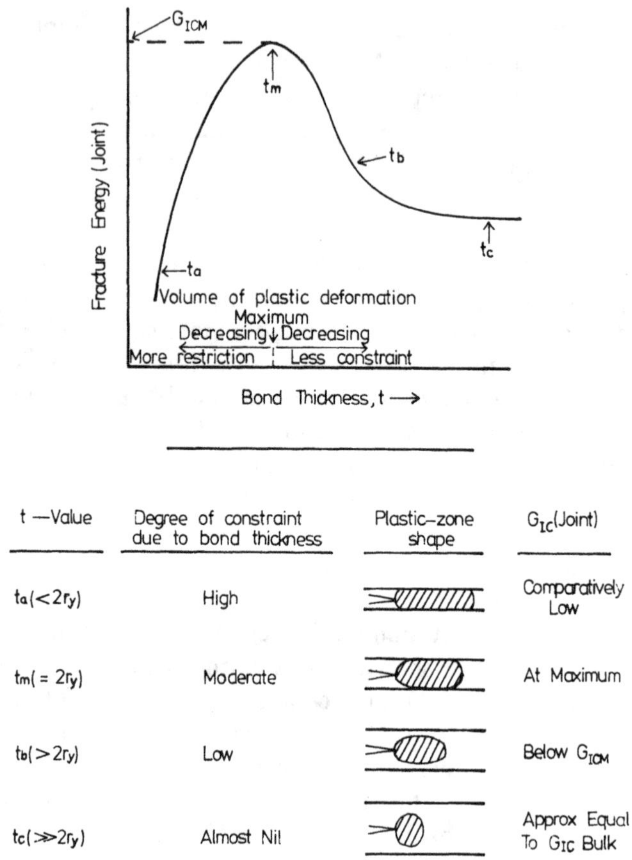

Figure 7.15 Crack tip plastic zone restriction/constraint mechanism.

terms of what can be described as a crack tip plastic zone restriction/constraint mechanism (Kinloch and Shaw, 1981). The theoretical basis of this mechanism is demonstrated in Figure 7.15. As shown, the volume of the crack tip plastic zone can be regarded as a main factor responsible for the bond thickness–adhesive G_{Ic} relationship. Due to the relatively large crack tip plastic zones which can develop with rubber-modified epoxies, a situation can exist whereby the dimensions of this zone will be of a similar magnitude to the bond thicknesses shown in Figure 7.14. Thus, at bond thicknesses below a critical value (equivalent to the plastic zone diameter), restriction of the development of this zone in the bond thickness direction is likely to result in a reduction in plastic zone volume. In addition, both finite element analysis (Wang *et al.*, 1978) and more recent experimental observations have shown that the degree of constraint imposed upon an adhesive layer can dictate to a large degree the distance over which the principal stresses responsible for the plastic zone exist. For example, increased constraint brought about by a reduction in bond thickness, has been shown to produce an extension in the 'length' of the crack tip plastic zone in a rubber-modified epoxy adhesive, which in turn will result in increased plastic zone volume.

Because of these two competing effects, the maximum volume of the plastic zone, and hence a maximum G_{Icm}, will occur when the maximum degree of constraint exists, at a given bond thickness, commensurate with the condition that no restriction on the development of the plastic zone from the metallic substrate exists. Studies conducted with a rubber-modified epoxy over a range of test rates and temperatures (Kinloch and Shaw, 1981) have shown that this condition generally occurs when bond thickness is equivalent to the plane stress plastic zone diameter, depicted in Figure 7.15 as $t_m = 2r_y$. At bond thickness below t_m, although a high degree of constraint will be imposed upon the adhesive layer thus tending to increase plastic zone volume by extension along the adhesive layer, the competing influence of plastic zone restriction in the bond thickness direction will also occur, resulting in an overall reduction in plastic zone volume and thus G_{Ic}. At bond thickness in the vicinity of t_m, the G_{Ic} of the adhesive layer will be greater than that obtained from its bulk counterpart due simply to an enhanced plastic zone volume brought about by the constraint effect. As bond thickness is increased beyond t_m, restriction in the bond thickness direction will no longer apply since $t > 2r_y$. At the same time constraint on the adhesive layer will continually decrease resulting in a reduction in plastic zone length and therefore volume, which in turn will cause G_{Ic} to fall until a bond thickness is reached whereupon G_{Ic} of the adhesive layer is similar to that obtained from bulk adhesive.

Adhesive joint fracture under mode II and mode I/II combinations has also been studied but to a far more limited extent (Bascom and Oroshnik, 1978; Mai and Vippond, 1978). Table 7.1 shows typical fracture energy values for similar unmodified and rubber-modified epoxy formulations obtained under mode I, mode II and combined mode I/II conditions. As expected, the unmodified adhesive provides the lowest fracture energy under all the loading conditions employed, particularly mode I. Although the introduction of a mode II component increases fracture energy by a

Table 7.1 Effect of loading mode on adhesive joint fracture energy

Adhesive	Fracture energy ($kJ \cdot m^{-2}$)				
	G_{Ic}	$G_{I,IIc}$ [a]			$GIIc$ [b]
		30°	45°	60°	
DGEBA-amine	0.07	0.15	0.13	0.15	1.45
CTBN-modified DGEBA	3.40	0.31	0.50	0.31	3.55

[a] Obtained from scarf-joint specimen; [b] obtained from independently loaded mixed-mode joint.

factor of approximately 2, it is not until virtually all of the mode I component is removed that a substantial improvement in fracture energy occurs. The rubber modified adhesive, as shown, once again exhibits more complex behaviour. The mixed mode I/II values are substantially lower than either mode I or II values with the two extreme loading conditions providing approximately equal fracture energy values. Of particular interest is an apparent maximum in the I/II fracture energy value at a taper angle of 45°; an observation which has been addressed by Bascom and Orshnik (1978).

7.7 Environmental effects

7.7.1 Introduction

A major problem with the use of structural adhesives concerns the adverse effects which certain environments can have on the mechanical properties of the joint. Notable amongst these include temperature extremes, stress and radiation. However, the most severe problem has been associated with the influence of atmospheric moisture. Experience over many years, gained from both accelerated tests and long-term exposure to warm/moist environments, has shown conclusively the potentially disastrous results which can result from the use of adhesively-bonded components in moist environments (Hockney, 1970, 1972, 1973; Butt and Cotter, 1976). To demonstrate the potential magnitude of this effect, Figure 7.16 shows the influence of water immersion at a temperature of 60°C on the strength of mild steel butt-joints. The adhesive formulation employed was prepared from a diglycidylether of bisphenol A (DGEBA) cured using a tertiary amine curing agent. Clearly this environment has a detrimental effect on the strength of this particular joint, 1500 hours immersion resulting in only a 15% retention of strength. Additionally, the immersion conditions indicated were also found to influence locus of failure with a pronounced change from cohesive within the adhesive layer to apparent interfacial occurring on increasing immersion time. Thermodynamic considerations can, in fact, be used to predict such interfacial behaviour as will be described shortly.

In this necessarily limited account an attempt will be made first to highlight the major parameters which, from experience, have been shown to influence joint durability, second to discuss the mechanisms which have been deemed responsible for the adverse effects of moisture and third to consider the various approaches which have been considered in an attempt to improve the environmental resistance of epoxy adhesive bonded joints. Finally, a brief discussion concerning other environmental factors such as elevated temperature performance will be given.

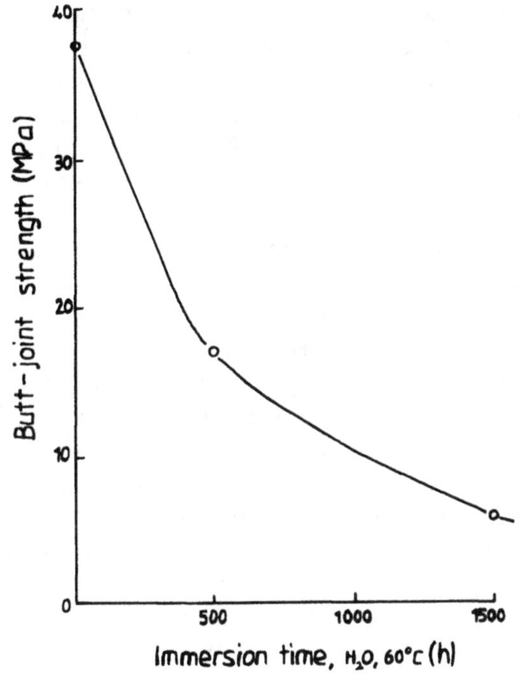

Figure 7.16 Influence of water immersion at a temperature of 60°C on the strength of epoxy-bonded mild steel butt-joints.

7.7.2 Moisture-related effects

As previously discussed, experience has shown that water generally present as atmospheric moisture, can exert a serious debilitating effect on epoxy-bonded joints. In addition, whilst the locus of failure of well prepared joints is usually cohesive within the adhesive layer, environmental exposure frequently focuses failure towards the interfacial zone between the adhesive and substrate.

Numerous factors have been shown to influence joint durability, including:

(i) *Environment.* As mentioned, water is by far the most hostile environmental agent which a structural adhesive joint can experience. Generally factors such as temperature and applied stress together with the amount of water available (e.g. relative humidity), usually exert a significant influence; increases in the magnitude of each generally resulting in an increased degree of environmental attack. Interestingly, a number of studies have indicated that, although exposure of adhesive joints to high humidity usually results in strength degradation, many joints appear able to withstand exposure to lower humidities for extended periods of time without any significant strength reductions. Although some researchers have argued that this signifies the existence of a critical concentration of water below which environmental attack will not occur (Gledhill *et al.*, 1980) others have suggested that this is merely a time-related effect.

(ii) *Influence of adhesive type.* It is now clear that the chemical nature of a structural adhesive can influence quite dramatically both the extent and rate of environmental damage. For example both vinyl-phenolic and nitrile-phenolic adhesives have been shown to exhibit long-term durability superior to that shown by epoxy-based adhesives (Bolger, 1973; Buck and Hockney, 1973). In fact one particular class of epoxy adhesive, the nylon-modified epoxies, have been shown to suffer serious degradation when subjected to warm/moist environments (Cotter, 1977) and are now generally considered inappropriate for humid service conditions. Althought relatively inferior in terms of durability, epoxy-based adhesives are nowadays preferred due to a combination of improved bonding characteristics (lower cure temperature/pressure) and superior peel strengths.

(iii) *Substrates.* Adhesive joints constructed from most substrate types, be they metallic or polymeric in nature, can and frequently do exhibit susceptibility to environmental attack. The mechanisms by which this degradation occurs can vary significantly. As will be discussed (section 7.3), these differences in behaviour influence the methods by which improvements in environmental stability can be imparted.

(iv) *Surface treatment.* Choice of surface treatment for a particular substrate has, in many cases, been shown to be of critical importance in imparting acceptable levels of durability with epoxy-based adhesives. In addition, the use of primers, of various types, has been found conducive to good moisture resistance as will be discussed.

7.7.3 Failure mechanisms

Water can influence the load-bearing capacity of an epoxy-bonded joint in several ways. These can include the following:

(i) It can affect the substrate, either chemically (observed with certain metallic alloys such as aluminium or titanium) or by physical modification. The latter has been shown to be applicable to various composite substrates such as carbon fibre-reinforced epoxies (CFRP).

(ii) It can exert an effect on the adhesive layer by chemical modification (e.g. hydrolysis) or by physical damage (e.g. microcracking), both of which can be regarded as irreversible. In addition a reversible plasticisation can occur, influencing in particular the glass transition temperature, T_g, and properties associated with it. As mentioned in section 7.1, the absorption of moisture into an adhesive layer can result in a not inconsiderable reduction in T_g which can itself result in a substantial reduction in load-bearing capacity at elevated temperatures.

(iii) Water can influence the interfacial region of the substrate and adhesive, resulting in an adhesive de-bond which in turn can cause a virtually complete loss of load-bearing capacity.

Although the effects outlined in (ii), namely bulk modification of the adhesive layer, can and indeed have been observed with epoxy-bonded joints, they are not usually responsible for incidents of environmentally-induced failure. Rather, it is the interfacial regions which are generally affected by atmospheric moisture, and upon which our attention must be focused. Two main mechanisms, which can be regarded as complementary, have been proposed to account for observations of environmentally-induced interfacial failure. The first, based upon an assessment of adhesive/metal oxide interface stability from a thermodynamic standpoint, has been shown to be applicable to steel-epoxy bonded joints. The second, concerned with elements of metal oxide stability, has been shown to be particularly pertinent to epoxy-bonded aluminium alloys.

7.7.3.1 Interface stability. The stability of an interface between, for example, a metal oxide and an epoxy adhesive, can be predicted on thermodynamic grounds. In a dry environment, the work needed to separate two phases of unit area in contact can be related to relevant surface free energy parameters by the Dupré equation.

$$W_A = \gamma_x + \gamma_y - \gamma_{xy} \tag{7.6}$$

where W_A is the thermodynamic work of adhesion, γ_x and γ_y are the surface free energies of the substrate and adhesive phases repectively and γ_{xy} the interfacial free energy. In the presence of a liquid environment (denoted by the suffix L), the work of adhesion is

$$W_{AL} = \gamma_{xL} + \gamma_{yL} - \gamma_{xy} \tag{7.7}$$

For adhesive metal oxide interfaces existing in an inert atmosphere, e.g. dry air W_A usually exhibits large positive values indicating thermodynamic stability of that interface. However, in water W_{AL} is generally negative indicating interface instability. Some typical values of both W_A and W_{AL} are shown in Table 7.2. For both the metal oxides indicated the change in work of adhesion from a positive to a negative value on going from a dry to a wet environment essentially predicts that water would act to displace the epoxy from the oxide surface thus promoting both a reduction in load-bearing capacity and an increase in the degree of apparent interfacial failure observed on failed surfaces. These locus of failure predictions have in fact

Table 7.2 Values of work of adhesion W_A (inert environment) and W_{AL} (in water) for various interfaces

Interface	W_A ($mJ.m^{-2}$)	W_{AL} ($mJ.m^{-2}$)
Epoxy/ferric oxide	291	−255
Epoxy/aluminium oxide	232	−137
Epoxy/CFRP	90	44

been confirmed, at least in the case of epoxy/mild steel joints, using modern surface specific analytical techniques such as Auger and X-ray photoelectron spectroscopy (Gettings *et al.*, 1977). The epoxy/CFRP interface is of particular interest in that both W_A and W_{AL} exhibit positive values indicating interface stability in both dry and wet environments. Indeed, as mentioned above, this stability is reflected in the locus of failure generally observed when CFRP joints are aged in hot/moist environments; the CFRP substrate and/or the bulk adhesive generally being influenced by water rather than the interface.

7.7.3.2 Metal oxide stability. Studies have shown that, with the surface pretreatments that have been and are currently employed with aluminium alloys, the surface oxide layer is generally regarded as the 'weak-link' in the presence of moisture, as opposed to the interface and adhesive. As described in section 7.3, the three usual methods for pretreating aluminium are chromic-sulphuric acid etching, chromic acid anodising and phosphoric acid anodising. Work conducted by Noland (1975) using X-ray photoelectron spectroscopy, showed that the oxide produced on aluminium alloys by the chromic–sulphuric acid etch was unstable in the presence of moisture changing to a weaker hydrated form. Furthermore, Noland observed that epoxy/aluminium joints pretreated by this method exhibited an apparent interfacial failure when the joints were exposed to hot/humid conditions. However, XPS showed that failure had occurred in fact in the weak hydrated oxide layer. Although using the thermodynamic arguments previously described, interfacial instability would be expected between aluminium oxide and an epoxy adhesive, it has been suggested that the apparent discrepancy between this thermodynamic prediction and the observed locus of failure can be reconciled by the existence of a mechanically interlocking component (Schmidt and Bell, 1986). The three surface treatments mentioned, particularly the anodising techniques, have been shown to produce surface oxide layers exhibiting topographical characteristics intuitively consistent with the ability to promote mechanical interlocking.

7.7.4 *Approaches to improved durability*

Two main approaches can be considered as a means of alleviating the undesirable effects of atmospheric moisture with epoxy bonded metal joints. These are:

(i) Improve the moisture resistance of the most vulnerable part of the joint, i.e. the interfacial zone.
(ii) Employ epoxy adhesive formulations which exhibit hydrophobic characteristics which would restrict the transport of large concentrations of water to the vulnerable areas of the joint (Shaw *et al.*, 1988).

Both of these approaches will now be considered.

7.7.4.1 Improving the moisture resistance of the interface. As will be apparent from the previous discussions, the approach to interfacial stability will clearly depend upon the precise location of the failure path. Experience has shown that, with epoxy-bonded mild steel joints, with the substrates being pretreated using a simple degrease/grit blast procedure, failure in the presence of moisture is usually interfacial in nature (Gledhill and Kinloch, 1974) the water literally causing a de-bond between the epoxy adhesive and the mild steel substrate. Thus, in this case it is necessary to focus attention upon the epoxy–oxide interface. Many of the studies conducted within this context have been concerned with the use of coupling agents employed as pre-bond primers.

7.7.4.2 Coupling agents. The majority of coupling agent studies have been devoted to one class of material, namely organosilanes. These materials were first recognised as having potential for use in adhesively-bonded structures following successful exploitation in glass-reinforced plastic (GRP) structures, where their incorporation between the glass reinforcement and polyester matrix was shown to provide substantially enhanced composite strength retention in moisture laden environments (Plueddemann, 1982). Following this recognition a number of investigations have confirmed this potential (Kinloch *et al.*, 1975; Gledhill *et al.*, 1990).

Most commercially available organosilanes are based upon the following generalised structure

$$R\text{–}Si\text{–}(X)_3$$

Where X is a hydrolysable group (considered necessary for interaction with a substrate) and R is an organo-functional group capable of some form of interaction with the intended polymeric component (Plueddemann, 1982).

Experience in the use of organosilanes has suggested that they perform best when applied as very dilute solutions. Thus areas which have received specific attention have included the influence of solvent type, solution age, film drying conditions as well as, of course, the chemical nature of the coupling agent. As far as the last is concerned, by far the majority of studies and indeed eventual use in service, has been devoted to one specific organosilane, namely γ-glycidoxypropyltrimethoxysilane, the structure of which is shown below.

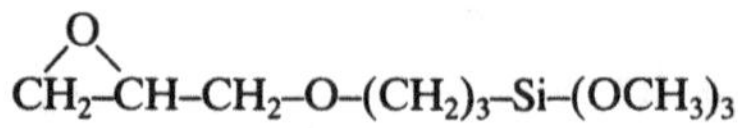

Before briefly describing some of these studies it is perhaps worth while to mention that this particular organosilane has been chosen for use with epoxies due primarily to the presence of the terminal glycidoxy group; the idea here being that reaction between this group and epoxide groups contained within the adhesive would result in a covalent bond linking the adhesive with the organosilane, which in turn could undergo reaction with the oxide surface via the methoxy groups. Such a 'covalent bridge' linking adhesive to substrate has been viewed by many adhesion scientists, the author included, as providing greater environmental stability to the interfacial region than one consisting purely of secondary force interactions. Whether such a mechanism exists in practice has been a subject of considerable debate for many years. We will return to this subject later in this section.

Returning to the experimental variables mentioned above, Figure 7.17

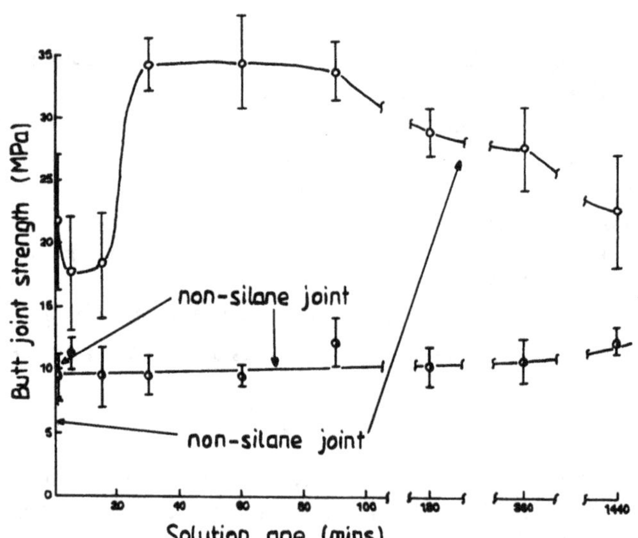

Figure 7.17 Effect of organosilane solution age and solvent type on the strength of epoxy bonded mild steel butt-joints following immersion in water at 60°C for 1500 hours (aqueous solution) and 1000 hours (ethanol-based solution).

shows the effect of two of the variables mentioned, namely solvent type and solution age on the durability characteristics of mild steel butt-joints bonded using a simple tertiary amine cured epoxy adhesive. Although rather complex, this figure demonstrates several interesting features.

First, for joints pretreated prior to bonding with a 1% v/v aqueous solution, the substantial improvement in moisture resistance over joints not treated with organosilane is clear; the treated joints exhibiting joint strength values of between 17.5 and 34.5 MPa following 1500 hours water immersion at 60°C, in comparison with 5.8 MPa for the untreated joints. Second, a maximum in joint strength is apparent over aqueous solution ages of between 30 and 90 minutes. Third, changing from an aqueous to an ethanol-based solvent system results in striking differences in behaviour. These can be itemised as follows:

(a) The ethanol based solutions exhibit no apparent dependence upon solution age.
(b) Joint strength values obtained from the ethanol system are substantially lower than their aqueous solution based counterparts.
(c) Joint strength values obtained from the ethanolic organosilane solutions are virtually identical to those obtained from the untreated joint.

Although such studies have provided clear evidence of the substantial improvements in moisture resistance which can be conveyed to epoxy-bonded joints by organosilanes it is clear that various experimental parameters are of vital importance so as to avoid misleading and pessimistically false impressions of their durability-enhancing characteristics. Our inability to predict precisely the appropriate optimum molecular and experimental parameters can be attributed, partly at least, to the fact that at present no universally acceptable mechanism has been proposed to describe their action. However, one particular theory, the chemical bonding theory, has generally received the most widespread support amongst adhesion specialists and thus a brief description here would not be out of place.

The *chemical bonding theory* is the oldest, best known and possibly the most easily understood of all the theories that have been proposed to account for coupling agent action (Plueddemann, 1982). The concept is strikingly simple, namely that in place of the predominantly secondary force interactions that would exist between a high energy substrate, such as a metal oxide, and an epoxy adhesive, a suitably employed coupling agent would act as a 'covalent bridge' linking together the organic and inorganic components of the joint. This covalent linking would be expected to provide far greater moisture resistance than mere secondary force interactions alone. The sequence of reactions which could be expected in the formation of an interfacial 'covalent bridge', starting from the initial organosilane, can be viewed in the following manner:

(a) Hydrolysis of the alkoxy groups on the organosilane molecule resulting in an organofunctional silanol.

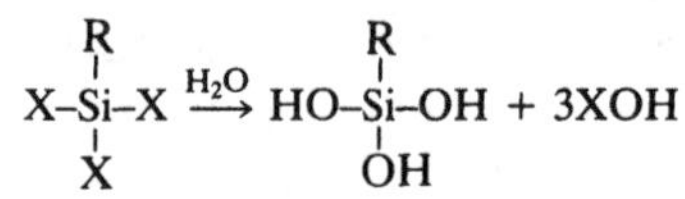

(b) Condensation of hydrolysed silane, i.e. reaction of organofunctional silanol hydroxyls with hydroxyl groups on the substrate surface (e.g. metal oxide) and interaction of silanol hydroxyls to form polymeric networks.

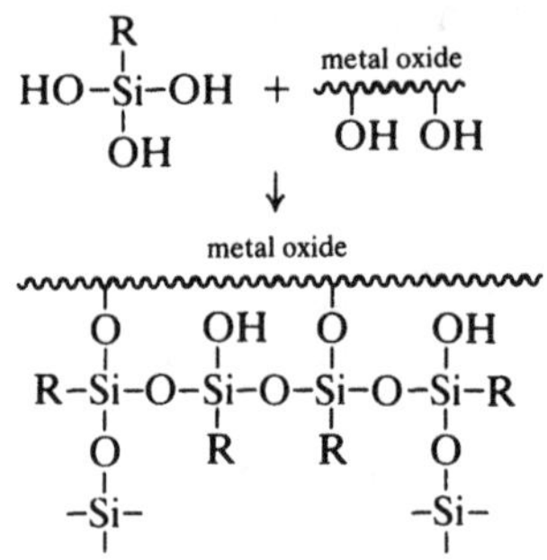

(c) Reaction/interaction of the organofunctional group, R, with the epoxy adhesive could then be envisaged as the next and final stage in the formation of the 'covalent bridge'.

A substantial amount of evidence has accumulated over the years in support of this theory. Gettings and Kinloch (1977) employed XPS and secondary ion mass spectroscopy (SIMS) to study the interactions between an organosilane film (derived from an aqueous-based solution of γ-glycidoxypropyl-trimethoxy silane) and a mild steel surface. Results obtained from SIMS analysis showed the existence of species such as $FeSiO^+$ radicals, which Gettings and Kinloch justifiably attributed to the presence of Fe–O–Si bonds within the interfacial zone; this clearly providing evidence for the existence of a chemical bond between the metal oxide and the organosilane layer. In addition, infrared spectroscopy has indicated evidence of silanol condensation between various organosilanes and high surface area particulate silica compounds (Plueddemann and Collins, 1975). Furthermore, Fourier transform infrared spectroscopy has made possible a direct study of silanol reactions with lower surface area glass fibres (Ishida and Koenig, 1978). Such studies have shown clear evidence of silanol condensation reactions between organosilanes and surface hydroxyls.

Covalent bonding between the organosilane and the polymer component has been more difficult to demonstrate, although considerable indirect evidence has indicated that such reactions can occur. For example in studies conducted as early as 1962, Plueddemann, whilst evaluating a wide range of coupling agents in glass-reinforced polyester and epoxy composites, found that the effectiveness of the organosilane paralleled the reactivity of its organofunctional group with the resin (Plueddemann *et al.*, 1962).

Although such evidence has indicated the existence of interfacial reactions and possibly 'covalent bridge' formation, a number of important observations have provided cause for doubt. Although organosilanes would appear capable of forming oxane bonds (M–O–Si) with various oxide surfaces where M=Si, Ti, Al, Fe, etc., it is not obvious that such bonds should contribute outstanding water resistance to the interface since oxane bonds between silicon and iron or aluminium, for example, are not resistant to hydrolysis. Yet, as demonstrated in this section, substantial evidence is now available indicating the benefits to be gained from organosilanes. In addition, the direct vulcanisation of rubber against organosilane treated glass has been shown to produce covalent bonds at the interface with, however, poor adhesion characteristics (Plueddemann and Collins, 1975). Also certain organofunctional silanes have shown good coupling abilities with inert polymers such as polyalkanes, where no chemical reaction could be demonstrated nor would be expected (Plueddemann, 1982). These discrepancies have resulted in other theories such as those based upon wetting and surface energetics, deformable and restrained interfacial layers. However, a variant of the chemical bond theory, the *reversable hydrolysable bond theory* has been proposed by Plueddemann (1982) which explains many of the discrepancies outlined above. With this theory, chemical bond formation is once again the main feature. However, the inherent vulnerability of oxane bonds to the effects of moisture is accounted for by the proposed ability of the hydrolysis reaction products to reform and reproduce oxane bonds, as shown in the following simple hydrolysis equilibrium reaction.

$$\text{M–O–Si} + H_2O \rightleftharpoons \text{M–OH} + \text{HO–Si–}$$

Clearly for an equilibrium condition such as this to be maintained, the adhesive contributing to the interfacial zone must have the ability to hold silanols at the interface. Equilibrium conditions would be lost if silanol groups resulting from hydrolysis were to be physically removed from the interface. Consequently the ability of an organosilane to promote increased moisture resistance would depend as much on the morphology it helps to control in the interfacial zone as its ability to develop primary covalent bonds across the adhesive–substrate interface. The development of a rubbery zone would result in its retraction from the interface as silanol bonds are released through hydrolysis. Water diffusing through the polymeric phase would then force the adhesive further from the substrate surface until, eventually, complete loss of adhesion occurred. Conversely, silanols formed at a rigid interface would not be capable of moving beyond the dimensions of molecular segments and would, therefore, be in a position to reform oxane bonds, with the original or adjacent active sites on the substrate surface.

Although the reversible hydrolysable bond mechanism still fails to explain all of the effects observed with organosilanes, it is the theory currently favoured by most adhesion scientists, the present author included.

Although much of the research described has been concerned with organosilanes employed as pre-bond primers, i.e. applied to substrate surfaces prior to epoxy adhesive application and bonding, it is fairly common commercial practice, at least with some manufacturers, to incorporate organosilanes into epoxy adhesives. Since, as expressed in this account, organosilanes are essentially 'interface active' materials, this formulatory approach would clearly rely on the ability of the organosilane molecules to diffuse to the interfacial zone and undergo the necessary interfacial reactions prior to cure of the adhesive. Although a number of researchers have harboured doubts as to the likelihood of such processes occurring in practice, it would seem unlikely that manufacturers would include coupling agents in adhesive formulations without evidence of a beneficial contribution to adhesive joint durability.

In addition to organosilanes, other types of coupling agent have been considered for use with epoxy adhesives. Most notable amongst these have been organotitanes and zirconates (Monte and Sugarman, 1980), although to date they have not achieved the prominence enjoyed by organosilanes.

7.7.4.3 Metal oxide stability. As previously mentioned (section 7.3.3.2), surface specific analytical techniques have shown that the relatively poor environmental (hot/wet) performance of aluminium joints pretreated prior to bonding using the chromic sulphuric acid etch, is due primarily to the poor hydration resistance of the surface oxide layer. Although this surface pretreatment has been shown to produce generally acceptable results when employed in conjunction with phenolic-based adhesives, the use of epoxy adhesives has resulted in relatively poor performance.

Two anodising surface treatments are now generally employed for pretreating aluminium alloys; namely phosphoric and chromic acid anodising (section 7.3). Both have been shown to provide superior hot/wet performance to the etching procedure. These differences in behaviour have been attributed to two main factors. Firstly, Noland (1975) using XPS and Ahearn *et al.* (1983) using ellipsometry, demonstrated a pronounced improvement in hydration resistance of the oxide produced by phosphoric acid anodising compared with the chromic-sulphuric acid etch. It has been proposed that this improved stability is associated with the presence of an aluminium phosphate layer on the anodised oxide (Venables, 1985). As mentioned in section 7.3.2, a mechanical interlocking contribution to the adhesion of epoxy aluminium joints has been proposed. Experiments described by Poole and Watts (1985) have, in a sense, confirmed this view as indicated in Table 7.3. As shown, a grit blast prior to chromic-sulphuric acid etching and chromic acid anodising results in a substantial reduction in apparent interfacial crack propagation.

Table 7.3 Wedge test crack propagation data obtained from 120°C curing epoxy-bonded aluminium alloy joints

Surface treatment	Crack growth (mm) during exposure to 50°C/96% RH			
	1 h	5 h	24 h	96 h
CAE/CAA	3.9	13.1	29.3	35.5
GB/CAE/CAA	2.3	3.7	6.2	6.9

CAE, chromic acid etch; CAA, chromic acid anodise; GB, grit blast.

The use of primers in addition to the aluminium anodising surface treatments is now regarded as providing optimum environmental resistance. Corrosion inhibiting primers comprising strontium chromate-epoxy combinations are possibly the most common.

However, there has been a desire in recent years to replace chromium-containing primers with more environmentally acceptable formulations. Recent work by Pike and co-workers (1988) has focused on inorganic primer systems based on amorphous inorganic aluminium oxides such as aluminium alkoxide. Promising results have been obtained allowing potential for eliminating the use of chromium-containing organic primers for corrosion protection.

The correct choice of surface pretreatment has also been shown to be of vital importance for achieving moisture resistance with epoxy-bonded titanium alloys. Figure 7.18 shows crack growth data obtained from epoxy-bonded wedge test specimens where the titanium substrates were subjected to the surface treatments outlined in section 7.3; i.e. alkaline/hydrogen peroxide etch, chromic acid/fluoride anodise and sodium hydroxide anodise. A simple grit blast treatment is also shown for comparison. The superior performance of all three chemical treatments in comparison to the grit blast is clearly apparent, with the sodium hydroxide anodise providing marginally superior behaviour to the chromic acid/fluoride treatment.

7.7.4.4 Reduction in moisture absorption. Since the previous discussions have focused attention on the debilitating effects of water on adhesive joint behaviour, particularly at the interfacial zone between substrate and adhesive, it seems intuitively reasonable that a means of preventing water reaching these critical regions in large concentrations would enhance moisture resistance. Indeed studies conducted in particular by Comyn and co-workers (1981) have demonstrated a linear relationship between joint strength and the total quantity of water absorbed by the adhesive.

One approach adopted by the current author and colleagues (Shaw *et al.*, 1988) has been to consider the concept of fluorination as a means of

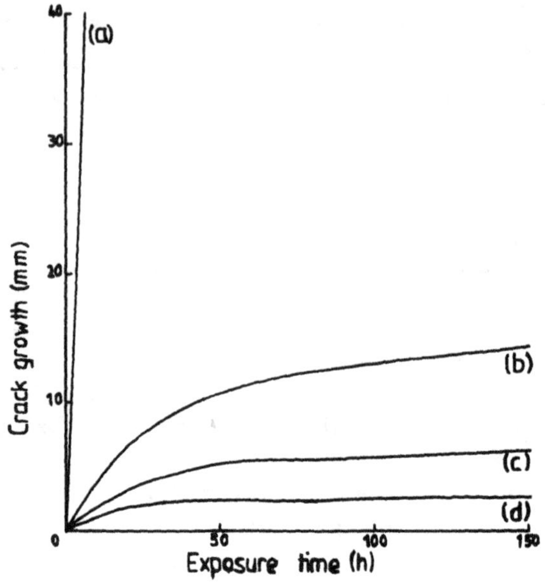

Figure 7.18 Influence of hot/humid atmosphere (50°C/96% Rh) on wedge test crack growth characteristics of epoxy bonded titanium joints. (a) Grit blast; (b) alkaline/hydrogen peroxide etch; (c) chromic acid/fluoride anodise; (d) sodium hydroxide anodise.

enhancing the hydrophobic characteristics of epoxy resins. In these studies the following type of epoxy structure was employed,

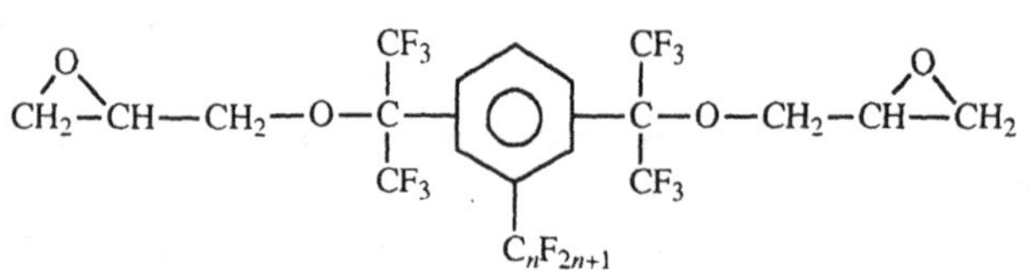

which when cured with suitable curing agents, resulted in equilibrium water concentrations as low as 0.2%. With regard to non-halogenated epoxies, the amount of water the cured adhesive will absorb will depend upon many factors including the type of resin and curing agent employed. In an excellent review, Wright (1981) has shown that water uptake may differ by a factor of ten between resin types and by a factor of three with a similar resin but different curative. In fact the amount of water absorbed by a cured epoxy adhesive can vary enormously from about 1% to 10% with 2–6% being fairly typical for a wide range of formulations. Fluoroepoxies of the type demonstrated above can therefore be considered as substantial improvements over the majority of 'conventional' epoxy systems with regard to hydrophobic character.

Further analysis of these systems, so as to determine whether the hydrophobic characteristics exhibited by bulk polymer are translated into highly moisture-resistant adhesive joints, has not been conducted. It would be of major interest to determine the existence or otherwise of such a correlation.

7.7.5 *Other hostile environments*

In addition to atmospheric moisture, other potentially harsh environmental conditions can include stress, low and high temperatures, radiation, and various liquid contaminants, e.g. fuels, de-icing fluids, hydraulic oils (i.e. those encountered in various aircraft). Although space does not permit discussions about each of these interesting areas, it is perhaps pertinent to discuss, very briefly, just one, namely epoxies as high temperature adhesives.

The majority of epoxy adhesives are formulated to yield high temperature limits only marginally greater than 100°C. This is indicated in Figure 7.19 which shows the retention of room temperature strength as a function of test temperature for a proprietary epoxy adhesive system (DICY-cured, rubber-modified formulation). As indicated, this adhesives load-bearing capacity is

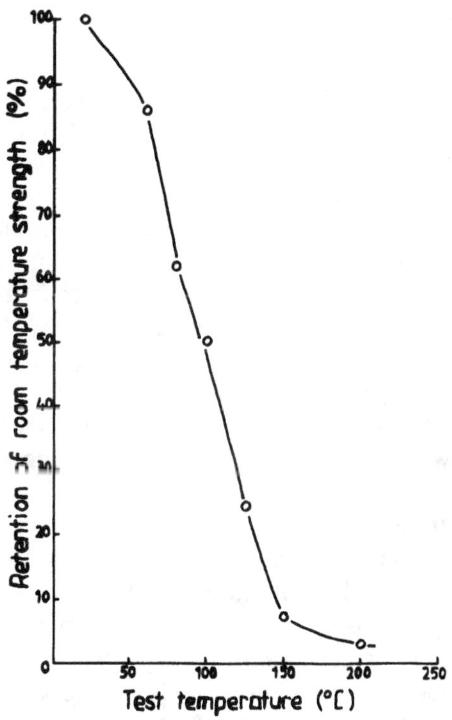

Figure 7.19 High temperature strength retention of a dicyandiamine cured, rubber-modified epoxy adhesive.

greatly diminished with increasing temperature, with less than 10% room temperature strength being retained at 140°C. This type of strength decline can usually be related to the onset of T_g. Indeed T_g can, to a first approximation, be considered as the high temperature limit for any structural adhesive.

Improvements in high temperature capability can be obtained by increasing the degree of cross-linking by a suitable choice of epoxy and/or curing agent. The use of polyfunctional epoxies and curing agents can produce cured products having T_g values in excess of 200°C, resulting in substantial retention of properties at temperatures in excess of 150°C. However, the propensity of epoxies to absorb, in some cases, considerable quantities of water, which in turn can reduce T_g, together with the somewhat brittle characteristics bestowed by the high cross-link density, has greatly diminished this approach to high temperature capability. In addition, due to their molecular structure, even the most highly cross-linked formulated epoxy adhesives are unable to withstand long-term use at temperatures in excess of approximately 175°C. For such applications other adhesive types must be considered, including for example, the epoxy-phenolic-based adhesives discussed in chapter 6 (section 6.2).

7.8 Applications

From humble beginnings in the 1940s and 1950s, epoxy adhesives are nowadays employed in a diverse range of applications associated with a number of important industries including aerospace, automobile, civil engineering and electrical/electronics including of course the domestic home market. Possibly the main impetus for their current popularity has been the aircraft industry which has spearheaded both the development and application of structural adhesives generally. Indeed, structural adhesives were first used in aircraft construction in the early 1940s with a vinyl-phenolic adhesive system being used to bond aluminium components on the de Havilland Hornet aircraft. Since this time various adhesive types have been used for a wide range of applications in both civil and military aircraft (Alberici, 1983), with epoxy adhesives now enjoying prominence at the expense of the older phenolic types.

Aircraft applications to date have included the bonding of reinforcing doublers and stiffeners to both fuselage and wing panels; the former being employed to reinforce holes for mechanical fasteners and window/door openings. In addition, epoxy film adhesives are frequently employed for the production of honeycomb sandwich structures where aluminium or Nomex honeycomb is bonded to aluminium or composite skins. Such structures have been employed to produce aircraft structures such as rudders, flaps, elevators, ailerons, doors, floor panels as well as engine structures such as nacelles and thrust reversers since they provide attractive combinations of strength, stiffness and weight in addition to offering excellent fatigue and sound insulating characteristics.

In addition to manufacture, epoxy adhesives, particularly cold cure systems, are also employed in the repair of aircraft components which have suffered damage due to accidents (ground impacts for example), lightning strike, hailstones and bird impact.

Epoxy adhesives have also been used in the construction of helicopter blades where their ability to confer enhanced fatigue resistance over the more traditional mechanical attachments has been amply demonstrated.

Epoxies have also been employed in slightly more esoteric aerospace applications such as spacecraft, where again they have been utilised for the bonding of stiffeners and doublers to skins and also for honeycomb structures. Additionally they have also been utilised in the manufacture of rocket motor components for guided weapon applications.

In contrast to acrospace, the car industry has approached the use of structural adhesives with a degree of caution. Although adhesive bonding is employed in a number of manufacturing processes, the all bonded car, where traditional joining techniques such as spot welding are totally replaced by adhesive bonding, is not yet a reality. Where it has been employed, the one-part heat-cured toughened epoxy pastes have predominated with applications including use either in cosmetically critical areas such as doors, boot lid or bonnet or in a combination with spot welds so as to reduce their number or to help provide an increase in stiffness of the vehicle.

Epoxy adhesives have also been employed in a range of civil engineering applications where their use over the past thirty years has increased considerably. Notable examples include the repair of cracks in concrete structures such as dams and bridges where the adhesive is poured, brushed or pumped under pressure into cracks. Epoxies have also been employed as gap filling materials for incorporation between precast segmental concrete units, as well as the bonding of external steel reinforcement to various concrete structures.

References

Anderson, G.P., Bennett, S.J. and De Vries, K.L. (1977) *Analysis and Testing of Adhesive Bonds*, Academic Press, New York.

Adams, R.D. and Wake, W.C. (1986) *Structural Adhesive Joints in Engineering*, Applied Science, London.

Ahearn, J.S., Davies, G.D., Sun, T.S. and Venables, J.D. (1983) *Adhesion Aspects of Polymeric Coatings*, ed. Mittal, K.L. Plenum, New York.

Albericci, P. (1983) *Durability of Structural Adhesives*, ed. Kinloch, A.J. Applied Science, London.

Allen, K.W. and Alsalim, H.S. (1977) *J. Adhesion* **8**, 183.

Allen, K.W. (1979) *Brit. Polym. J.* **11**, 50.
Arnold, D.B. (1981) *Developments in Adhesives – 2*, ed. Kinloch, A.J. Applied Science, London.
Bascom, W.D., Cottington, R.L., Jones, R.L. and Peyser, P. (1975) *J. Appl. Polym. Sci.*, **19**, 2545.
Bascom, W.D., Cottington, R.L. and Timmons, C.O. (1977) *Appl. Polym. Symp.* **32**, 165.
Bascom, W.D. and Orshnik, J. (1978) *J. Mats. Sci.*, **13**, 1411.
Bolger, J.C. (1973) *Adhesion and Adhesives* Vol. 3, ed. Patrick, R.L. Marcel Dekker, New York.
Buck, B.I. and Hockney, M.G.D. (1973) *Aspects of Adhesion – 7*, eds. Allen, K.W. and Alner, D.J. Transcription Books, London.
Butt, R.I. and Cotter, J.L. (1976) *J. Adhesion* **8**, 11.
Comyn, J. (1981) *Developments in Adhesives – 2*, ed. Kinloch, A.J. Applied Science, London.
Cotter, J.L. (1977) *Developments in Adhesives – 1*, ed. Wake, W.C. Applied Science, London.
Cotter, J.L. and Mahoon, A. (1982) *Int. J. Adhesion and Adhesives* **2**, 47.
Deryaguin, B.V. (1955) *Research* **8**, 70.
Deryaguin, B.V., Drotova, N.A., Karasser, V.V., Kirillova, Y.M. and Aleinikova, I.N. (1957) *Proc. 2nd Int. Congress on Surface Activity – III*. Butterworths, London.
Deryaguin, B.V. and Smilga, V.P. (1969) *Adhesion, Fundamentals and Practice*, McLaren, London.
Ellis, B. and Rashid, H.U. (1984) *J. Appl. Polym. Sci.* **29**, 2021–2038.
Gettings, M., Baker, F.S. and Kinloch, A.J. (1977) *J. Appl. Polym. Sci.*, **21**, 2371.
Gettings, M. and Kinloch, A.J. (1977) *J. Mats. Sci.*, **12**, 2511.
Gledhill, R.A. and Kinloch, A.J. (1974) *J. Adhesion* **6**, 315.
Gledhill, R.A., Kinloch, A.J. and Shaw, S.J. (1977) *J. Adhesion* **9**, 81.
Gledhill, R.A., Kinloch, A.J. and Shaw, S.J. (1979) *J. Mats. Sci.* **14**, 1769.
Gledhill, R.A., Kinloch, A.J. and Shaw, S.J. (1980) *J. Adhesion* **11**, 3.
Gledhill, R.A., Shaw, S.J. and Tod, D.A. (1990) *Int. J. Adhesion and Adhesives* **10**, 192.
Griffith, A.A. (1920) *Phil. Trans. Roy. Soc.* **A221**, 163.
Hockney, M.G.D. (1970) *RAE Tech. Rep.* 70081; (1972) *RAE Tech. Rep.* 72100; (1973) *RAE Tech. Rep.* 73016.
Huntsberger, J.R. (1967) *Treatise on Adhesion and Adhesives*, Vol. 1 ed. Patrick, R.L. Marcell Dekker, New York.
Irwin, G.R. (1964) *Appl. Mats. Res.* **2**, 65.
Ishida, H. and Koenig, J.L. (1978) *J. Coll. Int. Sci.* **64**, 565.
Kaelble, D.H. (1971) *Physical Chemistry of Adhesion*, Wiley, New York.
Kinloch, A.J., Gledhill, R.A. and Dukes, W.A. (1975) *Adhesion Science and Technology*, ed. Lee, L.H. Plenum, New York.
Kinloch, A.J. (1980) *J. Mats. Sci.* **15**, 2141.
Kinloch, A.J. and Shaw, S.J. (1981a) *Developments in Adhesives – 2*, ed. Kinloch, A.J. Applied Science, London.
Kinloch, A.J. and Shaw, S.J. (1981b) *J. Adhesion* **12**, 59.
Kinloch, A.J. and Young, R.J. (1983) *Fracture Behaviour of Polymers*. Applied Science, London.
Kinloch, A.J., Shaw, S.J. and Hunston, D.L. (1983) *Polymer* **24**, 1341.
Lee, H. and Neville, K. (1967) *Handbook of Epoxy Resins*, McGraw-Hill, New York.
Mai, Y.W. and Vipond, A.S. (1978) *J. Mats. Sci.* **13**, 2280.
Marceau, J.A. and Scardino, W. (1975). *Durability of Adhesive Bonded Joints*, ADML TR-75-3, Dayton, Ohio.
Marceau, J.A., Moji, Y. and McMillan, J.C. (1977) *Adhesives Age* **20**, 28.
May, C.A. (1988) *Epoxy Resins, Chemistry and Technology*, Marcel Dekker, New York.
McGarry, F.J. and Willner, A.M. (1968) *Toughening of an epoxy resin by an elastomeric second phase. Res. Rept. R68-8*, Massachusetts Institute of Technology.
McMillan, J.C. (1979) *Bonded Joints and Preparation for Bonding*, NATO/AGARD Lecture Series 102.
McMillan, J.C. (1981) *Development in Adhesives - 2*, ed. Kinloch, A.J. Applied Science, London.
Moji, Y. and Marceau, J.A. (1976) US Patent 3,959,091.

Monte, S.J. and Sugarman, G. (1980) *Proc. 35th Ann. Tech. Conf. SPI,* Paper 23-F.
Morgan, R.J. (1980) *Developments in Reinforced Plastics – 1*, Applied Science, London.
Mostovoy, S. and Ripling, E.J. (1966) *J. Appl. Polym. Sci.* **10**, 1351.
Noland, J.S. (1975) *Adhesion Science and Technology*, ed. Lee, L.H. Plenum, New York.
Parker, B.M. (1983) *Composites*, **14**, 226.
Partridge, I.K. and Bucknall, C.B. (1983) *Polymer,* **24**, 639.
Pike, R.A. and Lamm, F.P. (1988) *Adhesives, Sealants and Coatings for Space and Harsh Environments*, ed. Lee, L.H. Plenum, New York.
Plueddemann, E.P. (1982) *Silane Coupling Agents*. Plenum, New York.
Plueddemann, E.P., Clark, H.A., Nelson, L.E. and Hoffmann, K.R. (1962) *Mod. Plast.* **39**, 136.
Plueddemann, E.P. and Collins, W.T. (1975) *Adhesion Science and Technology*, ed. Lee, L.H. Plenum, New York.
Poole, P. and Watts, J.F. (1985) *Int. J. Adhesion and Adhesives* **5**, 33.
Poole, P. (1985) *Industrial Adhesion Problems*, ed. Brewis, D.M. Orbital Press, Oxford.
Potter, W.G. (1970) *Epoxide Resins*, Iliffe, London.
Sage, G.N. and Tin, W.P. (1982) *Composites,* **13**, 228.
Schliekelmann, R.J. (1979) *Bonded Joints and Preparation for Bonding*, NATO/AGARD Lecture Series 102.
Schmidt, R.G. and Bell, J.P. (1986) *Advances in Polymer Science* 75, ed. Dusek, K. Springer-Verlag, Berlin, pp. 33–71.
Sharpe, L.H. and Schonhorn, H. (1964) *Advances in Chemistry Series 43,* Gould, R.F. ed. American Chemical Series, Washington.
Shaw, S.J. (1984) *PhD Thesis*, City University, London.
Shaw, S.J., Tod, D.A. and Griffith, J.R. (1988) *Adhesives, Sealants and Coatings for Space and Harsh Environments*, ed. Lee, L.H. Plenum, New York.
Shaw, S.J. and Tod, D.A. (1989) *J. Adhesion* **28**, 231.
Sherriff, M. (1976) *J. Adhesion,* **7**, 257.
Smith, T. and Haak, R. (1979) *Treatment of Am 355 Steel for Adhesive Bonding*. SC5169.16FR.
Ting, R.Y. (1988) *Epoxy Resins, Chemistry and Technology*, Ch. 5 ed. May, C.A. Marcel Dekker, New York, pp. 551–601.
Titow, W.V. (1978) *Adhesion – 2*, ed. Allen, K.W. Applied Science, London.
Trantina, G.G. (1972) *J. Composite Mat.* **6**, 192.
Venables, J.D. (1985) *J. Mats. Sci.* **19**, 2431.
Voyutskii, S.S. (1963) *Autohesion and Adhesion of High Polymers*, Wiley, New York.
Wake, W.C. (1982) *Adhesion and the Formulation of Adhesives*, 2nd Edn, Applied Science, London.
Wang, S.S., Mandel, J.F. and McGarry, F.J. (1978) *Int. J. Fract.* **14**, 39.
Wenzel, R.N. (1963) *Ind. Eng. Chem.* **28**, 988.
Wright, W.W. (1981) *Composites* **12**, 201.
Young, T. (1805) *Trans. Roy. Soc.* **95**, 65.
Young, R.J. (1980) *Developments in Reinforced Polymers – 1*, ed. Pritchard, G. Applied Science, London.

8 Composite materials

F.R. JONES

8.1 Introduction

Composite materials technology exploits the high specific moduli and strength of reinforcing fibres to produce low density high performance structures. For engineering applications, the anisotropy of unidirectionally reinforced materials is of crucial importance. It is therefore necessary to develop techniques to align the fibres in the direction of the principal stresses. For many applications of epoxy resins, this involves stacking preimpregnated fibrous sheets (generally known as prepreg) to form laminates. The fibres can have any number of geometrical arrangements but a quasi-isotropic material might be obtained with the laminae arranged with the fibres in the different plies at these orientations: $0_2°/\pm 45°/90_4°/\pm 45°/0_2°$. The subscripts refer to the number of individual plies. The same laminate could be referred to as $[0_2/\pm 45/90_2]_s$ where s denotes a symmetrical lay-up.

The other way of providing quasi-isotropy is to use woven reinforcements or to combine differing fibre placement techniques. The most important fibres are those based on E-glass, carbon and aramid polymer and are available in various forms and with a wide range of properties. Over recent times, the use of textile technology to weave or knit preforms for resin impregnation has been developed. Short or discontinuous fibres are not generally employed in epoxy resins because of their poorer reinforcing efficiency compared with their continuous counterparts.

8.2 Fibre reinforcements

By comparing the mechanical properties of potential reinforcing fibres with matrix and bulk materials (Tables 8.1 and 8.2), the benefits of fibre composite technology become clear. Apart from steel, the fibres all have significantly enhanced strength and stiffness. This point is made even more pertinent by comparing their specific properties, in which strength and modulus are normalised to their respective densities. In Figure 8.1, the specific values of modulus and strength for each are plotted. Efficient structures can be fabricated from materials with the highest values and in Figure 8.1, it is seen that 'carbon fibre' and the intermediate modulus fibres

Table 8.1 Typical mechanical properties of reinforcing fibres

	Relative density	Young's modulus (GPa)	Tensile strength (GPa)	Failure strain (%)	Fibre diameter (μm)
E glass	2.55	72	1.5–3.0	1.8–3.2	10–16
S glass	2.5	87	3.5	4.0	12
S-2 glass	2.49	86	4.0	5.4	10
Carbon-mesophase pitch	2.02	380	2.0–2.4	0.5	10
Carbon-PAN					
Type A high strength	1.8	220–240	3.0–3.3	1.3–1.4	7
Type A high performance	1.8	220–240	3.3–3.6	1.4–1.5	7
Type A high strain	1.8	220–240	3.7	1.5–1.7	7
Type II high strength	1.8	250	2.7	1.0	7
Type I high modulus	2.0	330–350	2.3–2.6	0.7	7
Intermediate modulus	1.9	280–300	2.9–3.2	1.0	7
SiCO-continuous	2.5	200	3.0	1.5	10–15
SiC-whisker	3.2	480	7.0	–	1–50
SiC Ti C O continuous	2.35	200	2.8	1.4	8–10
Boron	2.6	410	3.4	0.8	100
α-alumina (FP)	3.9	380	1.7	0.4	20
β-alumina (Saffil)	3.3	300	2.0	0.5	3
δ/ϑ-alumina (Safimax)	3.3	250	2.0	–	3.1
η alumina (Safimax LD)	2.0	200	2.0	–	3.2
$Fe_{80}B_{20}$ metallic glass	–	100	3.6	–	–
H.C. steel	7.8	210	2.8	–	250
Polyphenylene terephthalamide					
Aramid – high modulus	1.47	180	3.45	1.9	12
Aramid – intermediate modulus	1.46	128	2.65	2.4	12
Aramid – low modules	1.44	60	2.65	4.0	12
Aramid – staple fibre (polymetaphenylene-isophthalamide)	1.4	17.3	0.7	22.0	12
Polybenzothiazole (PBT)	1.5	250	2.4	1.5	20
Polyamide 66	1.44	5.0	0.9	13.5	≃10
Polyethene (theory)	–	200	–	–	–
Solution spun	1.0	100–150	–	–	–
Drawn	≃1	6–80	–	–	–

Table 8.2 Typical mechanical properties of isotropic bulk materials

	Relative density	Young's modulus (GPa)	Tensile strength (GPa)	Failure strain (%)
Steel	7.8	210	0.34–2.1	–
Aluminium alloys	2.7	70	0.14–0.62	–
Glass	2.6	60	–	–
Resins:				
Phenolic	1.4	7	–	≃0.5
Epoxy	1.2	2–3.5	0.05–0.09	1.5–6
Polyester	1.4	2–3.0	0.04–0.085	1.25
Thermoplastics:				
Nylon 66	1.4	2.0	0.07	60
High density polyethene	0.96	1.3	–	–
Low density polyethene	0.91	0.25	–	–

in particular, together with the high performance polymeric fibres offer the best solution.

Where light weight is more important than the cost of materials, such as in aerospace and sports goods applications, these reinforcements dominate. However, where economic considerations are more important, in industrial applications, glass fibres still have much potential. For the polymeric fibres,

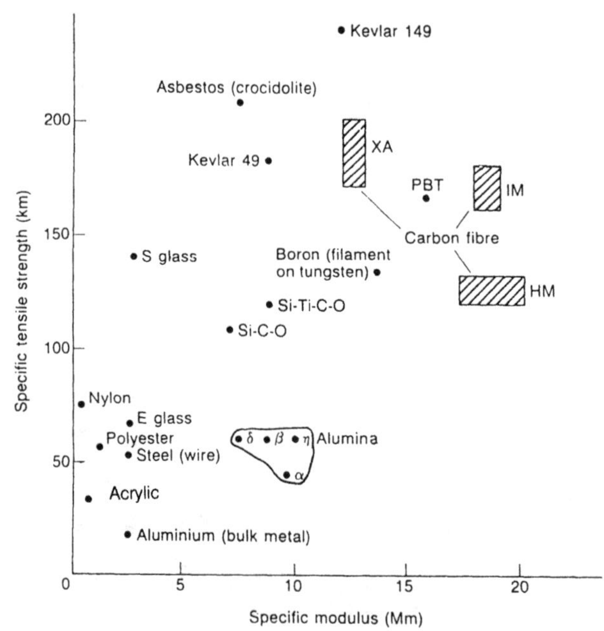

Figure 8.1 Specific properties of fibre reinforcements. XA, 1M, HM, refer to differing grades of carbon fibre. For other definitions see Table 8.1.

PBT is still only available in sample quantities for strategic applications and so aramid fibres such as Kevlar, Nomex and Twaron are the main contenders. However, differing fibre and composite performance are more important criteria for the selection of the grade of aramid or carbon reinforcement.

It is important to recognise that the strength of a material is determined by the presence of Griffith flaws which reduces the theoretical strength significantly. The fibre form reduces the preponderance of flaws so that filament strength is larger than that of the bulk material. However, the strength of a filament is still determined by the presence of defects and as a consequence the reported values are gauge length-dependent and are usually average values determined from either unimpregnated or resin-impregnated bundles. Individual filaments within the bundle will have a different strength so that the statistics of fibre strength contributes significantly to damage mechanisms of the composite. A typical distribution of filament strengths is given in Figure 8.2. A further consequence of the distribution of flaws on the fibre surfaces is the need for protection by a resin coating or so-called size during handling immediately after manufacture. For this application emulsified epoxy resins are commonly employed for carbon and glass fibres.

8.2.1 *Manufacture of carbon fibres from polyacrylonitrile (PAN) precursors*

The process consists of five major stages: (1) precursor spinning; (2) oxidation; (3) carbonisation; (4) graphitisation; and (5) surface treatment.

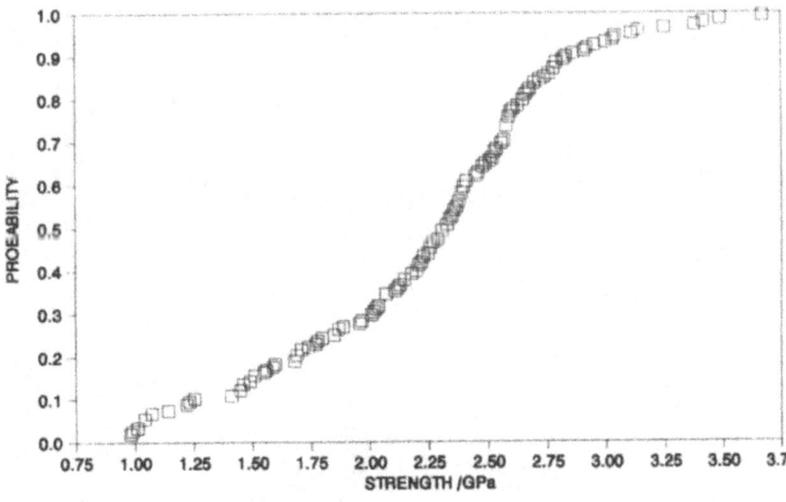

Figure 8.2 The probability of failure of single E-glass filaments under tensile stress showing the statistical aspects of strength.

The *special acrylic precursor* fibre is obtained by wet-spinning a carefully filtered polyacrylonitrile copolymer solution. The degree of orientation of the macromolecules during steam stretching has a profound effect on the properties of the final carbon fibre.

Oxidation of the precursor fibre is carried out under restraint in oxygen atmosphere at 200 to 220°C to maintain the molecular orientation induced during stretching. Polyacrylonitrile is the favoured precursor because of the self-induced thermal polymerisation of the pendent nitrile groups, to produce a ladder polymer. Because of the limited number of adjacent isotactic nitrile groups, the cyclised sequences involve approximately 5–15 groups. In addition, in areas where the molecules are not so well orientated, cyclisation reactions between syndiotactic nitrile groups will cause the ladder polymer to take on a curved rather than planar ribbon-like form. The cyclised rings of the thermally-modified PAN are also puckered. It is now known that these reactions precede the oxidation reaction and that oxygen aids the aromatisation of the cyclised polymer into a planar aromatic structure which facilitates condensation of the graphitic nuclei in the carbonisation and graphitisation stages, which are carried out in an inert atmosphere over the temperature range of 400 to 2500°C. In the original work at the Royal Aircraft Establishment (now DRA-RAE), Farnborough, UK, the tensile strength of the fibre reached a maximum at 1500°C, the so-called type-II or high strength (HS) fibre (Watt, 1985). Graphitisation at a temperature near 2500°C produced the type-I or high-modulus (HM) fibre. Improvements in precursor and spinning technology have led to improvements in fibre strength. This is illustrated in Figure 8.3 where clean

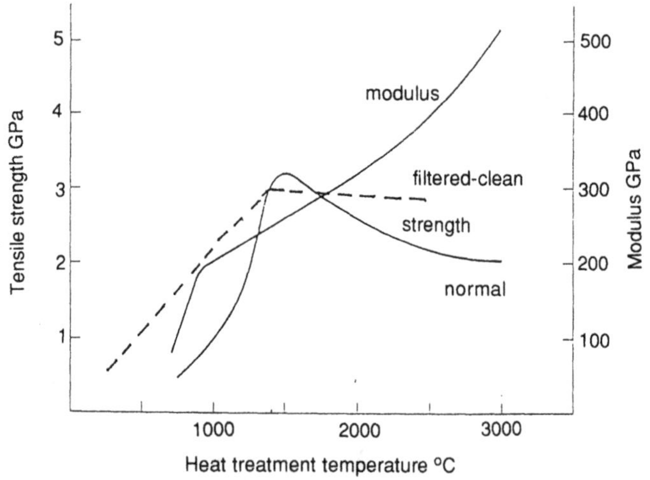

Figure 8.3 The effect of heat treatment temperature and clean room conditions on the strength and modulus of PAN-based carbon fibres. (Redrawn from data in Watt, 1985).

room conditions have reduced the number of strength-reducing flaws. A wide range of commercial fibres are now available of which the type-A fibre, which has been heated to only 1000°C is often preferred. More recently, intermediate modulus (IM) fibres have become available which combine high strength with an improved modulus. A modulus approaching 800 GPa (i.e. 80% theoretical) can be produced by hot-stretching techniques.

The microstructure of the carbon fibres is determined by the relatively small size of the graphite nuclei and the curved ribbon-like molecular structure of the oxidized acrylic fibre which yields a turbostratic graphite. A skin-core structure is most pronounced for the high modulus fibres, as a result of graphitisation at the high temperatures involved and the subtle chemical differences in oxidised PAN fibre which arise from the diffusion-controlled oxidation of the thermally cyclised polymer.

Surface oxidation is an essential post-carbonisation stage. This process can be carried out chemically, thermally, or electrolytically, with the last being the favoured method. The treatment is essential to the optimisation of the adhesion of the fibre to the resin matrix and the performance of the composite. If the surface treatment is too severe the composite material will behave like a brittle monolith. Figure 8.4 demonstrates the need for a carefully controlled surface oxidation to optimise the performance of the

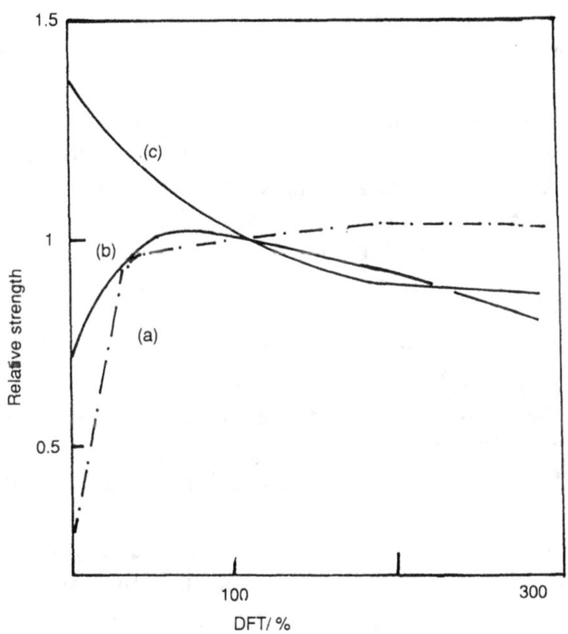

Figure 8.4 A schematic representation of the effect of electrochemical oxidative surface treatment on the properties of carbon fibre epoxy composites. (a) Interlaminar shear strength of a 0° laminate; (b) impact strength; (c) notched tensile strength of a $(0°/\pm 45°/0°)_s$ laminate. (Redrawn from data given by Dunsford *et al.*, 1981.)

composite. The interlamina shear strengths (normalised to the normal industrial treatment and denoted 100%) at different degrees of treatment are compared to the notched tensile strengths (also normalised to the value for 100% treatment) for varying surface treatments. The impact strength of a multidirectional composite reaches a maximum at commercial treatment levels.

The nature of the adhesive bond between the fibres and matrix is still the subject of controversy. This is largely because of the complicated surface microstructure and microchemistry which originates from the treatment and the techniques employed to assess the chemical differences and the true value of the interfacial shear strength of the matrix. Most of the controversy arises from attempts to deactivate an oxidised surface. Apparent removal of oxygen moieties has not always led to a reduction in the fibre–matrix interfacial shear strength. As a consequence some workers (Drzal *et al.*, 1982) have maintained that mechanical interlocking is more important. Significant increases in surface area are not achieved but the surface microporosity is increased (Denison *et al.*, 1988). Recent work has confirmed that the surface oxygen concentration of oxidised fibres increases with time after oxidation, probably by continued water adsorption. Furthermore, detreated fibres (Denison *et al.*, 1989) were shown to exhibit the same acidity as present before thermal desorption of oxygen species. This was interpreted as the presence of 'active' sites located in the micropores of the high modulus fibres, which could be at 'unsaturated' carbon atoms or other functional groups. Either the latter reform on a desorbed fibre surface or the former can react with the matrix resins, during composite fabrication. The micropores arise from erosion at intercrystallite boundaries on high modulus fibre surfaces to provide a location for these active sites. For Type-A fibres the edges of the basal planes probably run directly to the fibre surface to provide reactive sites (Oberlin, 1985).

On balance, the weight of evidence appears to favour the formation of chemical bonds to the resin matrix at the fibre surface. Most telling is the number of observations which have demonstrated strong adsorption of epoxy molecules. The functional groups most likely involved are amines, carboxylic and hydroxyl groups. For the surface-treated fibres the last two appear to be the most abundant species which account for ≃ 50% of the surface oxygen, with reactability to epoxy resins. At least 50% of the remaining surface oxygen appears to come from adsorbed water.

8.2.1.2 Sizing resins for carbon fibres. To protect the fibres from damage during handling and subsequent fabrication the fibres are dressed with an epoxy resin size. This is usually applied from an emulsion, immediately after the surface treatment. The role of these resins in the formation of the interfacial bond has not been studied in detail but it is now recognised that these may not have the same chemical structure as the resin matrix and the presence of an interphase region needs to be considered.

8.2.1.3 Carbon fibres from pitch. High-strength carbon-fibre is also manufactured commercially from petroleum pitch (Bacon, 1980; Rand, 1985). This involves the polymerisation of the condensed aromatic molecules to form a mesophase or liquid crystal solution at temperatures above 350°C for extrusion melt spinning into fibres for thermosetting in air and carbonisation at temperatures around 2000°C. The main difference between the two types of fibre lies in differing microstructures. The pitch fibres tend to have a radiating crystal structure. The differences arise because the crystal nuclei are formed prior to spinning and orientation of these nuclei occurs during spinning. This results in fibres with a lower transverse strength than for PAN-based fibres.

Commercial limitations have arisen from a variability in the pitch precursor and the high cost of producing the mesophase.

8.2.1.4 Polymerisation induced at carbon fibre surfaces. Walterson (1985) has observed complications during attempts to react epichlorohydrin (ECH) with C-fibre surfaces, to probe the reactivity to epoxy groups. Alexander *et al.* (1991) chose to adsorb ECH from the bulk liquid at 120°C to more closely emulate the reactions which could occur at the fibre surface in neat epoxy resin. The chlorine atom provides a highly sensitive element for surface analysis by X-ray photoelectron spectroscopy (XPS) and for imaging by time of flight secondary ion mass spectrometry (ToF SIMS). They showed that the ECH was adsorbed non-stoichiometrically but uniformly across the fibre surface. The ToF SIMS spectrum produced a series of secondary ions up to a mass of 1400 D. The fragmentation pattern was indicative of a polymer based on $(CH_2CH_2OCH_2)_n$. This demonstrated that polymerisation of ECH had been induced. It was also possible to show that hydroxyl groups were formed. The latter can result from the reaction of active hydrogen species (e.g. COOH) at the fibre surface. Because chlorine levels were lower than anticipated, the loss of hydrogen chloride was postulated with the induction of a polymerisation, which results in macromolecules growing from the surface. This would appear to be in contrast to the work of Denison *et al.* (1988) where the difunctional epoxy appeared to be bound at both ends.

8.2.2 Aramid fibres

These are based on poly 1,4-phenylene terephthalamide (**1**) (Kevlar, Twaron) and poly 1,3-phenylene isophthalamide (**2**) (Nomex) (Dobb, 1985). The fibres are produced by an analogous technology to pitch-based

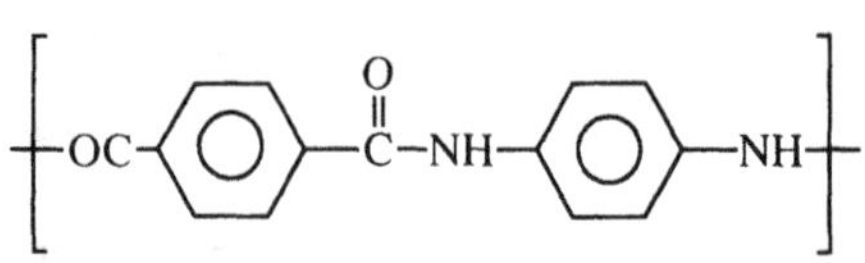

(1)

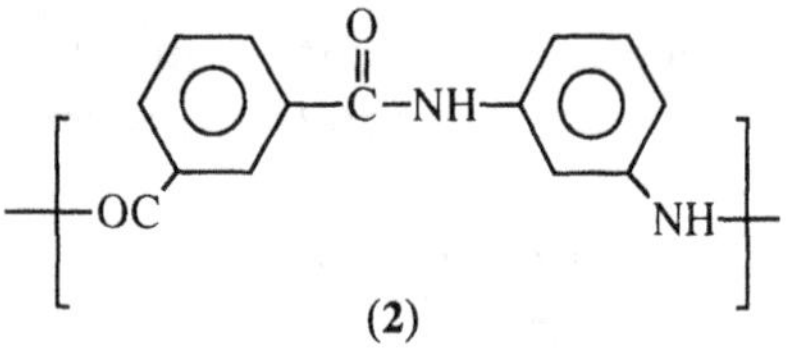

(2)

carbon fibres in that liquid crystal or nematic spinning dopes are employed. The fibres are wet spun from 99.8% H_2SO_4 at 90°C, into an aqueous coagulant bath and washed with a weak alkali to remove acidic impurities. Residual sodium sulphate is considered to be responsible for the observed variations in fibre-moisture absorption (3 ± 0.5%). The final morphology is achieved by a subsequent but brief heat treatment for a few seconds in an inert atmosphere at temperatures between 150 and 550°C when recrystallization occurs. For high-performance, heat treatments at temperatures of the order of 450°C lead to a perfection in the lateral order of the polymer.

The spinning conditions result in a radial crystallite structure similar to pitch-based carbon fibres. Because of the conformational arrangement of the polymer chains, a pleated structure is formed.

The lower-modulus, 1,4-aramid with less lateral order is primarily used for tyre cord and ropes. The higher modulus fibres which have a more perfect crystal structure are used for composites. The irregular chain conformation of the 1,3-aramid is not suitable for high stiffness, and has a lower modulus. These are generally used as staple fibres for honeycomb applications.

The highly aromatic molecular structure gives rise to high thermal stability to the polymer. However, ultraviolet (UV) light can sensitise the oxidative degradation process. Although the degradation products are self-screening, it is essential, in applications where UV environments are encountered, to employ protective films.

The polar nature of the molecule means that the equilibrium moisture content is significant and steps are required to dry the fibres and to prevent moisture ingress during use.

The aligned crystalline morphology is responsible for the poor fibre transverse strength and the tendency to fibrillate. Consequently, the fibres have poor compressive strengths. Composites also have relatively poor transverse properties because of the limited adhesion chemistry between fibre and matrix and between fibre crystallite planes. These properties can be utilised in applications where impact resistance is critical, such as ballistic protection.

Attempts have been made to improve the fibre-resin adhesion by using plasma-type surface treatments and increase fibre integrity by γ-irradiation cross-linking, but commercially-treated fibres are not generally available (Verpoest, 1987).

8.2.3 *Glass fibres* (Loewenstein, 1983; Jones, 1989)

A number of glass compositions are commonly used for the production of glass fibres. The most important is E-glass, whereas for specialist applications others can be used: for example, S-glass where a higher modulus is required, ECR glass for acid resistance, and AR glass for alkali resistance (i.e. cement matrices). C-glass is usually produced only in the form of veil for chemically-resistant GRP linings.

The fibres (other than C-glass veil) are prepared by drawing from the bushing below a molten glass reservoir. Since the strength of glass fibres is determined by the distribution and size of surface defects the most important and commercially sensitive stage in the process is the surface treatment or size which is applied by rubber roller prior to assembly into rovings. The rovings consist of strands, each containing 2000 filaments. The rovings can be either used directly, rewound onto a bobbin (a small twist being put in to improve handling) for further textile processing, or chopped into short fibres for the production of chopped strand mat.

The surface 'finish' is a critical stage in the process, ensuring compatibility with the resin or matrix and high-speed conversion into composite materials with minimum catenation or fluffing of the strands, fibre breaks, and damage. The 'finish' consists of an adhesion promoter called a coupling agent, a protective polymeric coating or size, binder, and other additives (e.g. fatty acid ester or amide) which ensure good handleability. It is applied in an aqueous medium and for epoxy resin matrices this will invariably be an emulsified epoxy resin. Some manufacturers use polyester/epoxy resin film formers for compatibility with a range of polyester and epoxy resins. Fibrous mats or reinforcing tissues will have an additional polymeric binder, which can be selected to control the wet-out of the fibres in the chosen fabrication process.

The adhesion promoter which is generally a silane coupling agent, is expected to generate a strong interfacial bond between the fibre and matrix. They are generally of the form

$$R\text{–}(CH_2)_n\text{–}Si(OR')_3$$

where R is a resin reacting or compatible group. As shown in Table 8.3 these are predominantly amines or epoxides. However, since some rovings may be of general use in epoxy and polyester resins, mixtures with vinyl, methacrylate or other type of silane may be employed. Typically n varies from 0–3. The $-Si(OR')_3$, with $R'=CH_3$ or C_2H_5, is hydrolysed in the aqueous medium (often in the presence of ethanoic acid) to $-Si(OH)_3$ for condensation with surface silanol groups to form a covalent bond. The statistics of silanol group condensation intra- and intermolecularly leads to varying degrees of oligomerisation/polymerisation and interfacial condensation. This leads to a complex interfacial deposit of $\simeq$ 100–200 molecular layers which is reported to involve an oligomeric or physisorbed component overlying loosely and chemisorbed layers. It is now accepted that the last two are really a graded polymeric network of variable hydrolytic stability. At the fibre surface the deposit has the highest network density and is strongly bound to the glass surface. As a consequence the mechanisms of adhesion are complex and may involve chemical covalent bonds to the matrix resin. Alternatively a semi-interpenetrating network may form to give a 'mechanical' bond at the molecular scale. Furthermore, a number of reports have highlighted that the presence of the oligomeric component can be detrimental to the interfacial bond strength. More recently, it has been reported that the glass substrate ions can be incorporated into the silane (Wang, 1992). These effects can also contribute to the hydrothermal stability of the interphase region. This has led to a number of theories on the structure of the silane interface (Rosen, 1978; Jones, 1989). For example, Plueddemann (1990) has proposed a reversible hydrolytic bonding mechanism which allows stress relaxation through reversible hydrolysis and condensation to account for the wet performance of the composites. It is therefore clear that an *interphase* will form between the fibre and the matrix. However, this is complicated by the presence of the polymeric size and binders which have to dissolve into the resin to form an interfacial region or modified interphase. Thomason and Morsink (1988) make the point that the coating constitutes up to 1% of the fibre weight of which $\approx$ 80% is acetone-extractable, representing $\simeq$ 5% of the resin matrix in the composite. They showed clearly that this can reduce the glass transition of the matrix significantly at a rate of 2 K per %. Since this will probably be unevenly distributed, the thermomechanical properties of the interphase will be complicated by the glass transitions of the coating and the variably plasticised matrix.

Table 8.3 Typical silane coupling agents for resin-based composites

Vinyl	$CH_2{=}CHSi(OCH_3)_3$
*Epoxy	$\overset{O}{CH_2CH}CH_2OCH_2CH_2CH_2Si(OCH_3)_3$
Methacrylate	$CH_2{=}C(CH_3){-}COOCH_2CH_2CH_2Si(OCH_3)_3$
*Primary amine	$H_2NCH_2CH_2CH_2Si(OCH_3)_3$
*Diamine	$H_2NCH_2CH_2NHCH_2CH_2CH_2Si(OCH_3)_3$
*Mercapto	$HSCH_2CH_2CH_2Si(OCH_3)_3$
Cationic styryl	$CH_2{=}CHC_6H_4CH_2NHCH_2CH_2NH(CH_2)Si(OCH_3)_3 \cdot HCl$
Cationic methacrylate	$CH_2{=}C(CH_3){-}COOCH_2CH_2{-}N^+((CH_3)_2)CH_2CH_2CH_2Si(OCH_3)_3\ Cl^-$
Cycloaliphatic epoxide	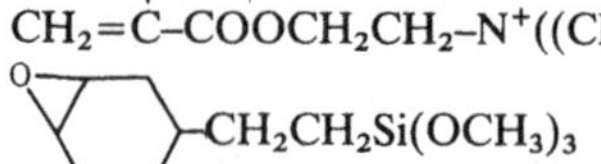

* Epoxy resin compatible.

8.3 Fabrication of composites

8.3.1 The reinforcement form

The form of the fibre reinforcement is dictated by the fabrication process, which in turn is an inherent aspect of the engineering design philosophy. Discontinuous fibres allow complex mouldings to be manufactured rapidly. With thermoplastics and thermoset moulding powders, the fibre length is generally less than 5 mm. As discussed elsewhere a critical fibre length exists below which the reinforcing efficiency is seriously undermined. Generally the critical length of fibres in thermosets is larger than for thermoplastics. Therefore for injection moulding, dough or bulk moulding compounds (DMC or BMC) containing longer 10–20 mm fibres are preferred where length retention in the artefact is improved. For longer discontinuous fibres (> 20 mm) they are bound together with a polymeric binder to form a random or chopped strand mat (CSM). Fibres of similar length are also incorporated into a preimpregnated sheet known as sheet moulding compound (SMC). The fibres can be present in various geometries in SMC. SMC employs thickened resins but with the wide range of epoxy resins, prepreg technology is employed.

For *epoxy resins*, however, the predominant fabrication processes incorporate continuous carbon, aramid or glass fibres and the techniques involve rovings directly, as woven cloth or as preforms. The woven cloth can take on several forms: unidirectional, plain, satin, triaxial, hybrid fibre fabric or woven tape. The more complex fabrics are generally employed where drape onto the former is crucial. The convenience of a complex reinforcement is often off-set by a reduction in average composite properties. These can be used directly or preimpregnated with uncured resin or as a 'prepreg'.

The fabrication of fibre composites with epoxy resins is dominated by those involving continuous fibres: lamination in an autoclave, pultrusion, filament winding, hot compression moulding. These techniques are reviewed in Table 8.4.

RRIM and Structural RIM generally employ polyurethane or polyamide matrices but for the latter, attempts to generate fast curing epoxy resins suitable for this application are under development (see section 8.3.3.2).

8.3.2 Prepreg mouldings

High-performance carbon-fibre and glass-fibre laminates are fabricated by prepreg moulding using autoclave or pressclave techniques. A 'prepreg' consists of collimated fibres or woven cloth preimpregnated with an epoxy resin (or other thermoset or high performance thermoplastic) modified with a thermoplastic or elastomeric toughening agent and containing a latent

Table 8.4 Fabrication processes for composites from epoxy resins

Processes	Reinforcement	Vol. (%)	Method	Typical application
1. Spray-up	Chopped rovings (G)	15–30	Direct application of chopped rovings and resin	Pipe linings
2. Autoclave	Continuous (G, C, A)	60	Forming of stacked prepreg in vacuum bag	Aerospace components
3. Filament winding	Continuous rovings, occasionally particulate fillers included (e.g. sand) (G, C)	~50	Wet resin carried by fibres onto mandrel at predetermined angles. Variants: prepeg tape, woven tape, dry preform manufacture	Pipes and containers, geodesic shapes
4. Pultrusion	Continuous fibres, occasional CSM and fillers (G, C)	60–80	Wet resin-impregnated roving or cloth formed through a heated die	I-beam or rods
5. Compression moulding	Chopped or continuous fibres or preforms and fillers (G, A, C)	10–60	Prepeg or wet resin/preform pressed to shape in a closed mould between matched dies	Medium-sized complex mouldings
6. Resin injection, resin (RTM) transfer	Continuous or discontinuous fibre plus preform polymer foam inserts (G, C, A)	25–30	Wet resin injected into woven, knitted or CSM preform within a closed mould	Panels
7. Reinforced reaction injection moulding (RRIM)	Short-milled glass fibre or fibre preform (G)	5 20	Fibre-resin dispersion and hardener mixed at rapid mixing head and injected into mould. Chemistry controlled to give rapid reaction	Large complex parts (e.g. auto parts for impact resistance but with low stiffness)
8. Structural RIM	Continuous fibre preform (G, A, C)	30–50	Similar to 6 with the more rapid RIM chemistry	Large structural parts. Rapid production possible

Fibres: G = glass; C = carbon; A = aramid.

hardener. The latent hardener such as the insoluble dicyandiamide (DICY) is micronised in the resin solution prior to wet-impregnation into the collimated fibres and woven cloth and drying. The curing temperature and the kinetics of cross-linking can also be adjusted by incorporating a promoter or accelerator (e.g. Monuron or Diuron). The resin viscosity is adjusted to give the prepreg the appropriate tack by adding thermoplastic polymers or 'advancing' the resin with partial reaction with the hardener.

The prepreg is stored between plastic release films in a sealed container in a freezer. After thawing and temperature equilibration to prevent moisture condensation and ingress, the unidirectional sheet is cut to appropriate dimensions and stacked at predetermined angles to produce a balanced laminate (e.g. $0_2°/\pm 45°/90_2°/\pm 45°/0_2°$; the subscripts define the number of prepreg lamina in each ply; the choice of stacking sequence is determined by the required laminate properties). The prepreg stack is laid up between release cloths and 'blotting paper or similar' bleed pack to soak up the excess resin in the prepreg (Figure 8.5). The bleeding of excess resin facilitates the removal of entrapped air and residual volatile impurities during consolidation of the prepreg into a void-free laminate. The curing schedule is also carefully controlled since the exothermic nature of the chemical reactions can lead to a temperature build-up in thick sections. Some of these processing difficulties can be overcome by using thermoplastic matrices but these have their own processing problems. Prepreg moulding techniques have a complicated technology, but offer the mechanical engineer the opportunity to orientate fibres in the direction of the principal stresses within complex structures.

8.3.3 *Matrices for fibre composites*

Plastic matrices for fibre composites generally can be divided into thermoplastics, chemosets, and thermosetting polymers. The last two broadly fall into the same category of thermosetting resins, but for the purposes of this discussion we differentiate between those requiring

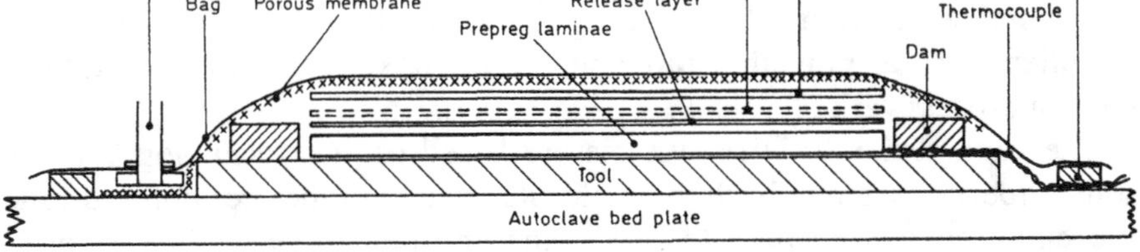

Figure 8.5 Autoclave moulding arrangement for laminates based on epoxy prepreg. (After Purslow and Childs, 1986).

additional hardeners, curing agents, or catalysts (the chemosets) and those that are quenched at a partially reacted state and continue to cure on heating. In both cases, curing involves the generation of a three-dimensional cross-linked molecular structure. Although it is generally assumed that the cross-linking process is homogeneous, there are notable examples where an abundance of indirect evidence points to a heterogeneous amorphous molecular structure. For the epoxy resins, it is more likely to result from the heterogeneous nature of some of the curing agents. The main types are epoxy or unsaturated resins such as polyester or vinyl esters. The latter are low polymers with $M_n < 2000$ g . mol^{-1} whereas the epoxides are generally of relatively low molecular weight. These two systems satisfy the requirements of an appropriate matrix.

1. Low viscosity for fibre impregnation
2. High reactivity on curing
3. Chemical control of cure without volatile formation
4. Good mechanical properties of the cured resin.

Criterion 3 is not readily achieved by the thermosets (e.g. phenolics, polyimides) since these involve condensation rather than step-growth or addition polymerisation. To prevent damage accumulation as a result of volatile formation, higher pressures and more complicated cure schedules are required. The bismaleimides (e.g. PMR-15) contain thermally polymerisable unsaturated groups and provide significantly higher temperature performance (up to 400°C) than the advanced epoxy matrices. Epoxies generally satisfy the above criteria because the curing occurs without evolution of volatiles, in a controllable manner with low shrinkage.

8.3.3.1 Epoxy matrix resins. The difunctional diglycidyl ether of Bisphenol A (e.g. Ciba–Geigy MY750) and related oligomers of degrees of polymerisation up to four, form the basis of the general-purpose resins for both adhesives and composites. The glycidyl amines have a higher functionality and potentially higher cross-link densities and glass transition temperatures and form the basis of the resins for aerospace composites. The latter are used as prepreg resins and are often mixtures based on the tetrafunctional MY720. Typical resins for laminates are given in Table 8.5. Judicial choice of resin blend and hardener is required to provide controlled cure and viscosity profiles during cure. Furthermore, thermoplastic modifiers such as polyethersulphone and phenoxy resins are also used to control process variables.

The Ciba–Geigy 913/914 systems are based on this technology, as are some Fiberite systems. Both the molecular weight of the thermoplastic and the end-groups contribute to processing and performance. For example amine end-group terminated polyether sulphones ensure chemical bonding with the epoxy resin. As will be discussed later, moisture absorption leads to

Table 8.5 Typical epoxy resins used for composite fabrication

Difunctional:

Diglycidyl ether of bisphenol A ($n = 0$), MY 750

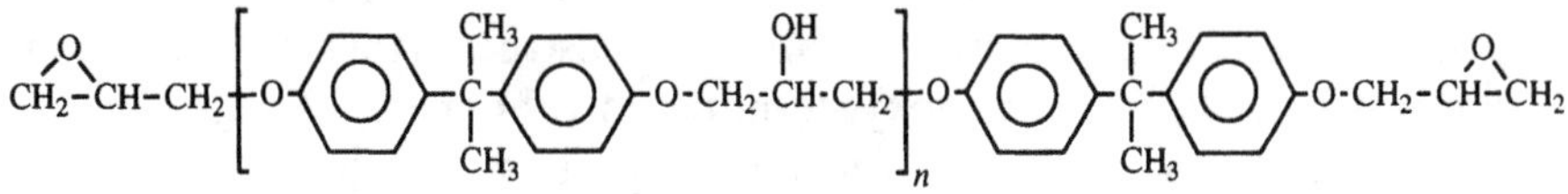

Trifunctional:

MY 0510

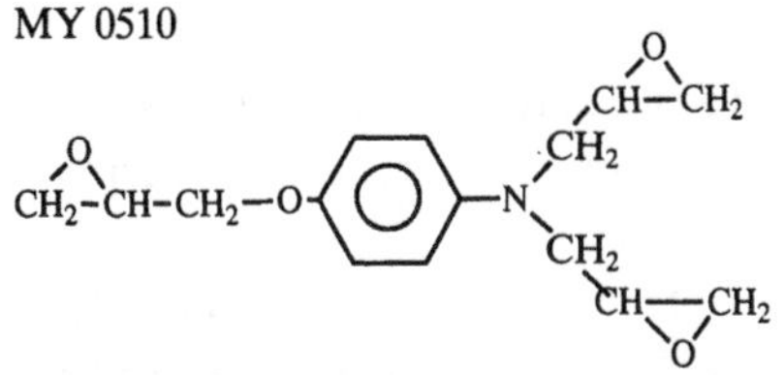

XD 7342.00L

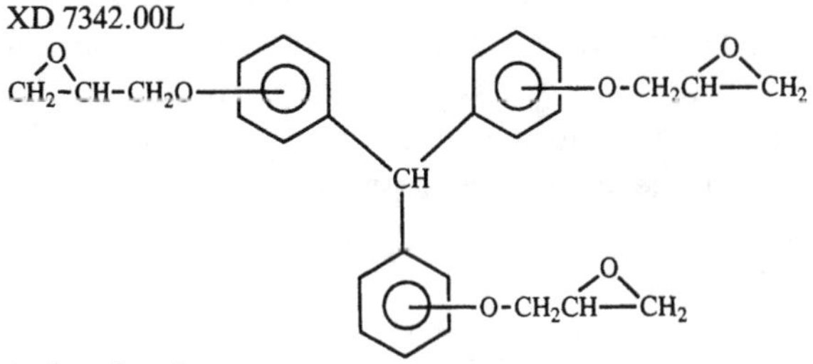

Tetrafunctional:

MY 720

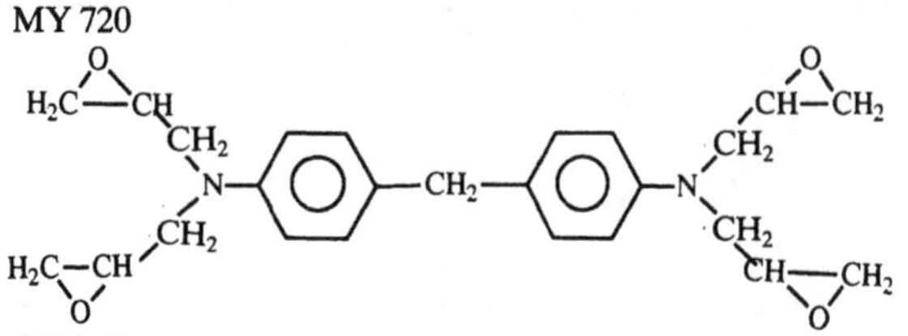

PGA-X

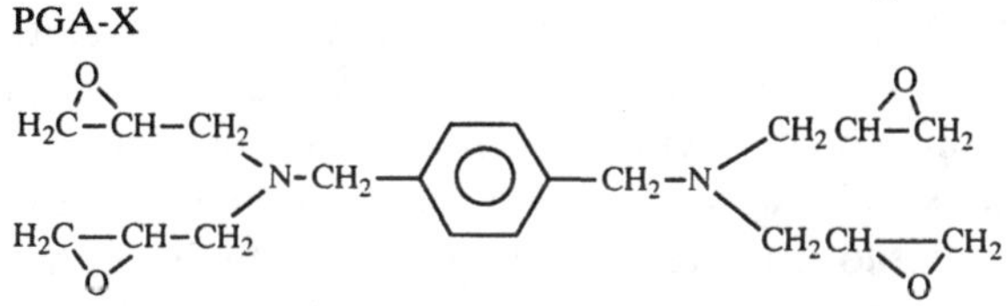

XD 9053.00L

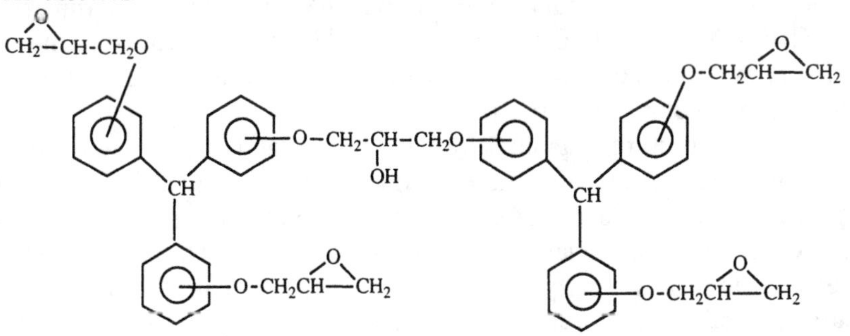

a significant reduction in glass transition temperature of the cured resin which seriously limits the maximum service temperature. To mitigate this effect highly cross-linked resins are preferred. However, to ensure that the microcrack resistance of the composite is maintained, toughening modifiers are required. The thermoplastic modifiers can also act in this way. However, in contrast to adhesives, carboxyl-terminated rubber (CTBN)-toughened epoxies are not generally employed for composites. This probably demonstrates that the advantage of modification with thermoplastics may lie predominantly with processing improvements.

Other important systems employ thermosetting modifiers such as bismaleimide or cyanurate resins to achieve high temperature performance with durability.

It is important to emphasise the difference between hardeners and catalytic curing agents. The latter are used in small quantities and primarily yield a cured resin with polyether network chains. The former are molecules with active 'hydrogen' atoms which can add to the epoxy group. There are two main types: the acid anhydrides and the amines and polyamides. They have a powerful influence on the cured resin properties as a result of the molecular structure of the network chains. More details are given in chapter 2.

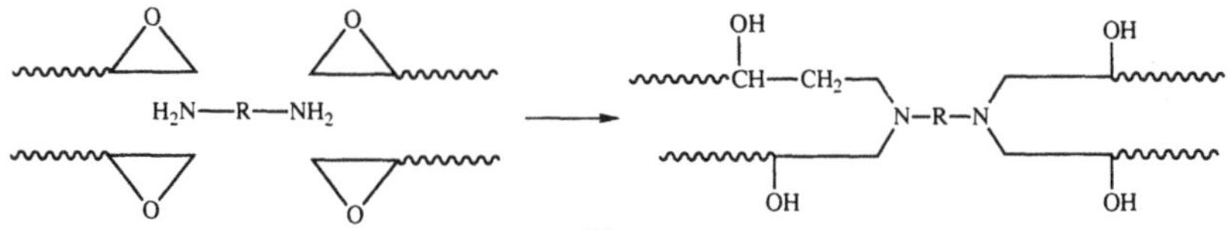

(3)

Amine hardeners react with epoxies to form hydroxy amines (**3**). At first sight, this might be expected to promote adhesion to the fibres through the hydroxyl groups, but the polarity promotes moisture sensitivity and hardening or catalytic systems which provide a less polar molecular structure are preferred. For industrial lamination, tertiary amine-promoted anhydride cure is often employed. Nadic methylenetetrahydrophthalic anhydride (NMA) is commonly employed with the diglycidyl ether (MY750) and tetrafunctional glycidyl amine (MY720) with benzyl dimethyl aniline (BDMA) or the similar Epikure K61b. This system provides a matrix with lower water absorption, which can be attributed to the 'polyester' copolymer molecular structure (**4**).

A polyether structure originates from the use of catalytic systems of which the cationic boron trifluoride amine complex is favoured (**5**). For advanced composites, BF_3 complexes, often in combination with diamino diphenylsulphone (DDS), act as latent hardeners becoming active at $\simeq$ 160°C. Therefore, the resin viscosity profile with temperature can be adjusted by a precure (sometimes referred to erroneously as B-staging) to optimise the

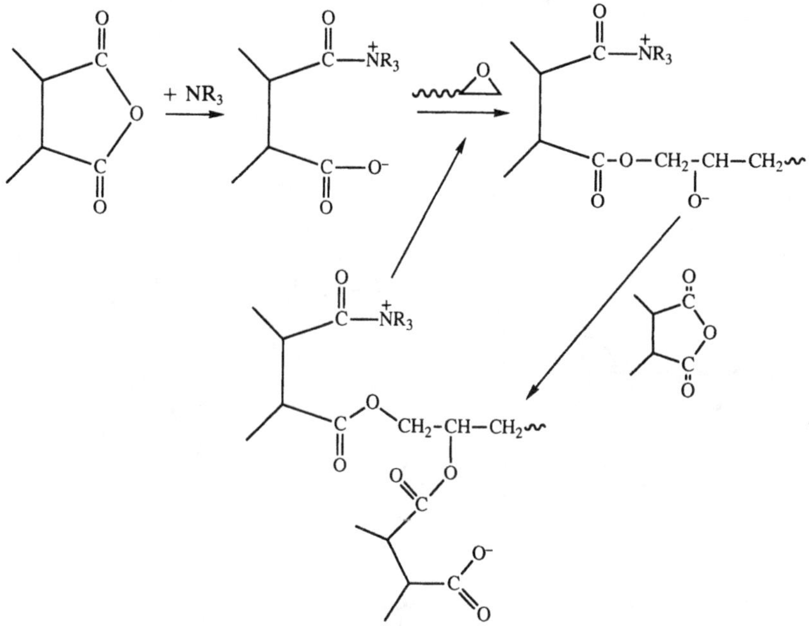

(4)

consolidation of the prepreg stack and removal of entrapped air in the fabrication of a laminate by vacuum bag/autoclave techniques, prior to the onset of cure.

Micronised dicyandiamide (DICY) also provides latent curing ability to prepreg with improved shelf-life. Both high and low temperature curing systems are available. Without the presence of 'Diuron' or 'Monuron' amine accelerator, the rate of curing is determined by the melting of DICY at 165°C at which temperature it dissolves into the epoxy resin. With 'Monuron' the curing temperature can be reduced to 120°C. While low

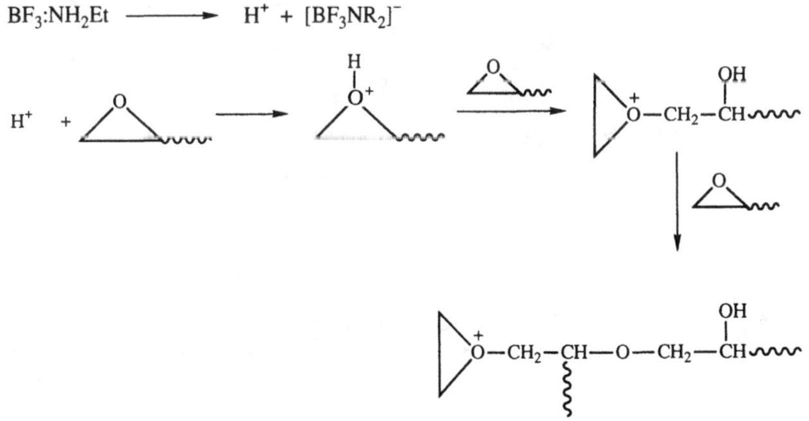

(5)

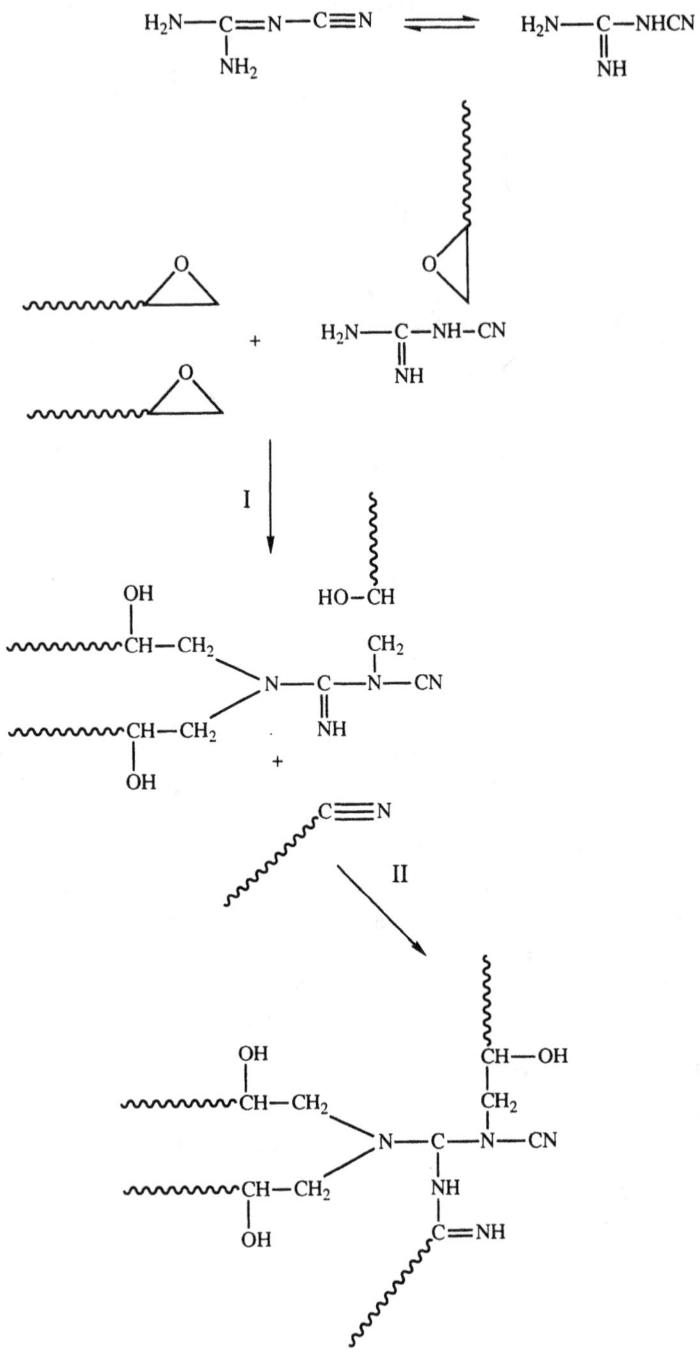

Scheme 8.1 Cure of epoxy resins with DICY. Alternative reaction schemes are discussed by Amdouni *et al.* (1990).

temperature systems have clear advantages, not least the control of the exotherm during cure, careful control of the curing schedule is required for the production of good quality void-free laminates. Thus a dwell time at 'minimum' viscosity has to be incorporated, usually $\simeq$ 90°C for these systems. The other disadvantage of the low temperature curing systems based on DICY is the increased probability of residual DICY hardener in the resin. Another observation is the inconsistent appearance of a microstructure in these cured systems. Much of the variability can be reduced by ensuring that the hardener is micronised to a uniform fine particle size prior to prepreg manufacture. However, microstructural variations can still occasionally appear but these could also be attributed to phase separation of the thermoplastic modifier.

DICY is used in catalytic quantities 4–6 phr but the curing chemistry is complex producing a structure expected of a polyamine as indicated in Scheme 8.1.

It should be pointed out that the complete reaction of the epoxy groups is not normally achieved except after long times at high temperatures so that higher-heat-distortion temperatures and chemical resistance can only be achieved by appropriate postcuring. The mechanical properties of epoxy resins are therefore determined by the hardener or curing systems and time–temperature profile. The time temperature transformation curing diagrams of Gillham (1986) are particularly valuable in this context. This is discussed in chapter 1. The effect of post-curing on mechanical properties can be generalised by appreciating that the heat distortion temperature and Young's modulus increase with cross-link density and the concentration of aromatic groups within the network. However, there are exceptions to this where a higher glass transition can arise alongside a lower modulus, since the glassy state is frozen in at higher molecular free volume.

8.3.3.2 Rapid curing resins for structural RIM. Much progress has been made recently in the development of rapid curing epoxy systems for alternative composites processing.

Stanford (Mortimer *et al.*, 1992) has reported gel times of < 25 s at 40°C using BF_3 diol complexes (**6**) of Dianol 240 and 320 to cure DGEBA (Dow Chemical DER332).

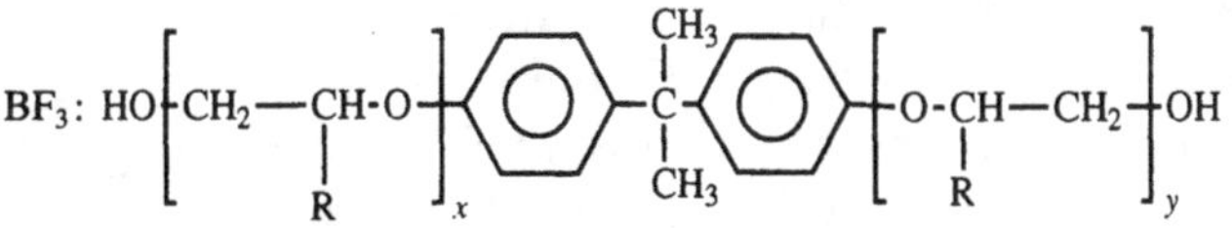

(**6**)

where with R = H (**6**) the average value of x and $y = 2.15$. With $R = CH_3$, $x = y = 1$.

To provide control over resin viscosity for RIM, processing co-monomers for the cationic polymerisation of the epoxy groups are included. Tetrahydrofuran yielded a resin of low T_g but with butyrolactone (**7**) a higher T_g of 103°C could be achieved. Gel times of 64 and 80 s were observed.

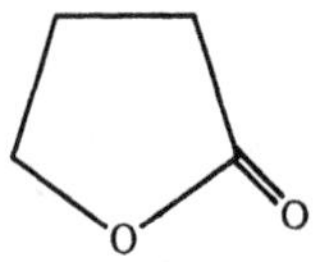

(**7**)

The shortest gel time of 35 s was observed with the BF_3 complex with diol (**8**). The maximum cure temperature was 70°C.

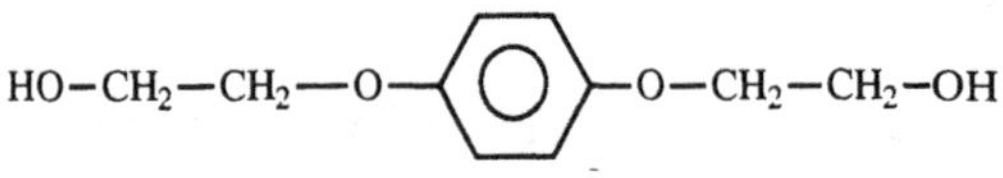

(**8**)

This appears to be a significant advance over the literature citations in which demould times were > 2 mins at the higher temperature of > 100°C.

8.4 Mechanical properties of unidirectional laminates

A unidirectional fibre composite is highly anisotropic so that in real laminates the fibres are arranged at angles to one another. Since transverse to the fibres, matrix and interfacial properties dominate and the statistical variability in fibre strength dominates in the fibre direction, the characteristic failure mode is one of damage accumulation, through debonding, microcracking and fibre fracture. However, under stress these events are comparatively innocuous under most service conditions but can act as nuclei for more severe failure processes such as stress corrosion cracking of GRP in acidic environments.

8.4.1 Longitudinal modulus, E_l

When a tensile load is applied in the longitudinal direction to the perfectly bonded continuous fibres the strain in the matrix, ε_m will be equal to the strain in the fibre ε_f. Since $E_f > E_m$ the stress carried by the fibres, σ_f, is greater than that in the matrix, σ_m. E_f and E_m are the Young's moduli of the fibre and matrix respectively. The stress in the fibres and matrix respectively is given by

$$\sigma_f = E_f \varepsilon_c \tag{8.1}$$

$$\sigma_m = E_m \varepsilon_c \tag{8.2}$$

where ε_c is the strain on the composite. The stress on the composite σ_c is given by equation 8.3.

$$\sigma_c = \sigma_f V_f + \sigma_m V_m \tag{8.3}$$

By equating the volume fraction of fibres V_f with the fractional fibre area and the volume fraction of matrix V_m with the fractional matrix area, the Young's modulus of the composite in the fibre direction, E_l, is given by,

$$E_l = E_f V_f + E_m V_m \tag{8.4}$$

and since $V_m = (1 - V_f)$

$$E_l = E_f V_f + E_m (1 - V_f) \tag{8.5}$$

This is known as the law of mixtures and assumes that both the fibres and matrix are perfectly elastic. At high V_f any deviations due to viscoelastic behaviour of the matrix are relatively insignificant. Similarly the additional stresses due to differences in the Poisson's ratios of the fibres and matrix only lead to an error of less than 2%. Experimental verification of equation 8.5 has been established in several studies.

8.4.2 *Longitudinal tensile strength, σ_{lu}*

Under a rising tensile stress parallel with the fibres, either the fibres or the matrix will fail depending upon the relative failure strains of the fibres (ε_{fu}) and matrix (ε_{mu}). Two cases need to be considered (Figure 8.6). Case 1 when

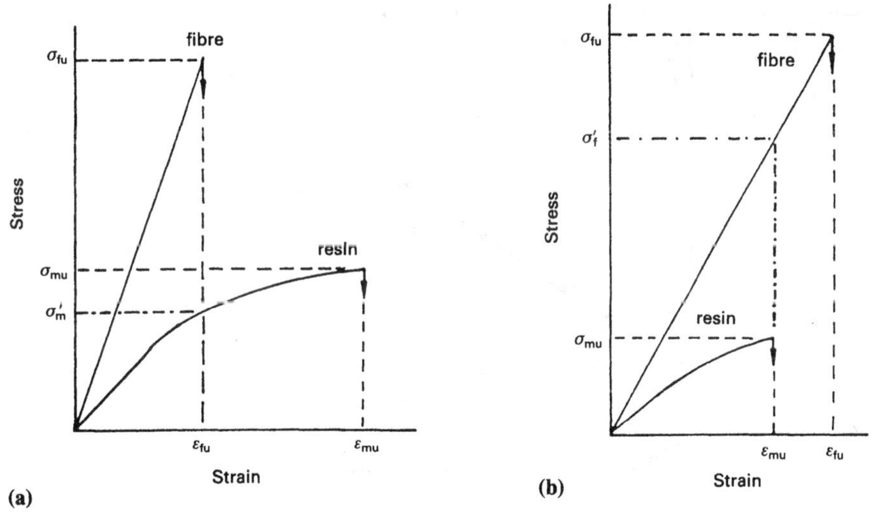

Figure 8.6 Individual fibre and resin stress/strain curves defining the failure processes described in Figures 8.7 and 8.8. (a) Case 1, $\varepsilon_{mu} > \varepsilon_{fu}$; (b) Case 2, $\varepsilon_{mu} < \varepsilon_{fu}$.

the failure strain of the matrix is greater than that of the fibres (a typical fibre-reinforced epoxy resin) and Case 2 when matrix failure strain is lower (a typical fibre-reinforced ceramic).

Case 1

$$\varepsilon_{mu} > \varepsilon_{fu}$$

With uniformly strong fibres at high volume fractions, when the fibres fail all the load is thrown upon the weaker matrix. The effective cross-sectional area is reduced and the composite will immediately fail. The strength of the composite σ_{lu} is given by

$$\sigma_{lu} = \sigma_{fu} V_f + \sigma'_m (1 - V_f) \tag{8.6}$$

where σ'_m is the stress in the matrix when the fibre breaks as shown in Figure 8.6.

At low volume fractions (typically less than 0.006–0.03 for GRP and CFRP) the matrix can carry the load, leading to multiple fracture of the fibres. The strength is given by

$$\sigma_{lu} = \sigma_{mu}(1 - V_f) \tag{8.7}$$

Equations 8.6 and 8.7 are plotted in Figure 8.7 where it is seen that a critical fibre volume fraction (V_{crit}) for reinforcement exists. The matrix contributes insignificantly to the failure strength of the unidirectional laminate in the fibre direction. The strength is given approximately by equation 8.8.

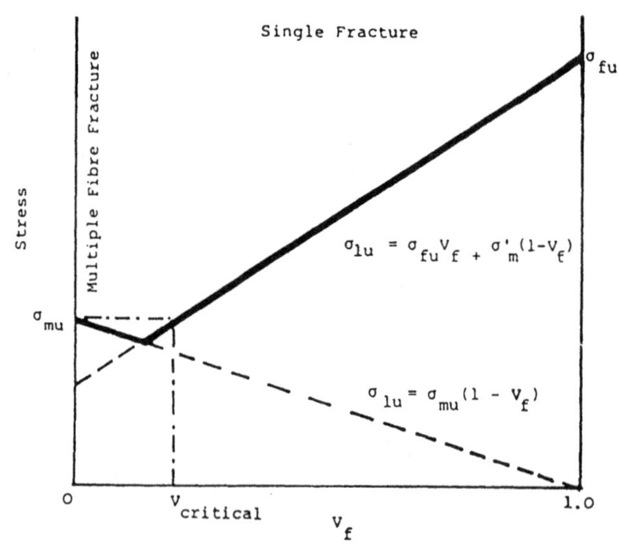

Figure 8.7 Fibre volume fraction dependence of fracture strength for a unidirectional composite, σ_{lu}, manufactured from a resin of failure strain higher than that of the fibres ($\varepsilon_{mu} > \varepsilon_{fu}$).

$$\sigma_{lu} = \sigma_{fu} V_f \tag{8.8}$$

However, as shown in Figure 8.2 the fibres have a statistical failure strength and strain, and in practice a cumulative weakening process occurs with fibre breaks at random. The first fibre break may lead to immediate failure if the flaw has sufficient stress intensity to cause the crack to propagate. This situation only arises when the fibres are very strongly bound to a matrix of low fracture toughness. Generally fibre-resin debonding occurs to a limited degree at the fibre ends so that stress transfer through shear enables the fibre damage to accumulate until the multiple fractured fibres reach their ineffective lengths. The probability of fracture of adjacent fibres is increased. This leads to a sequential failure and the production of a Griffith flaw which can result in composite failure. The generation of a critical number of adjacent fibre breaks can be inhibited by dispersing a higher failure strain fibre into the bundle to increase the ultimate strength in what has been termed the hybrid effect.

The tensile fracture surface has a brush-like appearance with varying degrees of fibre fracture and pull-out making meaningful strength predictions difficult. For further discussion of this problem the reader is referred to the texts by Hull (1981) and Kelly (1973).

Case 2

$$\varepsilon_{mu} < \varepsilon_{fu}$$

Under these circumstances the matrix fractures first, throwing all additional load onto the fibres (Figure 8.6). At low V_f as illustrated in Figure 8.8, the

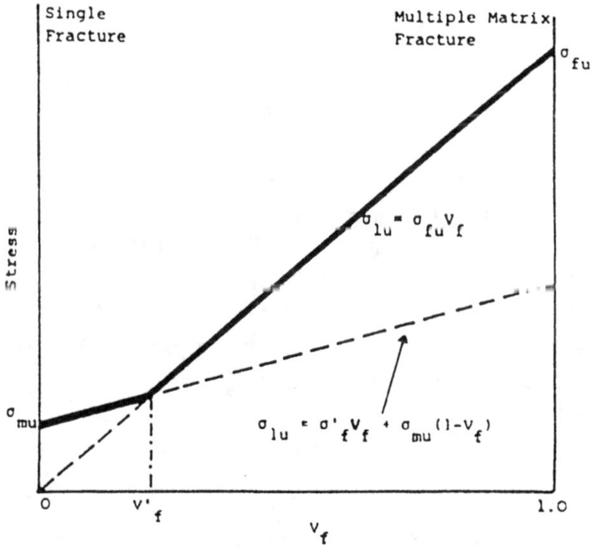

Figure 8.8 Fibre volume fraction dependence of fracture strength of a unidirectional composite manufactured from a matrix of failure strain lower than that of the fibres ($\varepsilon_{mu} < \varepsilon_{fu}$).

fibres are unable to support the additional load and the composite fractures. The strength is given by

$$\sigma_{lu} = \sigma_f' V_f + \sigma_{mu}(1 - V_f) \tag{8.9}$$

where σ_f' is the stress on the fibres when the matrix fractures (see Figure 8.6). At higher V_f when $V_f > V_f'$ $(= \sigma_{mu}/\sigma_{fu} - \sigma_f' + \sigma_{mu})$ the fibres are able to support the extra load when the matrix fails. Additional loads cause multiple cracking of the matrix. Further details are given by Aveston and Kelly (1973). The strength of the laminate is clearly determined by the average strength of the fibres $\bar{\sigma}_{fu}$ and

$$\bar{\sigma}_{lu} = \bar{\sigma}_{fu} V_f \tag{8.10}$$

In practice the distribution of values of σ_{fu} means that some fibres may fail before the onset of matrix failure and continual damage accumulation of matrix cracks and fibre fracture occurs.

8.4.3 *Transverse modulus,* E_t

A simple model for predicting the transverse Young's modulus of a unidirectional composite assumes a continuity of stress across the fibre-resin interface such that

$$\sigma_c = \sigma_f = \sigma_m \tag{8.11}$$

thus

$$E_t = (E_f E_m)/[E_f(1 - V_f) + E_m V_f] \tag{8.12}$$

or

$$1/E_t = V_f/E_f + V_m/E_m \tag{8.13}$$

The rule of mixtures equation 8.13 is a poor estimate of the actual transverse modulus and a number of other attempts have been made. The most rigorous approach is that of Hashin and Rosen (1964) in which upper and lower bounds may be calculated. The Halpin and Tsai (1969) equations are generally applicable and allow for variations in packing geometry and regularity with the factor ξ. These rules are compared in Figure 8.9 where it is clearly seen that transverse fibres are inefficient reinforcing elements.

$$E_t = E_m(1 + \xi\eta/V_f)/(1 - \eta V_f) \tag{8.14}$$

where

$$\eta = [(E_f/E_m) - 1]/[E_f E_m + \xi]$$

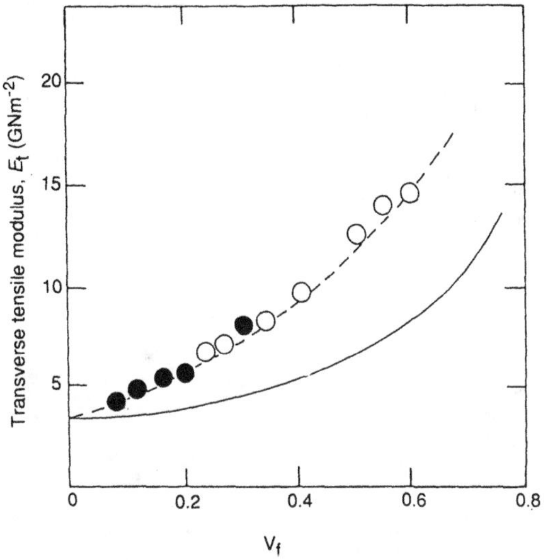

Figure 8.9 The effect of fibre volume fraction, V_f, on the transverse tensile modulus of a unidirectional composite. To accommodate the range of V_f both epoxy resin (V_f = 0.25–0.6) and an unsaturated polyester (V_f = 0.08–0.3) resins have been used. (●), Glass–polyester; (○), glass-epoxy; (—), Law of mixtures (equation 8.12); (---), Halpin–Tsai equation (equation 8.14) which gives the best fit to the data. (Redrawn from Bailey and Parvizi, 1981.)

8.4.4 *Transverse strength, σ_{tu}*

The transverse strength of a unidirectional composite will be determined by (a) the transverse strength of the fibres, (b) the interfacial bond, or (c) the matrix, whichever is weakest. This simple model fails to account for stress concentrations in the matrix.

The composite will therefore fail at a stress σ_{tu} which is significantly less than σ_{mu} such that

$$\sigma_{tu} \simeq 0.5\ \sigma_{mu} \tag{8.15}$$

The failure strain of the matrix in a composite, ε_{mu}, is significantly reduced by strain magnification as a result of the stiff fibre. Thus for a stress continuity, the strain in the matrix resin will be proportionally higher than the strain in the composite by the ratio of the moduli E_m/E_f.

The strain magnification theory of Kies (1962) modified by Garrett and Bailey (1977) gives $\varepsilon_{tu} = \varepsilon_{mu}\{1 + 2[(\pi/V_f)^{1/2} - 2]^{-1}\}$, where

$$V_f = \pi r_f^2/(D + 2r_f)^2 \tag{8.16}$$

However, as shown in Figure 8.10 this is still only an approximate solution since it does not take into account additional stress intensities at debonded fibres or other defects.

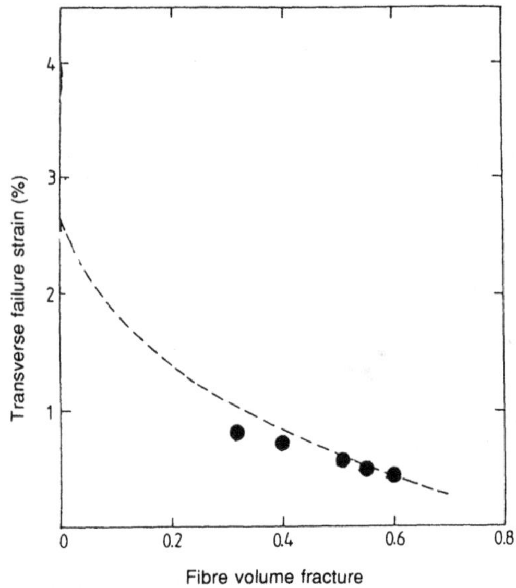

Figure 8.10 The effect of fibre volume fraction, V_f, on the transverse failure strain of a unidirectional E-glass–epoxy composite. (—), predictions of the Kies theory.

The discrepancy between theory and experimental therefore reflects the presence of strength-reducing flaws. In this case, they can occur at specimen surfaces but also at partially debonded fibres as a result of poor wet-out or other mechanism. The former aspect can be reduced by measuring the first transverse cracking strain of a crossply laminate but compensation for residual stresses has to be carried out.

Table 8.6 gives typical properties of 90° and 0° composites from carbon, glass and aramid fibre epoxy where the poor composites transverse properties are confirmed. The transverse strength is a complex function of the fibre, matrix and interfacial strength. In the case of aramid fibres, the aligned crystalline structure and poor inter-crystallite bonding also contribute to the low transverse strength. Fibre surface treatments are common for glass and carbon fibres but as discussed previously they have proved less effective for aramid fibres. However, these fibres impart high impact strength to their composites because of the energy absorption

Table 8.6 Mechanical properties of unidirectional epoxy resin composites

Fibre	V_f	E_l (GPa)	E_t (GPa)	σ_{lu} (GPa)	σ_{tu} (GPa)	ε_{lu} (%)	ε_{tu} (%)	ν_l*	ν_t
Carbon	0.62	127	8.3	1.7	0.039	1.16	0.48	0.29	0.019
Glass	0.55	42	14	0.92	0.056	2.2	0.5	0.27	0.09
Aramid	0.62	83	5.6	1.31	0.039	–	–	0.35	–

* ν is the Poisson ratio in the longitudinal (l) and transverse (t) directions.

associated with transverse fibre failure and fibre-matrix debonding. Design should employ a careful selection of fibres to provide the required performance.

Fibre surface treatments and coatings are currently receiving more attention as a means of improving and optimising interlaminar shear and transverse properties.

8.4.5 *Off-axis properties*

The modulus of a unidirectional fibre composite with fibres at an angle θ to the applied stress is given by:

$$1/E_\theta = \cos^4\theta/E_1 + (1/G_1 - 2\nu_c/E_1)\cos^2\theta\sin^2\theta + \sin^4\theta/E_t \qquad (8.17)$$

where θ is the angle of the fibres to the applied stress and ν_c is the Poisson ratio for the specimen. However in an off-axis test failure can occur by transverse, longitudinal or intralaminar shear. In the *maximum stress theory* (Tsai and Hahn, 1980; Tsai, 1985), the stresses reach a critical value for fracture in the principal material directions. Thus the criterion for failure is

$$\begin{aligned} \sigma_l &= \sigma_{lu} \\ \sigma_t &= \sigma_{tu} \\ \tau &= \tau_u \end{aligned} \qquad (8.18)$$

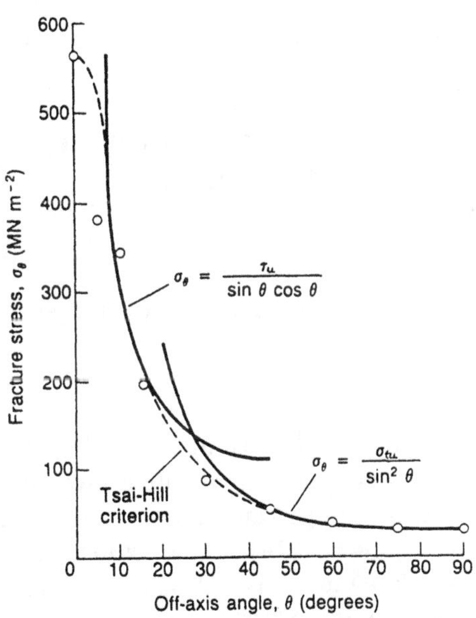

Figure 8.11 Effect of angle of test (θ) on the strength of a unidirectional carbon fibre (high strength) epoxy composite (V_f =0.5) showing the predictions of the Tsai–Hill model. (Redrawn from the original given by Hull, 1981.)

For angles of θ to fibres between 0° and 90° the applied stress (at angle θ), σ_θ will have the following components and the failure will be determined by the weakest:

$$\begin{aligned}\sigma_\theta &= \sigma_{lu}/\cos^2\theta \\ \sigma_\theta &= \sigma_{tu}/\sin^2\theta \\ \sigma_\theta &= \tau_u/\sin\theta\,.\cos\theta\end{aligned} \tag{8.19}$$

This is illustrated in Figure 8.11, where these equations are compared for epoxy-based CFRP. At angles of 0–5° the failure is dominated by longitudinal fracture, from 5–18° by intralaminar shear failure and from 45–90° by transverse fracture. In the region from 18–45° a mixed mode failure occurs. The actual angles at which these transitions take place will clearly be dependent upon the values of σ_{tu}, σ_{lu}, τ_u. The Tsai and Hill criterion which is based on maximum work theory however gives a better prediction of the off-axis strengths in this region (Halpin and Tsai, 1969; Tsai and Hahn, 1980; Hull, 1981; Tsai, 1985).

8.5 Failure processes in laminates

Fibre composites are constructed with fibres at angles to one another to maximise resistance to multiaxial stresses. For advanced composites, stacking of unidirectional plies (e.g. prepreg) at appropriate angles is commonly employed. Balanced constructions are required to prevent warpage. The nomenclature describes the number of plies of each type: for example a simple 8-ply crossply laminate, $0_2°/90_4°/0_2°$ or more complex angle ply laminate, $0_2°/90_2°/\pm 45°/0_2°/\pm 45°/90_2°/0_2°$. From the foregoing it will be appreciated that the failure processes are highly complex. In particular the free edge stresses in angle ply laminates can significantly modify the sequence of failure of individual laminate. The detailed discussion of these failure mechanisms by Hull (1981), Tsai (1985) and Tsai and Hahn (1980) are recommended. However, the failure of model crossply laminates illustrates the micromechanics of these types of composites.

8.5.1 *Crossply laminates*

A model 0°/90°/0° crossply laminate is illustrated in Figure 8.12. A typical stress-strain curve for an epoxy-glass laminate is given in Figure 8.13 where the first knee in the response of this laminate coincides with the observation of stress whitening (Figure 8.14). This coincides with fibre-matrix debonding and a cure-dependent viscoelastic response in the epoxy resin matrix. But these debonds initiate transverse cracking at an applied strain of 0.35%. Because the transverse cracks are pinned by the 0° plies (i.e. delamination at the ply interface does not occur) the stress builds up again in

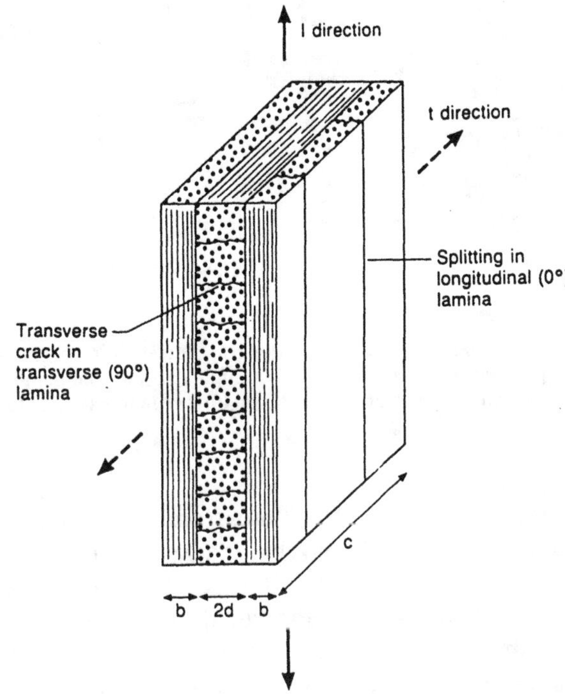

Figure 8.12 A 0°/90°/0° crossply laminate showing transverse cracks which progressively form under a rising tensile stress. Longitudinal splits in the 0° laminate also occur at high composite strain.

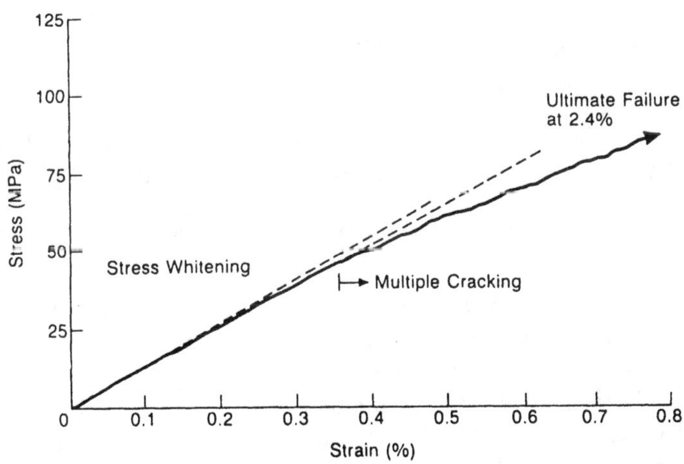

Figure 8.13 Low strain part of a stress/strain curve for a 0°/90°/0° laminate showing the onset of cumulative damage beginning with fibre-matrix debonding (stress whitening) and transverse cracking (see Figures 8.12 and 8.14) (after Manders *et al.*, 1983).

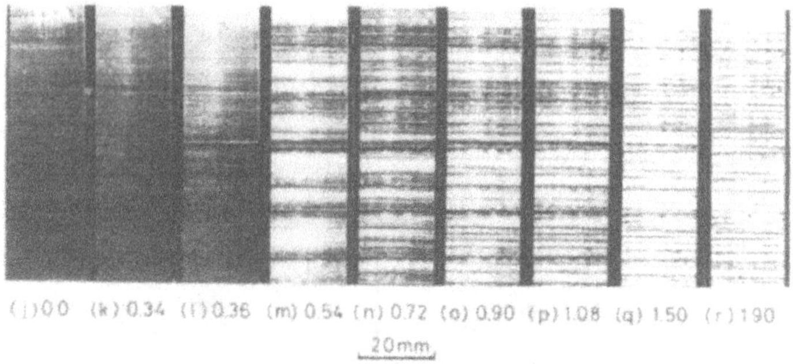

Figure 8.14 Progressive multiple cracking in the transverse ply of a glass-epoxy crossply laminate. Note the presence of longitudinal splits at higher strain. The numbers represent the strain at which the photograph was taken (after Manders *et al.*, 1983).

the 90° ply at a transfer distance of $\simeq$ 2 mm from the crack. This results in a black band either side of the transverse cracks as shown in Figure 8.14. Multiple cracking of the inner ply continues as the load on the specimen is increased until saturation has occurred when the shear-lag stress transfer can no longer exceed the transverse strength of the remaining elements.

Under continued loading Poisson strains are induced which cause the longitudinal ply splitting shown schematically in Figure 8.12. The onset of ply splitting is also seen in the photographic record of the cracking sequence in Figure 8.14.

Parvizi *et al.* (1978) used a modified shear lag analysis to calculate the additional stress transferred back into the transverse ply at a distance from the crack, y. These authors assumed that the cracks would form midway between the previous ones. If $\Delta\sigma_o$ is the additional stress required to load the transverse ply to its cracking strain then

$$\Delta\sigma_o = \varepsilon_{tlu} E_t d/b[1 + \exp(-\Phi^{1/2}t) - 2\exp(-\Phi^{1/2}t)/2)]^{-1} \tag{8.20}$$

where $\Phi = E_c G_t(b + d)/E_l E_t b d^2$ and d is the semi-90° ply thickness, b is the 0° ply thickness, t is the crack spacing and σ_a is related to $\Delta\sigma_o$ according to

$$\Delta\sigma_o = \sigma_a(b + d)/b - E_l\sigma_a/E_c \tag{8.21}$$

The nature of the accumulation of transverse cracks dictates that the crack spacings predicted by these equations are a stepped curve. However, it has been suggested that since the transfer distance y is small by comparison with the crack spacing that the first cracks form statistically at the weakest points in the 90° ply (see for example Manders *et al.* (1983) and Peters (1984)). Kimber and Keer (1982) showed that the average crack spacing is related to the lower bound of the shear lag analysis.

The statistical aspects of crossply cracking can arise from variations in fibre dispersion within the laminae, and could be worse when a memory of the individual tows remains. Peters and Chou (1987) have analysed the process in terms of an increasing element strength of a weak link theory and have argued that the statistics can be extrapolated to obtain the true transverse strength of the 'defect-free' ply.

8.5.2 *Constraint cracking*

Practical laminates make use of the fact that the transverse cracking strain increases with decreasing 90° ply thickness as a result of the suppression of transverse crack propagation by energy constraint. Parvizi and Bailey (1978) used equation 8.22 to predict the minimum transverse ply failure strain

$$\varepsilon_{\mathrm{tlu}}^{\mathrm{min}} = [2bE_1\gamma_t\Phi^{1/2}/(b + d)E_1E_c] \tag{8.22}$$

where γ_t is the fracture surface energy of the transverse ply. This equation provides a good correlation with experimental values at ply thickness less than 0.5 mm. Complete suppression occurred at $2d < 0.1$ mm. At large ply thicknesses factors which have been discussed above control the transverse ply failure strain. Therefore laminates should ideally be constructed from finely dispersed plies, to control damage accumulation. However, it should be recognised that debonding may still occur and that environmental factors might dictate a different laminate construction. Furthermore the growth of transverse cracks is a prime mechanism of stiffness reduction during fatigue and energy constraint may not be a good mechanism of crack suppression under these circumstances. Therefore prepreg technology utilising fine, well dispersed tows provides the most appropriate fabrication route for advanced composites. In many ways this effect has discouraged the development of cheaper, larger tow carbon fibres. The developments of interphases and higher failure strain matrices are required to encourage these developments.

8.5.3 *Epoxy resin matrix failure strain*

A study of equation 8.15 shows that the transverse cracking strain is directly proportional to the matrix failure strain. The increased toughness generally associated with the matrix failure strain can also restrict propagation and the transverse cracks take on a diffuse character.

A matrix of increased failure strain could have a larger expansion coefficient but since this could also result in an increased residual thermal strain a lower transverse cracking strain may still result.

8.5.4 *Thermal strains in crossply composites*

The expansion coefficients of a unidirectional lamina in the transverse and longitudinal directions α_t and α_1 which are dominated by those of the matrix and fibres α_m, α_f respectively, can lead to the induction of a significant thermal strain in the laminate. A large difference exists between α_m (60–100 × 10^{-6} K^{-1}) and α_f (glass = 5, HMS carbon (axial) −0.54, Kevlar 49 (axial) −2 × 10^{-6} K^{-1}). The equations of Schapery (1968) have been shown to predict the expansion coefficients of the composite reasonably accurately.

$$\alpha_l(T) = (E_m\alpha_m V_m + E_f\alpha_f V_f)/(E_m V_m + E_f V_f) \qquad (8.23)$$

$$\alpha_t(T) = \alpha_m V_m(1 + \nu_m) + \alpha_f V_f(1 + \nu_f) - \alpha_l \nu_c) \qquad (8.24)$$

For the crossply laminate, the higher longitudinal stiffness constrains the large transverse contraction on cooling from the post-curing temperature putting the plies into tension in the transverse direction and into compression in the longitudinal direction. The thermal strain in the longitudinal direction of the transverse ply, ε_{tl}^{th}, is given by

$$\varepsilon_{tl}^{th} = \frac{E_l b(\alpha_t - \alpha_l)(T_1 - T_2)}{(E_l b + E_t d)} \qquad (8.25)$$

where T_1 is the stress-free temperature. $(\alpha_t - \alpha_l)(T_1 - T_2)$ can be obtained directly from the curvature of an unbalanced 0°/90° coupon in a manner analagous to a bimetal strip.

$$(\alpha_t - \alpha_l)(T_1 - T_2) = (b + d)/2\varrho + \frac{E_l b^3 + E_t d^3}{6\varrho(b + d)} . [1/E_l b + 1/E_t d] \qquad (8.26)$$

The radius of curvature ϱ is related to the displacement of the coupon from flat δ by

$$1/\varrho = \frac{2}{L}\cos^{-1}[1 - \tau/\varrho] \qquad (8.27)$$

where L is the length of the beam and T_1 can be experimentally determined by recording the temperature at which the beam becomes flat (i.e. when $\delta = 0$).

The transverse ply cracking strain, ε_{tlu}, is consequently reduced by the magnitude of ε_{tl}^{th} and transverse ply failure strain ε_{tu}^{h} is given by equation 8.28.

$$\varepsilon_{tu} = \varepsilon_{tlu} + \varepsilon_{tl}^{th} \qquad (8.28)$$

Typical values of ε_{tl}^{th} can range from 0.2% to >1.0% depending on the stress free temperature and the difference in the longitudinal and transverse expansion coefficients ($\Delta\alpha$). It is generally larger for carbon fibres in equivalent resins since α_f along fibre is slightly negative. Since ε_{tu} can be typically 0.6–0.8%, at worst the propensity to crack during cooling can be high for crossply laminates manufactured from resins with high T_g and cured at temperatures in excess of T_g while at best the first cracking strain, ε_{tlu}, can be reduced to a low level.

8.5.5 *Poisson-generated stresses and longitudinal splitting*

In GFRP, the Poisson-induced strains can cause longitudinal splitting of the 0° plies. The effect of thermal strains is similar to that described above. However, longitudinal splitting has not been observed in CFRP prior to failure even though the Poisson strains are greater because of the lower overall failure strain of the composite.

8.5.6 *Angle ply laminates*

For laminates in which the fibres are arranged at angles to one another the micromechanics are complicated by the presence of interlaminar shear stresses at the free edges. Delamination of off-axis plies can be initiated at the free-edge or at damage in an infinitely wide specimen. From a consideration of the stresses occurring in symmetrical angle ply laminates at $\pm\,\theta$ to the uniaxial tensile stress it is clear that the shear stresses will be maximised at approximately ± 45°. At larger angles, transverse cracking will dominate the failure process although the actual value will largely depend upon the relative magnitude of the transverse, longitudinal and shear strengths. At smaller angles, shear dominated failure occurs.

The interactions between the stresses in the individual plies and the viscoelastic nature of the matrix resin makes the prediction of the stress–strain curves complex. However, this type of laminate is often subject to biaxial stresses and under these conditions, transverse cracking dominates the failure of each laminae at all angles.

In practice, complex laminates with balanced symmetrical configuration, e.g. $[0°/\pm 45°/90_2°]_s$ are used to provide load-bearing ability. The stacking sequence is selected to minimise the free edge stresses but the damage mechanisms will follow the general phenomena described in the previous sections, with fibre debonding developing into transverse cracks in the 90° plies, which in turn can initiate damage in the ± 45° plies. Debonding in ± 45° plies, where there is no adjacent 90° ply can also occur at higher strains leading to an extensive pattern of cracks. On further loading, delamination at ply interfaces, in-plane longitudinal cracking of the 90° plies and longitudinal splitting of 0° plies occurs. To optimise the performance, the 90° plies are normally finely dispersed and those with a large Poisson's ratio (i.e. ± 45° plies) should be located in the laminate mid-plane. A further variant on ply geometry is the use of woven fibrous prepreg whose drape properties enable complex curved surfaces to be fabricated more easily, but with the introduction of a more complex micromechanics involving fibre cross-over and resin-rich areas. In service, microcracking is likely to occur so that design practice needs to take account of the presence of benign cracks in composite structures.

8.5.7 Discontinuous fibre composites

A variety of discontinuous fibre composites is available ranging from the relatively long aligned fibre sheet moulding compound SMC through to the in-plane random dough moulding compound DMC and the short, random, reinforced fibre reinforced thermoplastics FRTP and thermosets FRTS. The performance of these materials differs from continuous fibre composites in several respects:

1. Lower volume fraction of reinforcing fibres (V_f)
2. The reinforcing efficiency (i.e. the aspect ratio of the fibres)
3. The orientation distribution of the fibres
4. The effects of flow-induced orientation leading to a heterogeneous fibre distribution.

The major advantage of short fibre materials comes mainly from fabrication economies and they do not play a major role in epoxy resin-based composites. However, the modulus of discontinuous in-plane random fibre composites can be given by a modified law of mixtures equation which includes orientation (η_θ) and length (η_1) factors:

$$\bar{E}_c = \eta_\theta \eta_1 E_f V_f + E_m(1 - V_t) \qquad (8.29)$$

or for $\eta_1 = 1$, by integrating over range of orientation angles

$$\bar{E}_c = 2\pi \int_0^{\pi/2} E_\theta \, d\theta \qquad (8.30)$$

where E_θ is defined by equation 8.17.

In practice the following simplified expression is more useful

$$\bar{E}_c = 3E_1/8 + 5E_t/8 \qquad (8.31)$$

The length efficiency factor is complicated by the nature of the transfer of load from the matrix by shear at the interface of the fibre. The strength of short fibre composites is dominated by these mechanisms which have been discussed in detail elsewhere (Cox, 1952; Kelly and Tyson, 1965; Folkes, 1982; Bader, 1985). The length/diameter ratio of the fibre has a critical value but for 99% efficiency, the fibres need to be 10 times this value. The minimum efficient reinforcing length of glass fibre in an epoxy resin is therefore in excess of 10 mm and nearer 100 mm for the higher modulus resins. As a consequence, short fibre reinforced epoxy resins do not find extensive applications.

8.6 Effect of moisture on the performance of epoxy resins

Most epoxy resin composites are employed in applications where humid environments are encountered. Since these are often in aerospace, the effect of moisture absorption and desorption on the mechanical properties of the laminates is crucial to the design procedures. The selection of a resin system is often determined by these factors.

8.6.1 *Moisture absorption kinetics*

In comparison to heat transfer, mass transfer of the water molecule is slower by a factor of $\approx 10^6$. Thus moisture equilibrium of a 12.5 mm thick carbon fibre epoxy laminate at 350 K can take $\approx$ 13 years whereas thermal equilibrium only requires 15 s. Further, thermal equilibration can be ignored when defining the moisture problem. Moisture absorption is generally also considered to be independent of concentration. Therefore the diffusion of moisture into cured epoxy resin is generally considered to obey Fick's law:

$$\frac{\partial c}{\partial t} = D_x \frac{\partial c^2}{\partial x^2} \tag{8.32}$$

where D_x is the diffusion constant in the direction normal to the surface, c is the moisture concentration, x is the distance from the edge of the plate normal to the surface, t is the time.

For fibre composites, diffusivities perpendicular (t) and parallel (l) to the fibres can differ and

$$D_x = D_1 \cos^2 \alpha + D_t \sin^2 \alpha \tag{8.33}$$

where α is the orientation angle of the fibre to the x direction (for in-plane fibres $\alpha = 90°$). D_x is related to the diffusivity of the matrix D_m,

$$D_x = D_m [(1 - V_f) \cos^2 \alpha + (1 - 2\sqrt{V_f/\pi}) \sin^2 \alpha] \tag{8.84}$$

and

$$D_1 = (1 - V_f) D_m \tag{8.35}$$

$$D_t = (1 - 2\sqrt{V_f/\pi}) D_m \tag{8.36}$$

D can be determined from the moisture absorption kinetics since it can be shown that:

$$M = \frac{4M_\infty}{h} \sqrt{\frac{t}{\pi} D} \tag{8.37}$$

where

$$D = D_x \left(1 + \frac{h}{l} \sqrt{\frac{D_y}{D_x}} + \frac{h}{n} \sqrt{\frac{D_z}{D_x}} \right)^2 \tag{8.38}$$

where h, l, n are thickness, length and width respectively of the coupon. M is the moisture concentration at time t and M_∞ is the equilibrium (i.e. maximum) moisture content at infinite time.

For a homogenous resin casting $D_x = D_y = D_z$ equation 8.38 reduces to

$$D = D_x\left(1 + \frac{h}{l} + \frac{h}{n}\right)^2 \tag{8.39}$$

The distribution of moisture through the thickness of a composite can be computed from these equations. Figure 8.15 illustrates that the moisture is unevenly distributed throughout the laminate for a significant proportion of its service life. The effect of thermal excursions on the moisture distribution profile can be appreciated from the effect of temperature on D. Since D increases with temperature, the moisture profile tends to even out as a result, as shown in Figure 8.15.

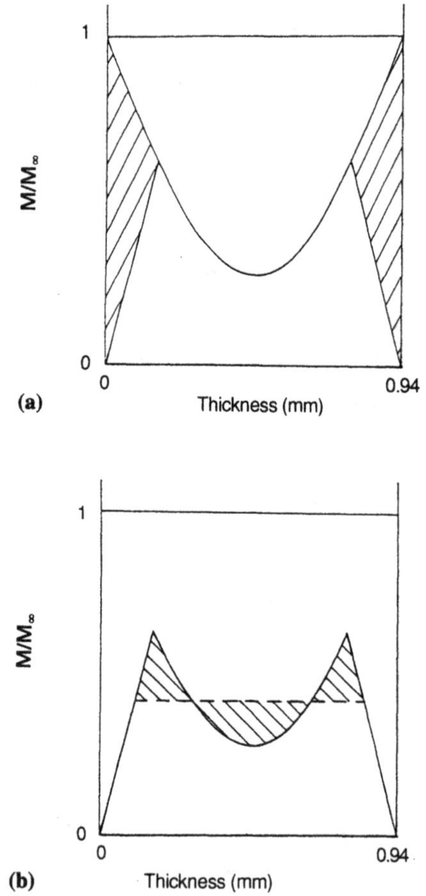

Figure 8.15 Theoretical moisture distribution within a unidirectional bismaleimide epoxy (Narmco 5245C) laminate, (a) subjected to 96% RH at 50°C for 3 days; (b) above following a rapid thermal excursion (thermal spike) to 150°C.

8.6.2 *Effect of resin structure*

The value of M_∞ is strongly dependent on the polarity of the cured resin structure. Generally the more polar, the higher the equilibrium moisture concentration. As a result, hardeners which yield 'polyester' or 'polyether' molecular structures rather than hydroxy amines are preferred for matrices.

Absorbed water acts as an efficient plasticiser for cured epoxy resins and as a consequence the glass transition temperature of the resin is reduced. As shown in Figure 8.16 and as a general rule for most resin systems, the T_g is reduced by $\simeq$ 20 K/1% moisture (Wright, 1981).

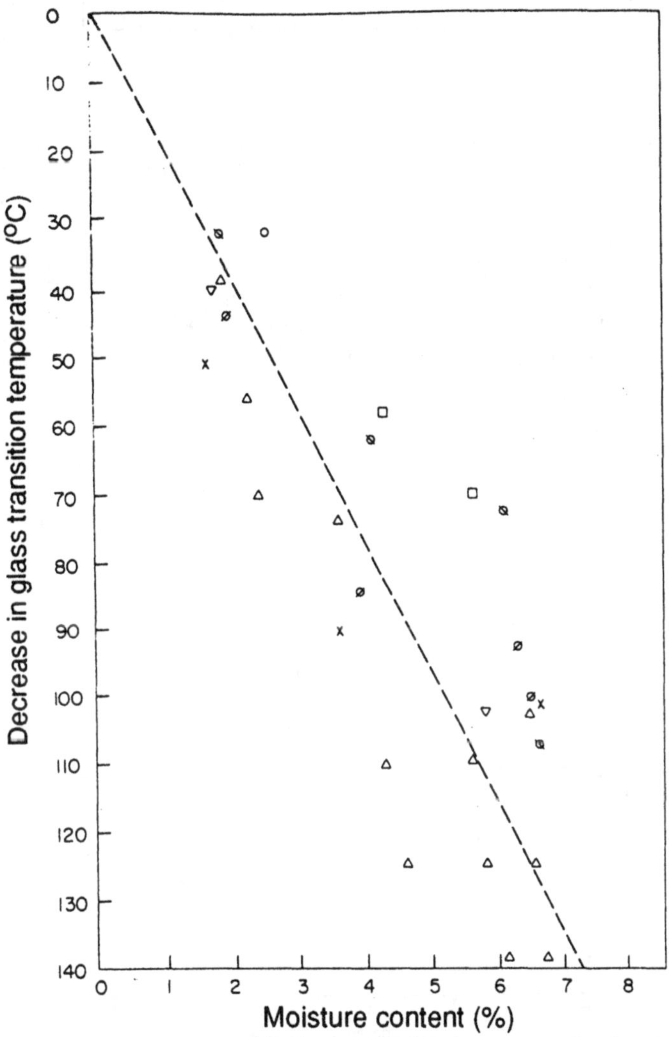

Figure 8.16 The effect of moisture on T_g of a series of differing epoxy resins showing generality of the effect of moisture plasticisation. Each symbol represents a different resin system. (Redrawn from Wright, 1981).

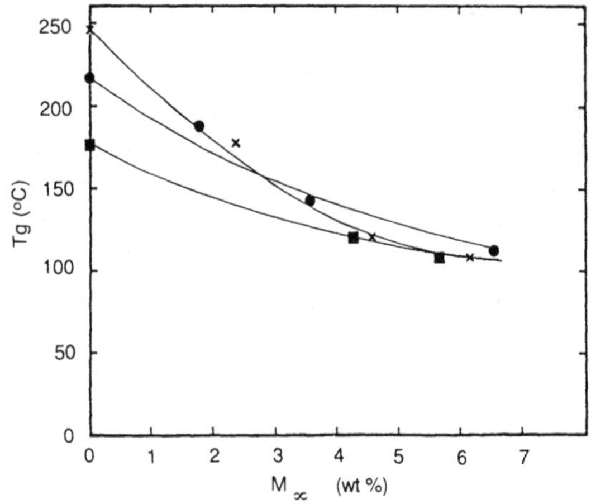

Figure 8.17 Change in T_g of Narmco 5208 epoxy resin as a function of equilibrium moisture content. Each symbol represents the results of different authors where variations in cure give differing 'dry' values. The wet T_g however tends to the same value in all work.

Figure 8.17 indicates the consistency of this observation with the variation of T_g with differing moisture concentrations for nominally the same resin system, as given by different workers (Mulheron, 1984).

The major consequence of this result is illustrated in Figure 8.18 where it is seen that not only the glass transition temperature but also the modulus is reduced. Since for most composite matrices, the average moisture absorption at equilibrium leads to a reduction in T_g by about 50°C, 150–

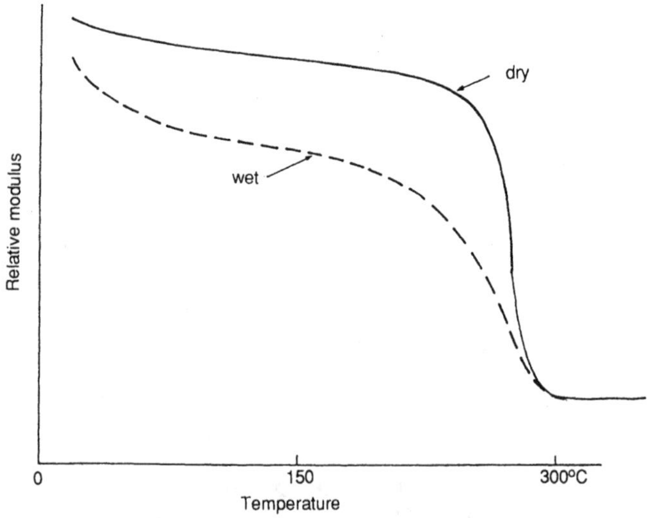

Figure 8.18 The effect of moisture on the temperature dependence of the relative matrix modulus of Narmco 5208 epoxy resin.

200°C resins can only be used where service temperatures are limited to 100–150°C. This has led to the development of more highly functional resins which have a higher cross-linked density, T_g and service temperature in humid environments.

While simple resin systems appear to obey Fickian diffusion kinetics under absorption and desorption conditions, the compounded resins employed for advanced composites may show irreversible effects. For example, the presence of residual DICY can increase the value of M_∞ significantly. Moreover, on immersion in water, blister formation may occur as a result of osmotic effects accompanied by leaching (Jones *et al.*, 1987). As shown in Figure 8.19 the presence of a second polymeric component may lead to moisture-induced changes in the matrix-dominated expansion coefficients and this can have severe effects on microcrack development in laminates during thermal excursions (section 8.5.4) (Jacobs and Jones, 1991).

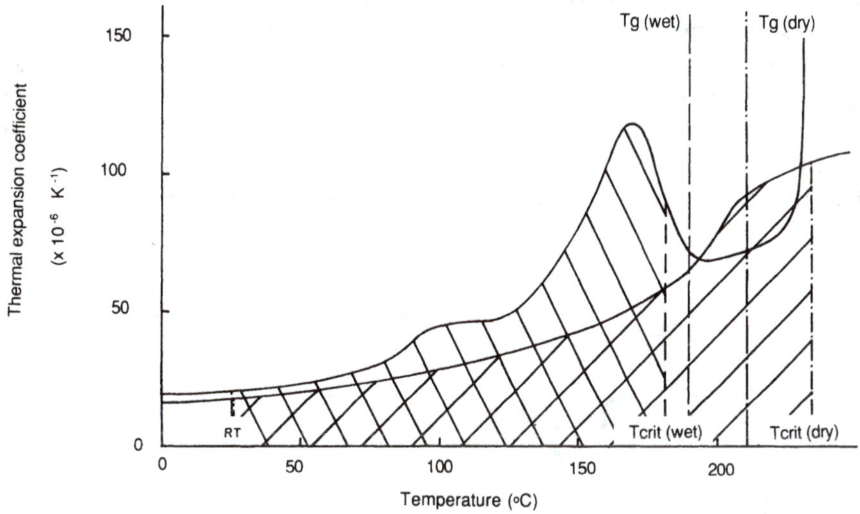

Figure 8.19 The temperature dependence of the transverse thermal expansion coefficients of the bismaleimide-modified epoxy resin based carbon fibre composite (Narmco 5245C). Dry (lower curve) and wet (upper curve). The cross hatching indicates the relative thermal strain induced on cooling from T_{crit} (dry) and T_{crit} (wet) (lower and upper curves respectively). T_{crit} is the temperature to which a crossply laminate needs to be heated and then cooled to induce a value of ε_{tl}^{th} which exceeds ε_{tu} and cause thermal cracking. Since T_g (dry) is below T_{crit} (dry) thermally-induced cracking does not occur but the reverse situation operates for the *wet* laminate and thermally-induced cracking does occur (Jacobs and Jones, 1991).

8.6.3 Effect of moisture on thermal residual strains

Since absorbed moisture is accommodated in the molecular structure of the resin through dipole–dipole and hydrogen bond interactions, this leads to swelling. Much debate has centred around the extent of this effect and the accommodation of water in the free or unoccupied volume of the dry glass. However, matrix swelling negates the restrained contractions which had already occurred on cooling through the strain-free temperature from the post-cure temperature after fabrication. As a result, the thermal strain is reduced by moisture ingress as shown in Figure 8.20.

For other systems the effect is bigger and, for example with Fibredux 913G, the residual stresses are reduced to a very low level as a result of moisture absorption.

8.6.4 *The combined effect of humidity and thermal excursions*

In service, composites often undergo thermal excursions and since under most conditions moisture absorption continues at the same time, the combined effects are important. For a moisture-equilibrated laminate, in which the thermal strains have been reduced, excursion to a higher temperature could lead to an enhanced level of thermal strain than present in the as-fabricated material. This occurs when $\alpha_m(\text{wet}) > \alpha_m(\text{dry})$ while the reduction in strain free temperature T_1 (equation 8.25) which is related to T_g, is insufficient to offset the corresponding increase in $(\alpha_t - \alpha_l)$. The effect is illustrated in Figure 8.19 for a bismaleimide-modified epoxy. In this case since the magnitude of α_l is such that it can be ignored, then ε_{tl} is directly proportional to the area under the curve. It can be seen that the enhance-

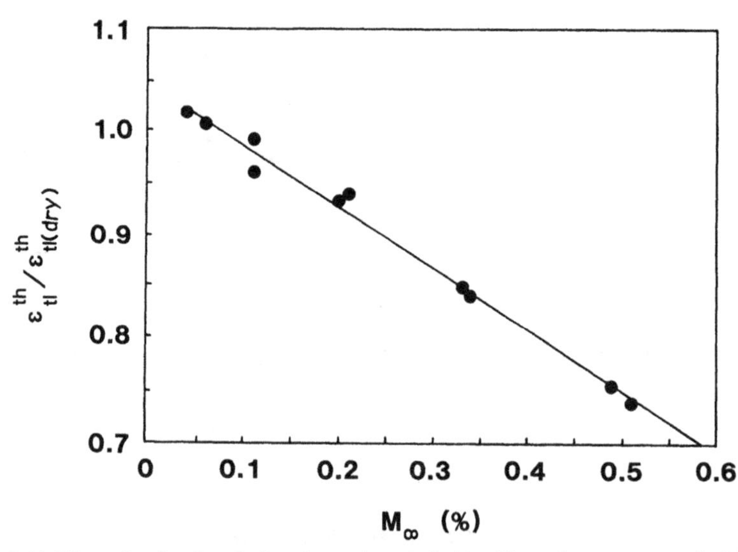

Figure 8.20 The reduction in relative thermal strain induced into the transverse ply (ε^{th}_{tl}) of an epoxy-based carbon fibre crossply laminate with differing equilibrium moisture levels. A typical value for a dry $0_2/90_2/0_2$ laminate is 0.63%.

ment in $(\alpha_t - \alpha_l)$ more than offsets the decrease in $(T_1 - T_2)$. Furthermore, the transverse cracking strain is also reduced as a result of moisture ingress, so that whereas a thermal cycle to T_{crit}(dry) for the dry laminate does not lead to microcrack formation, a thermal cycle to T_{crit}(wet) does. Thus a wet laminate subjected to a series of thermal cycles can lead to significant levels of damage. A corollary is that moisture absorption by the prepreg can also lead to significantly enhanced levels of thermal strain in the as-fabricated laminate which may cause process difficulties.

8.6.5 Thermal spiking

Since the timescale for moisture equilibration is long, the effect of thermal excursions on the kinetics of diffusion is also of importance. This is especially relevant in aerospace applications where for example a supersonic dash by an aircraft can raise the laminate temperature momentarily by ≈ 100°C. There have been several reports of enhanced moisture absorption as a result of a thermal spike. A typical example is shown in Figure 8.21 where the isothermal moisture absorption is compared to that which occurs under non-isothermal conditions.

During the thermal spike, moisture is lost but on reimmersion in the humid environment, the moisture climbs to a higher value. The exact

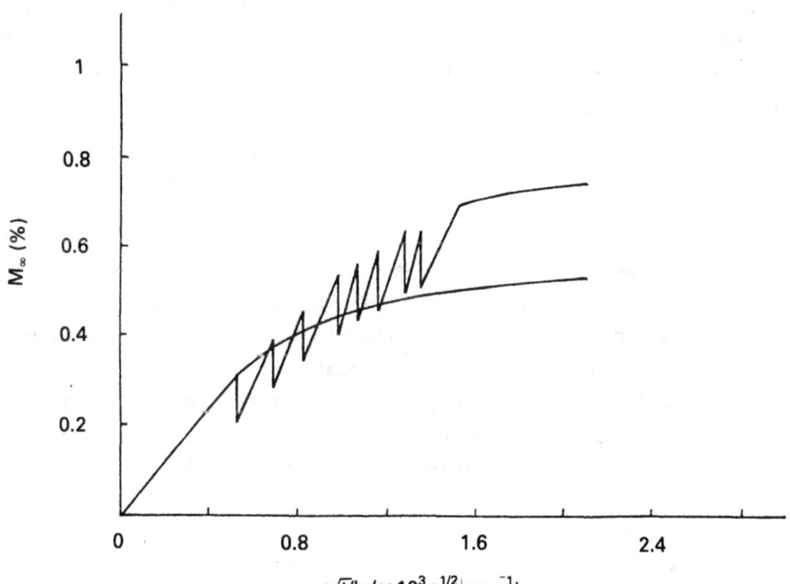

Figure 8.21 Comparison of moisture absorption by a 0° carbon fibre epoxy laminate (Narmco 5245C) under isothermal conditions (continuous line) and subject to intermittent thermal spikes to 150°C (saw-tooth). The environment was 96% RH at 50°C. The saw-tooth illustrates partial drying during thermal cycling and the subsequent enhanced water gain.

mechanism of this phenomenon is not fully clear but may involve the re-equilibration of the network molecular structure, since the critical thermal spike temperature above which this phenomenon is observed appears to be related to the glass transition of the wet resin.

8.7 Selection principles

Epoxy resin-based composite materials generally find application in high performance applications where high specific strength and stiffness dominate the requirements (aerospace structures or sports goods). In most of these cases, therefore, carbon fibre reinforcements are preferred. The grade of fibre employed is determined by a high strength or high stiffness criterion. For the former, Type A fibres and for the latter HM or IM fibres are employed. On a cost per equivalent strength, carbon fibre composites compare favourably with the cost of aluminium, its natural competitor. However, cost savings during fabrication can make the composite solution beneficial. On similar arguments glass fibre-reinforced epoxy is highly competitive and economic at $\approx$ 1/3 the material cost. In comparison to other plastics the cost of epoxy glass composites is low (except for polyester-based laminates) but where the fabrication rate and the complexity of the moulding dominates, other plastics solutions win out.

On a cost for similar stiffness basis, glass fibre laminates still compete effectively with high performance metals such as titanium but the carbon fibre solution is expensive and fabrication economies or lower running costs (in the case of airlines) need to be considered in the design.

It is often advisable to use hybrid systems which combine the benefits of glass and carbon reinforcements, as in the case of the prop-shaft for a motor vehicle and the rotor blade of a helicopter.

Aramid fibres offer higher performance than glass fibres but the off-axis properties are somewhat poorer. Therefore aramids find application where high stiffness and strength are combined with impact performance. Where high retained compression strength after impact is required, as for structural aerospace components, aramids are generally unsuitable.

Toughened epoxy resin textile fibre (nylon) reinforced films (e.g. Redux 330) are used as adhesives for joining composite laminates. These represent an additional application of epoxy resins in composites materials engineering.

8.8 Conclusions

Epoxy resins form the basis of the advanced composites industry as the preferred matrix for carbon fibres. These systems are designed to give good process viscosity profiles and involve latent hardeners for use in prepreg technology. The micromechanics of laminate systems is strongly dependent on the properties of the matrix. However, the absorption of moisture and the maximum achievable glass transition temperature can limit the maximum service temperature of the composite to < 200°C.

8.9 Glossary of symbols

8.9.1 Laminates

D	Interfibre spacing
$E(\bar{E})$	Young's modulus (average value)
G	Shear modulus
L	Length of composite coupon
T_2	Measurement temperature in equation 8.25
T_1	Strain-free temperature
V	Volume fraction
b	Outer ply thickness
c	Specimen width
d	90-ply semi thickness
d_f	Fibre diameter
r_f	Radius of fibre
t	Crack spacing
y	Shear lag transfer length
α	Linear expansion coefficient
γ_t	Fracture surface energy of the 90° ply parallel to fibres
δ	Displacement of 0°/90° beam from flat
ε	Strain
σ	Stress
ϱ	Radius of curvature of a 0°/90° beam
τ	Shear stress
ξ	Tsai–Halpin geometry and dispersion factor
η_θ, η_l	Orientation (θ) or length (l) factors for random discontinuous composite
ν	Poisson ratio

Diffusion theory

D	Diffusion constant or diffusivity
M	Moisture content at time t (in %)
M_∞	Moisture content at $t = \infty$ (in %)
C	Moisture concentration (g . mm^{-3})
t	Time

x	Distance into the specimen normal to the flux
α, β, γ	Orientations of the fibres with respect to x, y, z dimensions
h	Specimen thickness
l	Specimen length
n	Specimen width

Subscripts

a	Applied stress (σ_a) or strain (ε_a)
f	Fibre (e.g. σ_f stress in fibre)
m	Matrix (e.g. ε_m strain in matrix)
c	Composite (or L_c critical length for a short fibre in a composite)
l	Longitudinal or 0° direction of a unidirectional ply
t	Transverse or 90° direction of a unidirectional ply
u	Ultimate value at failure
fu	Ultimate value for failure of a fibre
mu	Ultimate value for failure of the matrix
lu	Ultimate value for failure of a longitudinal composite
tu	Ultimate value for failure of a transverse composite
tlu	Ultimate value for transverse cracking of transverse ply
tu	Ultimate value for failure of a transverse ply
tl	Property of transverse ply in longitudinal direction
lt	Property of longitudinal ply in transverse direction
ll	Property of longitudinal ply in longitudinal direction
Crit	Critical value

Superscripts

th	Thermal
′	Property in the component when other fails
P	Poisson
Min	Minimum value of property

References

Alexander, M.R. and Jones, F.R. (1991) In *Composites, Design Manufacture and Application*, Vol. 1, eds Tsai, S.W. and Springer, G.S. Sampe, Corina, USA. Chapter 11, paper G.

Amdouni, N., Sautereau, H., Gérard, J-F. and Pascault, J.P. (1990) *Polymer* **31**, 1245–1253.

Aveston, J. and Kelly, A. (1973) *J. Mat. Sci.* **8**, 352.

Bacon, R. (1980) *Phil. Trans. R. Soc., London* **A294**, 437.

Bader, M.G. (1985) In *Handbook of Composites*, Vol. 4, ed. Kelly, A. and Mileiko, S.T. Elsevier, North Holland, p. 177.

Bailey, J.E. and Parvizi, A. (1981) *J. Mat. Sci.* **16**, 649.

Cox, H.L. (1952) *Br. J. Appl. Phys.* **3**, 72.

Denison, P., Jones, F.R. and Watts, J.F. (1988a) *Surface and Interface Analysis* **12**, 455.
Denison, P., Jones, F.R. and Watts, J.F. (1988b) In *Interfaces in Polymer, Ceramic and Metal Matrix Composites*, ed. Ishida, H. Elsevier, New York, pp. 77–86.
Denison, P., Humphreys, P., Watts, J.F. and Jones, F.R. (1989) In *Interfacial Phenomena in Composite Materials*, ed. Jones, F.R. Butterworths, Guildford, pp. 105–110.
Dobb, M.G. (1985) In *Handbook of Composites*, Vol. 1, eds Watt, W. and Perov, B.V. Elsevier, Amsterdam, pp. 673–7.
Dunsford, D.V., Harvey, J., Hutchings, J. and Judge, C.H. (1981) *RAE Technical Report TR81096*, HMSO, London.
Drzal, L.Z., Rich, M.J. and Lloyd, P.F. (1982) *J. Adhesion* **16**, 1.
Folkes, M.J. (1982) *Short Fibre Reinforced Thermoplastics*, Research Studies Press, Letchworth, UK.
Garrett, K.W. and Bailey, J.E. (1977) *J. Mat. Sci.* **12**, 157.
Gillham, J.K. (1986) *Encyclopaedia Polym. Sci. Eng.* 2nd edn, Wiley, New York, pp. 519–524.
Halpin, J.G. and Tsai, S.W. (1969) *AFML-TR67*, 423.
Hashin, Z. and Rosen, B.W. (1964) *J. App. Mechanics Trans. ASME* **31**, 223.
Hull, D. (1981) *An Introduction to Composite Materials*, Cambridge University Press, UK.
Jacobs, P.M. and Jones, F.R. (1991) In *Composites, Design, Manufacture and Application*, Vol. 2, Ch. 16. eds Tsai, S.W. and Springer, G.S. SAMPE, Corina, USA, paper G.
Jones, F.R. (1989) In *Interfacial Phenomena in Composite Materials* ed. Jones, F.R. Butterworths, Guildford, pp. 25–32.
Jones, F.R., Shah, M.A., Bader, M.G. and Boniface, L. (1987) *Sixth Int. Conf. on Comp. Materials, 2nd European Conf. Comp. Materials* ICCM VI/ECCM2 Vol. 3, ed. Matthews, F.I. *et al.* Elsevier, London, pp. 443–436.
Kelly, A. (1973) *Strong Solids*. Clarendon Press, Oxford.
Kelly, A. and Tyson, W.R. (1965) *J. Mech. Phys. Solids* **13**, 329.
Kies, J.A. (1962) *US Naval Research Report* No. 5752.
Kimber, A.C. and Keer, J.G. (1982) *J. Mat. Sci. Letters* **1**, 353.
Loewenstein, K.L. (1983) *The Manufacturing Technology of Continuous Glass Fibres*. Elsevier, Amsterdam.
Manders, P.W., Chou, T-W., Jones, F.R. and Rock, J.W. (1983) *J. Mat. Sci.* **18**, 2876.
Mortimer, S., Ryan, A.J. and Stanford, J.L. (1992) *Proc. 5th Int. Conf. FRC '92*, paper 2 Plastics and Rubber Inst, London.
Mulheron, M.J. (1984) *PhD Thesis*, University of Surrey, UK.
Oberlin, A. (1985) In *Developments in the Science and Technology of Composite Materials* eds Gunsell, A.R., Lamicq, P. and Massiah, A. Elsevier Applied Science, London. pp. 15–20.
Parvizi, A. and Bailey, J.E. (1978) *J. Mat. Sci.* **13**, 2131.
Parvizi, A., Garrett, K.W. and Bailey, J.E. (1978) *J. Mat. Sci.* **13**, 195.
Peters, P.W.M. (1984) *J. Comp. Mat.* **18**, 845.
Peters, P.W.M. and Chou, T-W. (1987) *Composites* **18**, 40.
Plueddemann, E.P. (1990) *Silane Coupling Agents* 2nd edn. Plenum, New York.
Purslow, D. and Childs, R. (1986) *Composites* **17**, 127.
Rand, B. (1985) In *Handbook of Composites, Vol. 1* eds Watt, W. and Perov, B.V. Elsevier, Amsterdam, pp. 495–575.
Rosen, M.R. (1978) *J. Coatings Tech.* **50**, 70.
Schapery, R.A. (1968) *J. Comp. Mat.* **2**, 380.
Shen, C.H. and Springer, G.S. (1977) *J. Comp. Materials* **11**, 107.
Thomason, J. and Morsink, H. (1988) In *Interfaces in Polymer, Ceramic and Metal Matrix Composites*, ed. Ishida, H. Elsevier, New York.
Tsai, S.W. (1985) *Composites Design*, USAF Mat. Lab. – Think Composites, New York.
Tsai, S.W. and Hahn, H.T. (1980) *Introduction to Composite Materials*, Technomic, Westport.
Verpoest, I. (1987) In *Proc. Europ. Symp. on Damage Development and Failure Processes in Composite Materials*, eds Verpoest, I. and Wevers, M., Leuven, Belgium, pp. 70–77.
Waltersson, K. (1985) *Comp. Sci. Tech.* **22**, 223.
Wang, D., Jones, F.R. and Denison, P. (1992) *J. Adhesion Sci. Tech.* **6**, 79.
Watt, W. (1985) In *Handbook of Composites*, Vol. 1. eds Watt, W. and Perov, B.V. Elsevier, Amsterdam, pp. 327–387.
Wright, W.W. (1981) *Composites* **12**, 201.

Further reading

Chawla, K.K. (1987) *Composite Materials – Science and Engineering*, Springer-Verlag, New York.

Hull, D. (1981) *An Introduction to Composite Materials*, Cambridge University Press, Cambridge.

Jones, R.M. (1973) *Mechanics of Composite Materials*, McGraw-Hill, Washington DC.

Kelly, A. (1973) *Strong Solids*, Clarendon Press, Oxford.

Kelly, A. and Rabotnov, Yu.N. (1985) *Handbook of Composites*, Vols 1–4, Elsevier Science, Amsterdam.

Tsai, S.W. (1985) *Composites Design 1985, Think Composites*, Dayton, Ohio.

Tsai, S.W. and Hahn, H.T. (1980) *Introduction to Composite Materials*, Technomic, Westport, Connecticut.

Donnet, J.B. and Bansal, R.C. (1984) *Handbook of Fiber Science and Technology*, Vol. 3, Marcel Dekker, New York.

Plueddeman, E.P. (1982) *Silane Coupling Agents*, Plenum, New York.

Weatherhead, R.G. (1980) *FRP Technology, Fibre Reinforced Systems*, Applied Science, London.

9 Coatings and other applications of epoxy resins

X.M. CHEN and B. ELLIS

9.1 Introduction

The current use pattern for epoxy resins is similar to that suggested by Lee and Neville in 1967 who estimated consumption as coatings 55%, composite matrices 20%, castings 10%, adhesives 5% and miscellaneous 10%. These figures may be compared with the 1990 and 1991 estimates given in chapter 1, Table 1.1, even though the categories are not identical. Lee and Neville (1967a) pointed out that there are difficulties in obtaining exact consumption estimates. However, the main difference is that coatings have a slightly lower share of the epoxy market. Epoxy resins are usually more expensive than their rivals, such as phenolic resins for coatings or laminates. Therefore, epoxy resins find application because of their superior properties, which include both processing and those of the cured resin. The processing is convenient since it is possible to formulate compositions with the required rheological properties, such as low viscosity, and there is also a wide choice of hardeners so that it is possible to cure at ambient as well as elevated temperatures. Because epoxy resins can be cross-linked without the formation of low molecular weight products, volatiles are not evolved during cure, and the resins have only a relatively low shrinkage during curing. Their mechanical and electrical properties are superior to other resins and they have good heat and chemical resistance.

There is a large range of epoxy resins available commercially and some of the more important types are listed in Table 9.1. Exact specifications can be obtained from resin suppliers and many mixtures are offered by compounders. Selection of a resin-hardener system can be made by a critical assessment of processing and end-use requirements, e.g. the electrical properties, or resistance to UV radiation. It is possible to enhance a desired parameter simply by altering the cure schedule without a change in the initial composition of the resin-hardener system. Thus, it is possible to increase the glass transition temperature (T_g) of the cured resin by post-curing. Such effects can be appreciated by study of the time – temperature–transformation diagram, TTT, given in Figure 1.2 and the detailed discussion of cure given in chapter 3. Extensive data compilations of the properties of epoxy resins are available and Table 9.2 is a selection of the 'tensile' mechanical properties and also an estimate of the heat distortion temperature. The Young's modulus of epoxy resins is somewhat lower than

Table 9.1 Summary of commercially available epoxy resins[a]

	E.e.w.	Viscosity (250°C) cp	Applications
1. Bisphenol A-epichlorohydrin resins			
a. Liquid resins			
i. Wide ranging characteristics	95–450	5000–30 000	Table 9.3
ii. Chain extension	185–200	2000–7000	Contain specific catalysts for conversion of these resins
iii. Lower viscosity resins			Contain reactive diluents, e.g. glycidyl ethers of *iso*-octanol or butane-1,4-diol
b. Solid resins			
i. So. pt. 60–180°C	500–6000		Coatings, especially powder coatings Manufacture of epoxy resin esters for coatings With amino- or phenolic-resins; high chemical resistance for tube coating primers and stoving enamels Castings Prepregs
ii. Mixtures and solutions			Solvents include butoxy(ethoxy)ethanol, xylene, acetone, methyl isobutyl ketone and combinations
2. Epoxy-novolacs			
Derived from bisphenol A-epichlorohydrin resins			
i. Phenol, *liquid*	170–190	30 000–90 000	Coatings, encapsulators, laminating, moulding, adhesives
ii. Phenol, *solid* (So. pt. 90–100)	190–220		Moulding powder
(So. pt. 90–95)	275–330		Adhesives, coatings
iii. Cresol, *solid* (So. pt. 35–90)	200–240		Adhesives, coatings, mouldings
3. Polyhydric phenols			
i. Bisphenol F			
Diglycidyl ether of bisphenol F	158–175	5000–7000	Additive for BPA resins in coating and flooring, e.g. Araldite G7 281

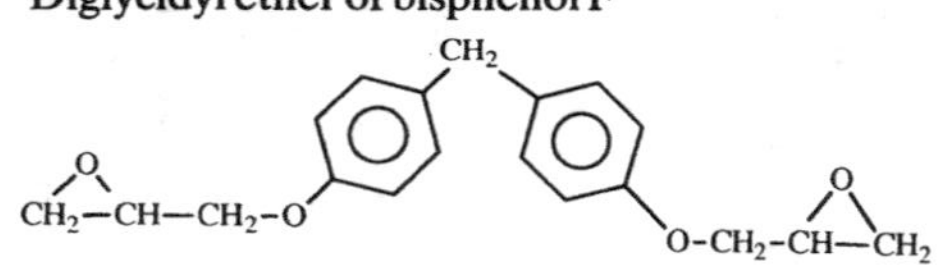

ii. Others include: Tetraphenol ethane derivatives			Powder coating Polyglycidyl ether of tetraphenol ethane: Shell 1031
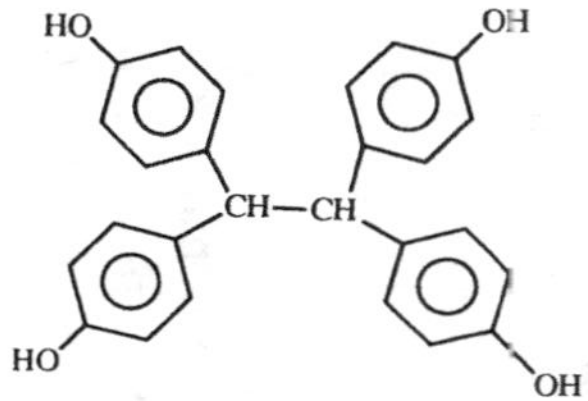			
4. Amines Reaction products of epichlorohydrin with amines such as:			
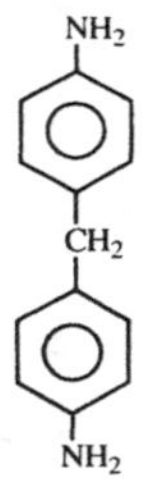 DDM  Aniline 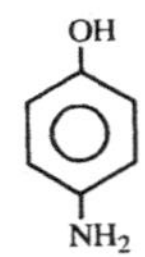 *p*-amino-phenol			
Triglycidyl *p*-amino phenol			Union Carbide ERL-0500
5. Halogen-containing epoxides			Flame retardancy applications: general purpose laminating, moulding
6. Cyclo-aliphatic resins Range of resins lower viscosities than BPA resins			Good UV light stability, electrical properties: general purpose casting resin, filament winding. Acid scavenger, plasticizer
3,4-epoxycyclohexylmethyl 3,4-epoxy-cyclohexane carboxylate	131–143	350–450	Union Carbide ERL-4221
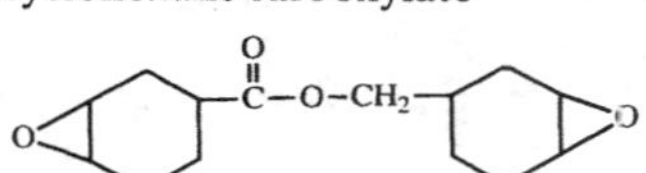			

So. pt. = softening point, °C. [a] For a fuller list see Tanaka (1988).

Table 9.2 Mechanical properties and range of heat distortion temperatures of cured epoxy resins

	Mechanical properties			
	Young's modulus ($GN.m^{-2}$)	Tensile strength ($MN.m^{-2}$)	Elongation at break (%)	Heat distortion temperature (°C)
Badge type:				
Flexible	0.1–1	10–70	5–10	–
Rigid	1.5–3.5	25–90	2.5–4	100–200
Epoxy-novolac	1.5–3.5	70	2	150–250
Cycloaliphatic epoxy	3.5	90	2–10*	Up to 200
Phenol-formaldehyde	2–4	20–60	1.5	150–170
Unsaturated polyesters	1.5–3.5	40–90	<2	Up to 200

* With up to 30 phr of a polyol flexibilizer, elongations at break may be up to 30%.

that of phenol formaldehyde (P/F) and unsaturated polyester resins but the tensile strengths are comparable and may be higher. This is because the extensibility, that is the fracture strain, of the epoxy resin is much higher than that of the 'brittle' resins. It is possible to select from a range of flexibilizers (see chapter 4) to 'tailor' the mechanical properties to suit the application requirements. For instance the 'tensile strength' can be significantly affected by cure treatment and the fracture behaviour of epoxy resins should always be considered when an application demands optimum strength. There is a detailed discussion of fracture behaviour in chapter 5 which is also relevant to the adhesive (chapter 7) and composite (chapter 8) applications of epoxy resins.

With standard BPA–EpiCl resins the heat distortion temperatures are similar to those of P/F and polyester resins, but with suitable hardeners and a post-cure at elevated temperature it is possible to attain higher glass transition temperatures and hence higher use temperatures. For instance a BADGE-type resin cured with trimellitic anhydride and cured for 1 hour at 120°C followed by 2 hours at 150°C may have a heat distortion temperature of up to 250°C.

Cyclo-aliphatic resins have very good outdoor ageing properties. This is because they do not contain aromatic rings which absorb the UV radiation which leads to degradation processes. Also, the chlorine, and other halogen, content of the cyclo-aliphatic resins is essentially zero and hence these resins have very good electrical properties.

Data compilations such as that given in Table 9.2 are indicative of the properties that may be achieved with epoxy resins. However, by comparison with the P/F and unsaturated polyester resins it would not be possible to assess the superior technical properties of the epoxy resins. Their processing properties offer considerable advantages. The range of hardeners available is a major asset since it is possible to formulate a curing system to suit almost

Table 9.3 Some typical curing agents and applications for BADGE-type resins

Hardener[1]	Physical state at ambient	Concentration (phr)[2]	Typical cure schedule	Approximate heat distortion temperature (°C)	Suggested applications[3]
1. Aliphatic amine					
Triethylene tetramine	Liquid	13	a. 25°C, 7 days b. 25°C, 1 day plus post-cure at elevated T	a. – b. Up to 120	General purpose Co, A, Ca, M, F, S and FW
Jeffamine D–400	Liquid	55	110–120°C, 30 min	~30	Flexible resin Co, A, Ca, F and S
2. Aromatic amine					
DDM	Solid	27	1 h at 150°C +3 h at 180°C	Up to 150	Glass cloth laminate
DDS (or DAS)	Solid	25	5 h at 125°C +1 h at 200°C	Up to 175	A, Ca, M, E, S and FW
3. Polyamides	Liquid	Depends on type	a. Ambient temperature 7 days b. 1 h at 100°C	Up to 100 Up to 100	a. Co, A, Ca, F, M and S b. Ca and A
4. Anhydrides					
Methyltetrahydro-phthalic anhydride (MTHPA)	Liquid	80	2 h at 20°C +2 h at 150°C	130	Liquid hardener processing A, Ca, M, S and FW
Trimellitic anhydride (TMA)	Solid	35	1 h at 120°C +2 h at 150°C	250	High heat distortion temperature A, Ca, M, S and FW
5. Miscellaneous					
Dicyandiamide (DICY)	Solid	4	1 h at 180°C	150	A, Ca, M, E, S and FW
BF_3–Monoethylamine (BF_3–MEA)	Solid	3	1 h at 120°C +2 h at 170°C	170	A, Ca, M, E, S and FW

[1] See chapter 2 for fuller details. Many hardeners are mixtures and/or derivatives and are not included in this list. Many other hardeners may be used. Shell list 30 curing agents for Epon 828. This table is only illustrative.
[2] phr = parts per hundred of resin by weight.
[3] *Code*: Co = coatings; A = adhesives; Ca = castings; E = electrical laminates; F = flooring; M = mouldings, S = structural laminates; FW = filament winding.

every specific requirement. The published tables can often be very misleading. For instance, in such data compilations the water absorption is often reported in terms of the w/o absorbed by $\frac{1}{8}$ inch thick specimen in 24 hours so that the value will not give either an estimate of the amount or even an indication of the rate of absorption. Also, the heat resistance and especially the solvent resistance of the resins are not adequately assessed. The epoxy resins have very good solvent resistance, which is a major advantage in many applications.

It is important to note that there is continual improvement and innovation in the resins, hardeners and other additives available. Resin manufacturers and converters have a policy of upgrading their products and offering speciality resins for particularly demanding applications. Laboratory samples and experimental products are usually available from resin manufacturers. The range of applications for a BADGE-type resin with different curing agents is given in Table 9.3.

In the present chapter it is not possible to discuss each application of epoxy resins in any detail, its purpose is to give an overview of the range of applications that currently provide a market for epoxy resins. It is important to note that many of the compositions that are used industrially contain many components, which are often not disclosed. Thus, full technical information on a product should be obtained from the resin manufacturer, compounder or supplier.

9.2 Surface coatings

9.2.1 Introduction

Surface coatings may be primarily applied for protective or decorative purposes but usually they have dual or more functions, such as exterior automotive coatings which are both protective and decorative. They may have specialized functions, such as anti-fouling paints for ships' bottoms or the protection of food and beverages in metal cans. There are also 'high-tech' applications in the electronic and related industries such as coatings for fibre optics, printed circuit panels and many other applications (Burkhart, 1988; Emerson *et al.*, 1990; Bonneau and Burkhart, 1991). Coatings contain a polymeric film-forming material, pigment, special additives and may be dissolved in a suitable solvent mixture or dispersed in water. Powder coatings do not contain a solvent. There is a range of film-forming materials which may be classified as either non-convertible or convertible (Martin, 1972). The former includes shellac, cellulose nitrate or acetate, cyclized and chlorinated rubber and vinyl and acrylic resins, which form films when a solvent evaporates. Convertible materials include the drying oils, varnishes, thermosetting resins such as alkyds, phenolics, epoxies and polyurethanes in

Table 9.4 Film forming materials for coatings

Application	Epoxy resins	Epoxy resin mixtures: Acrylic	Alkyd	Polyester	P/F[a]	A/F[b]	Coal tar modified	Other thermosetting resins[c]: Alkyd	Polyester	A/F[b]	P/F[a]	Polyurethane
Can and drum coatings	√	√		√	√	√		√			√	√
Heavy duty plant maintenance coatings	√	√	√	√				√	√	√	√	
Automotive – primers, topcoats and respray	√	√	√	√				√		√	√	
Pipe coatings	√				√		√					
Appliance finishes	√	√	√	√				√	√	√	√	
Marine coatings, primers and top coats	√				√		√	√			√	√
Coil coatings	√							√			√	
Metals, general primers and top coats	√	√	√	√				√			√	

[a] P/F is a phenol-formaldehyde resin; [b] A/F includes urea and melamine formaldehyde resins; [c] Non-convertible coatings include drying oils and cellulose ethers and esters.

which the film former becomes cross-linked either by reaction with oxygen or suitable curing agents or hardeners. A general text on surface coatings is that of Paul (1986).

There are many different coating methods which can be adopted for specific applications, depending on the coating thickness, coating speed and type of finish required. A review of 25 methods with outline descriptions is given by Scharenberg (1985), spray coating is discussed by Coeling (1985) and powder coating by Richart (1982, 1985). A listing of the polymer systems used for specific industrial coatings is given in Table 9.4. Epoxy resin coatings are discussed in detail by Somerville and Smith (1973), Potter (1970), Lee and Neville (1967) and Tess (1988a). Resin manufacturers offer advice on the use of epoxy resins for the fomulation of coating compositions, Ciba–Geigy (1988) and Shell (1991) offer a range of epoxy resins for surface coatings, including Epon 828, with an epoxy equivalent-weight of 185-192. This resin and also the Ciba–Geigy equivalent GY-250 have been formulated as a heat-cured coating for strengthening glass (Chen *et al.*, 1992). Epoxy resins can be formulated to have much shorter cure times (1 to 30 minutes at 240–135°C) than equivalent polyurethanes or polyesters (10 minutes at 200°C) and have excellent adhesion to most substrates, provided the surface has had an appropriate pre-treatment.

9.2.2 *Surface preparation and primers*

It is a primary requirement that coatings, when cured, should form a continuous film. This depends not only on the rheological properties of the epoxy formulation when applied, but also and in many cases more importantly, on the interfacial tension of the coating formulation and the surface that is to be coated. Any surface features which may hinder film formation must be treated prior to the coating operation. For the surface to be readily wetted by the epoxy formulation, a critical condition is that the contact angle between the formulation and the solid surface is low. Simple cleaning of the surface by appropriate solvents may be sufficient surface preparation when there is only surface contamination and non-adherent scale is absent. For example iron or steel should be rust free. On the other hand, if the coating is designed to do more than just cover the surface, specific treatments will be required. For example, the application of a strengthening coating to glass surfaces requires exposure to water vapour prior to coating to optimize hydrolytic durability (Ellis *et al.*, 1991). Other surface preparation methods include mechanical abrasion, chemical etching, and flame treatment, to name but a few. These are treatments used to modify different surfaces in order to improve the adhesion of coatings. In chapter 8, there is a discussion of surface preparation to achieve good adhesion and the same principles are involved here. Misev (1991a) gives details of the surface treatment of aluminium, steel and galvanized steel prior to the application of a powder coating.

A primer may be used to improve the 'wettability' of the surface and also to protect a cleaned surface so that it will not become contaminated before application of the coating. The primer has to adhere well to the substrate surface and offer good wetability and bonding of the coating. Silane and other coupling agents can be used to improve bonding, especially to hydrated surfaces. Typical coupling agents have two types of reactive groups, one reacts readily with inorganic substrates and the other with organic groups. Thus, coupling agents are often used in applications where inorganic surfaces are to be bonded to an epoxy resin, and they may be incorporated either into a primer or the coating. Coupling agents are discussed in chapters 8 and 9.

9.2.3 *Solution coatings*

An important class of epoxy coatings are those in which the rheological properties of the composition are adapted for the specific applications by the use of a suitable solvent or solvent mixture. The rheological properties will depend on the resin, its type and molecular weight, the pigment and its concentration and the solvents. The apparent viscosities are normally within the range of 0.05–1.0 Pas (0.5–10 poise), but it should be noted that viscosities are highly temperature-dependent. Factors which determine the selection of solvent include its vapour pressure and hence its rate of evaporation and flash point, its toxicity as well as its effect on the flow properties of the composition. However, the paramount condition for selection of solvent is its miscibility with the epoxy resin together with the hardener and any other additives present. Solutions should be stable for the required shelf-life of coating composition and during application.

Selection of a solvent or solvent mixture which is miscible with epoxy resins is aided by matching the Hidebrand's solubility parameters, δ (see Barton, 1983a), of the polymer and solvent. This is defined as

$$\delta = C^{1/2} = \left(-\frac{U}{V}\right)^{1/2} = (\Delta_l^g U/V)^{1/2}$$

where C is the cohesive pressure, or cohesive energy density (cohesive energy per unit volume), U the molar internal energy, that is the molar potential energy of the material relative to the ideal vapour pressure at the same temperature, V the volume, the $\Delta_l^g U$ is molar vaporization energy, that is the energy required to vaporize a mole of the liquid to its saturated vapour. For volatile liquids the solubility parameter (Grulke, 1989) is more readily calculated from the molar enthalpy of vaporization, $\Delta_l^g H$,

$$\delta = [(\Delta_l^g H - RT)/V]^{1/2}$$

where R is the gas constant and T the absolute temperature. For the use of this equation the vapour phase must obey the ideal gas law. For epoxy resins the value may be determined from swelling measurements or by calculation useing additive group cohesive energy constants (see Barton, 1983).

The simple matching of solubility parameters is only suitable for an initial selection of a solvent because other factors such as polar attraction force and hydrogen bonding complicate solvent-solute interactions. Tess (1988b) gives useful diagrams for solvent selection and Barton's (1983) solubility maps for epoxy resins can also be consulted. Solvents for epoxy resins include xylene, methylethyl ketone (MEK), methyl isobutyl ketone (MIBK) and proprietary solvents, such as oxitol. Mixed solvents are often used so that not only solubility but the rate of evaporation and also the rheological properties of the solution are optimized for the specific application. Solvent mixtures include toluene with acetone, MEK, MIBK and also xylene, MIBK and 2-ethoxyethanol (Lee and Neville, 1967).

After the selection of suitable solvents and subsequent formulation of a coating solution, other factors need to be considered carefully to ensure successful coating application. These include the condition of the surface prior to coating, concentration of the solution, curing technique, pot-life of the solution, etc. It should be noted that the coating solution is a dynamic system within which chemical reaction (cure) progresses with time. Therefore care should be taken in its storage. Also, the rate of evaporation of the solvent in the system depends not only on the vapour pressure of the solvent, but also the curing method used. Evaporation will be faster at higher temperatures, but then the rate of cure is higher and the film may harden before all solvent molecules diffuse out of the coating. The trapped solvent molecules will affect the properties of the film which may be impaired and may lead to poor adhesion. Thus, it is always important to consider all the factors in order to establish a set of optimized application conditions.

The presence of pigment will have a considerable effect on the flow properties of a coating solution. Nielsen (1977) discusses the effect of fillers on viscosity and suggests a generalized relationship for the viscosity of a dispersion, related to that of the fluid η,

$$\eta/\eta_0 = \frac{1 + AB\phi_2}{1 - B\psi\phi_2}$$

where $A = k_E - 1$; k_E is the Einstein coefficient, $B = 1$ for rigid fillers,

$$\psi = 1 + \left(\frac{1 - \phi_m}{\phi_m^2}\right)\phi_m$$

ϕ_2 is the volume fraction of pigment and ϕ_m is the maximum packing fraction,

$$\phi_m = \text{(True volume of the filler)/(apparent volume of the filler)}.$$

However, the flow properties of dispersion systems are often very complicated and require measurements over an appropriate range of rates of shear (Utracki, 1988). They may be Bingham bodies with a yield stress, pseudoplastics or thixotropic.

9.2.4 *Dip coats*

As probably the oldest and easiest method to apply coatings, dipping has intrinsic advantages over other coating techniques. For example, compared with spraying, dipping produces minimum material waste and provides, in a single dip, a coating layer covering all the immersed parts. Furthermore, in the dipping process, objects to be coated do not need to be spun to ensure full coverage, which is inevitable for many spraying operations. The minus side of dipping is also obvious. Firstly, since the object has to be dipped into the fluid epoxy coating solution or emulsion, the size of the container (dip tank) has to be large enough to accommodate a reservoir of material plus the volume of the immersed part. This means that larger objects which can be sprayed with ease would be more difficult to coat by dipping. Secondly, since the container has to be constantly open during continuous operation, evaporation of the carrier (solvent or water) will cause a change in the concentration of the system in the dipping tank. Some overflow arrangement may be installed in order to overcome such problem as well as providing agitation.

In the actual coating process, the following points deserve careful and combined consideration for successful applications.

- It is important to minimize the evaporation surface of the liquid. This requires well thought out tank design.
- The material with which the tank is made must not react with the coating solution or emulsion.
- Anti-contamination measures should be considered, depending on the type of application.
- Degree of agitation should be determined to suit the application. For example, continuous agitation may be required for some emulsion systems which are likely to agglomerate.
- When the formulation is light sensitive, such as in the case of epoxy acrylate-based UV curable systems, a light shield should be installed to prevent unnecessary initiation of cure of the system in the tank.
- Dipping and withdrawal speed are the factors which need careful monitoring for control of the coating thickness. The concentration of the coating solution and the temperature difference between the liquid and to-be-coated objects are also factors which affect coating thickness.

9.2.5 *Epoxy emulsions and other water-based coatings*

With the ever growing environmental concern over pollution by chemical wastes it is expected that, in the not too distant future, tougher environmental legislation will be introduced. This will certainly change current industrial practice. As a direct consequence, many industrial operations that are common today will have to be up-graded when chemical wastes require further treatment before discharge into the environment, and some may be banned all together. In either case, substantial capital investment will be inevitable to ensure on-going industrial production. Volatile solvents are prime candidates for legislation, but at present the rate of change is uncertain. Installation of solvent extraction and/or recovery systems will involve additional costs and could mean that the operation loses its commercial viability. The long-term solution may lie in changing the carrier from organic solvents to water wherever possible. Research has been carried out recently to extend the application of such water-based coating systems.

To use water instead of solvent as carrier, the immediate difficulty is that most epoxy resins do not form stable solutions in water. Thus, unlike organic solvents, the system formulated will be heterogeneous at the colloidal level when the resin is dispersed or emulsified with water as the continuous phase. However, for many coating operations heterogeneity up to a certain scale will not affect the properties of coatings. Water-based epoxy emulsion type coatings are now readily available and also water soluble epoxy resins have been produced by the introduction of hydrophilic groups.

Epoxy resin emulsions are produced by mechanical shearing to reduce the particle size of the resin and addition of a suitable surface active agent to stabilize the emulsion. One example, of the many of the emulsifiers available that can be used to stabilize epoxy emulsions is

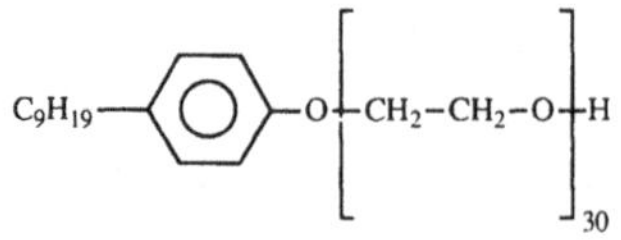

The hydrocarbon 'end' is soluble in the resin and the polar part of the molecule is hydrophilic where the length of the ethylene oxide chain can be chosen to suit the application. There are very many such emulsifying agents available commercially, and mixtures may be used to ensure that an emulsion is stable (Ash and Ash, 1980–1983). It is essential that the particle or droplet size of the emulsion is less that one micrometre to ensure that the emulsion does not settle out. Also, the emulsifying agent should be chosen to ensure that the droplets do not aggregate or even worse, coalesce. The stability of an emulsion is affected by shearing and hence the surface active agent should form a protective layer around each droplet. Changes in temperature also affect the hydrophile-lyophile balance (HLB) of the surfactant (Shinoda and Friberg, 1986).

To recapitulate from this discussion it can be easily appreciated that two basic criteria for the selection of emulsifiers must be satisfied.

- Reduce the interfacial tension between water and resin
- Form protective films around the resin droplets.

Although analytical methods can assist in the selection of emulsifier, a direct screening of the available emulsifiers is essential in practice to ensure that all requirements are satisifed.

There are three sets of conditions that an epoxy coating emulsion must comply with:

1. Preparation, stability and shelf-life
2. Application of the coating
3. Film formation.

The essential factors for the first set have already been outlined. For the second set the rheological properties of the emulsion must be such that the emulsion can be applied either by dipping, spraying or electrodeposition. With dipping the film thickness is determined by 'drainage' of the emulsion as the coated product is removed from the emulsion. The 'quality' of the film formed depends on even drainage, such quality involves control of the coating thickness as well as the avoidance of surface blemishes. It is particularly important that there is even coverage, which is difficult to achieve by spray application at places adjacent to corners, which should be avoided if possible by alternative product design. The drip that forms in the region of the 'neck' as the product leaves the coating bath can be troublesome. With spraying it is essential that fine droplets are formed since when large they splash where they hit which can lead to a 'more or less grainy surface' which eventually leads to a film with imperfect specular reflection. The rheology of the emulsion will affect size reduction in the spray equipment and also the flow of the emulsion on the coated surface which can lead to the same type of defects that occur with dip coating, except for a 'neck' blemish. Design of equipment for the support of the product during spraying is required so that the uncoated region adjacent to suspension hooks or grips is unimportant or can be subsequently treated appropriately.

Film formation involves the evaporation of water which requires more heat than for the removal of an organic solvent. The rate of diffusion of solvent or water through a polymer film is also important since it is essential that all of the low molecular weight species are removed prior to curing. With evaporation of water, a stage will be reached when the emulsion inverts and the epoxy resin becomes the continuous phase. Such a phase inversion will depend on the concentration of water present in the film and the temperature as well as other factors. The phase inversion temperature (PIT) is discussed by Shinoda and Friberg (1986). Evenness and film quality may be improved by the use of an organic solvent to reduce the viscosity of the resin prior to cure. The flow properties of the uncured film will depend on the pigment, its type, concentration and particle size distribution as well as the presence of other additives. The flexibility of the coating can be improved by the addition of alkyd, polyester or acrylic resins, polymeric amines and polyamides may be used as curing agents. A can coating has been formulated from a higher molecular weight epoxy resin (M.wt ~15–20 000) which was initially dissolved in solvent and then dispersed in water and cured with either a melamine or urea formaldehyde resin. The baked films were 'flexible' and resistant to the pasteurization conditions required for beer cans. Water-based epoxy resin coatings can be formulated for electrodeposition, both anodic and cathodic (Brewer, 1985). These coatings were developed for the applicance and automobile industries with cathodic systems being replaced by the anodic coatings because higher corrosion resistance can be produced. With electrodeposition (ED) the loss of paint due to overspray is eliminated, so that the process offers advantages compared with spraying since with ED it is possible to obtain more uniform coatings even in recessed regions. Very many formulations have been proposed for water-based coatings for dipping, spraying or electrodeposition.

It is possible to build hydrophilicity into an epoxy resin and an example is the reaction product of an epoxy resin with dibasic acids or anhydrides, and many different combinations have been evaluated. The synthesis of hydrophilic epoxy resins may be illustrated by the following reactions with both epoxy and hydroxyl groups,

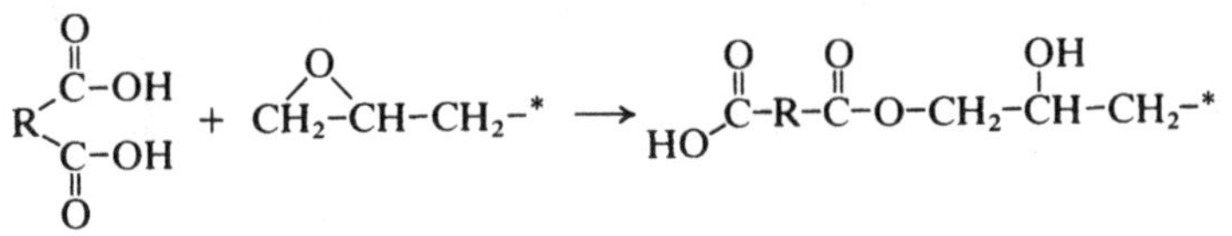

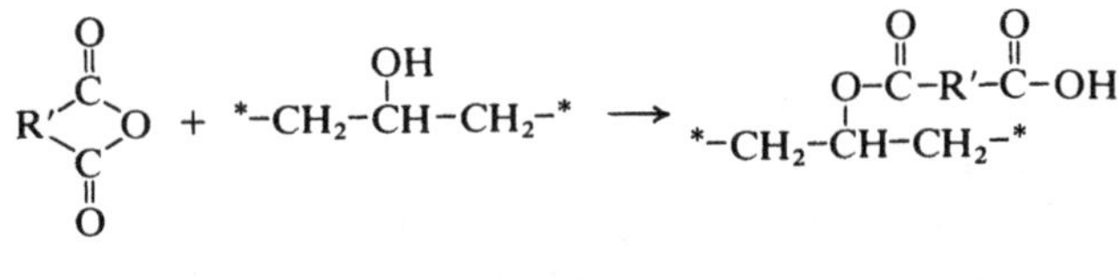

The carboxyl acid groups are, when neutralized, hydrophilic and the higher the acid content the more readily dispersible these resins are in water containing amines or inorganic bases. The degree of hydrophilicity is controlled by the extent of neutralization of the pendant acid groups.

9.2.6 *Powder coatings*

Powder coatings with 100% solids content do not contain a solvent or water carrier. They are blends of liquid and solid resins with curing agents plus other additives. Thus pollution, fire and explosive hazards are avoided in the coating plant. Also thick coatings can be applied in a single operation, with high application rates. The technology of powder coating is the subject of a recent text (Misev, 1991). There are many varied formulations and several different methods of application may be used. To formulate powder coating systems, high molecular weight (usually over 1000) epoxy resins are compounded with hardeners and other additives such as a flow-control agent, pigments, fillers and catalysts. The exact combination depends on requirements of the specific application. Fusion and dry blending are commonly used techniques in producing appropriate powder formulations. It should be noted that some dry-blending in a Z-blade mixer may require heating the components to a relatively 'high' temperature, thus formulations with reactive hardeners must be avoided. Latent hardeners may be used or two part mixes, with the hardener and catalyst kept separately until the coating is applied. The compounded resin is ground in a ball-mill or other comminuting machine so that it is fine enough to pass a 45 mesh, with a particle size less than 350 μm. It may be finer for specific applications. The hardeners that may be used include dicyandiamide, amine adducts, polyesters and phenolic resins. Very fast cures of 30 seconds with temperatures up to 250°C can be obtained with acid anhydrides catalysed with stannic octoate. The anhydride and catalyst can be mixed into separate portions of resin to avoid premature cure. Also, this type of system is sensitive to water, which will be absorbed during storage unless suitable precautions are taken. Water absorption reduces the hardener efficiency which leads to poorly cured coatings.

Methods of application which are frequently used include fluidized-bed, electrostatic spray, flock spray, and electrostatic fluidized-bed (Richart, 1982, 1985).

Powder coatings have excellent electrical resistance and thermal stability. These are properties attractive to electrical and electronic industries as coatings for vital components which are solvent sensitive and where contact with water must be avoided. Coating underground pipes is another important application due to the good resistance to corrosive soil and cathodic disbonding of such coatings. It can also be applied to many general objects as protective and decorative coatings. The composition of powder coatings can be varied to suit the requirements of specific applications.

The adoption of powder coating may involve expensive investment especially when changing from another type of coating line, colour matching may also be difficult. However, powder coatings have many advantages over other coating methods which include, little pollution or fire hazard problems, readily automated process technology, wide range of coating thickness possible in one application with exceptional film quality.

9.3 Industrial and related applications

9.3.1 Tooling

Epoxy resin tooling was initially used in the aircraft and car industries but has now been extended to mould and tool making for many different end-products, such as ship and boat building, domestic equipment and pottery. A combination of techniques similar to those used for casting and laminating are used for the production of moulds and foundary patterns. The resins are usually formulated with a liquid resin cured with a room temperature hardener. The properties are modified by the use of additives which not only improve specific properties but also usually reduce materials' costs. Tools can usually be fabricated in a fraction of the time required to machine metal and the costs of epoxy tooling are much lower than conventional metal tools. Epoxy resins are especially advantageous for the construction of prototypes and tooling for short runs. Lee and Neville (1967) have a chapter on epoxy resin tooling in which much detailed information is given which is still relevant. Special resins for tooling applications can be formulated and details are available from resin manufacturers. Some of the more important tooling applications of epoxy resins include:

1. Moulds for various purposes such as form master moulds for vacuum forming and injection moulding.
2. Jigs and fixtures to assist accurate positioning and checking shapes and dimension of components for such operations as drilling, cutting, and welding.
3. Metal-forming tools to replace the much more costly steel tools for prototype and short run production. Examples of such tools are stretch blocks, press tools, drop-hammer punches and dies. A rubbery resilient facing on one of the tools of the set may be required, e.g. for a drop hammer.
4. Foundry patterns ranging from master moulds from the original pattern, duplicate patterns made from the master mould to core boxes.

The benefits of using epoxy resins instead of traditional steel or aluminium alloy as tooling materials are numerous. In addition to facile forming of complex shapes the resins have low cure shrinkage and close tolerance can be maintained. Also, the inert nature of epoxy resin to both heat, when operational temperature is below its glass transition temperature, and chemical attack, which may corrode steel, ensures shape stability. Other important advantages include lower costs than for metals, lighter weight, also the shape of the tool can be readily adjusted, and lead-times are shorter.

Epoxy resin formulations for tooling applications are normally room temperature curable systems made up with liquid BADGE-type resins with polyamines (normally aliphatic) as hardeners. Depending upon specific application, diluents, plasticizers, fillers can also be introduced into the system for the purpose of reducing cost and improving required physical and chemical properties. In cases where special toughness is required, laminates of epoxy resin with various fibres can also be used. Open mesh 'tooling fabrics' are available and higher stiffness can be attained by the use of carbon fibre reinforcement (chapter 8).

9.3.2 Civil engineering

In the field of civil engineering the consumption of conventional materials such as concrete, wood, metal and glass is astronomical. Epoxy resins exhibit excellent adhesion to these materials and therefore have been used widely for surfacing, coatings and for repairs (Maslow, 1979). The most frequent application of epoxy resins in civil engineering include:

1. Flooring
2. Road and bridge coatings
3. Concrete bonding and repair.

9.3.2.1 Flooring Epoxy flooring can be formulated to be 100% non-volatile and 100% reactive so that one inch thick coatings may be applied in a single application. Floors of industrial premises such as chemical plants and warehouses are subject to constant and severe abuse of various forms. Typical examples are chemical spillage, harsh detergent cleaning, and heavy frictional wear by vehicles such as forklifts etc. Since floors of such industrial buildings are usually constructed with materials such as concrete, which under adverse service conditions are likely to gradually and then rapidly disintegrate, epoxy resins have been found to be very useful in aiding the maintenance of such floors, due to their chemical resistance, high mechanical strength and facile application.

In formulating a flooring system, the epoxy resin is usually mixed with hardeners, fillers and pigments according to service requirements. Colouring materials can also be added to suit different purposes. Prior to application of the resin the surface has to be thoroughly cleaned and pre-treated with an epoxy or other primer. Depending on the viscosity of the formulated system, it can be applied either by brush, roller, squeegee or trowel. For high viscosity formulations such as those primarily for wear resistance where large quantities of solid filler are added, the material has to be screeded by trowelling. On the other hand, low viscosity formulations for covering continuous floors with normal wear conditions can be applied with a squeegee or plastic comb. Self-levelling epoxy flooring compositions are formulated so that their viscosity is low. Epoxy flooring may be classified into three types, seamless flooring, industrial or epoxy concrete and terrazzo flooring. Seamless flooring has a 'low' viscosity due to the use of a reactive diluent and can be applied in thickness of 10–60 mm. They can be used to coat Portland cement floors, wood or metal after a specific surface preparation before application of the resin. Industrial epoxy flooring or epoxy concrete contains 75–85% particulate filler, usually sand (10–100 mesh). With high filler loadings the viscosity is high and the composition is applied by trowelling with thickness of 3–25 mm ($^1/_8$ to 1 inch). These compositions have low shrinkage and may be used to repair cement floors. Epoxy terrazzo floors contain coloured stone or marble chips and a fine particulate filler, such as calcium carbonate, and are trowelled to give 6–12 mm ($^1/_4$–$^1/_2$ inch) thick coatings. After setting overnight they are wet-ground to a polished terrazzo surface which is readily maintained.

The advantages of epoxy flooring may be summarized as follows:

1. Ease of seamless application, room temperature cure with low shrinkage
2. Excellent adhesion to many substrates
3. Chemical resistant and durable
4. Tough, flexible and abrasion-resistant
5. Dust-free, skid-resistant and readily maintained.

9.3.2.2 Road and bridge coatings. Epoxy resin coatings have been used to protect and repair road and bridge surfaces. Spalling is a commonly encountered problem on the surfaces of roads and bridges. It is caused by the combination and accumulation of many different factors such as freeze-thaw cycles, spillage by vehicles, road gritting and de-icing operations.

Epoxy resins that are used for the treatment of these conditions are usually liquids cured by amine hardeners at ambient temperatures. Coal tar is used as extender and often in high proportions in their formulation. When coated onto the structure and cured, the resin formulation forms a film which prevents the spalling of concretes on the road surface and bridge structure. Furthermore, the use of such formulations to seal cracks can significantly slow down their growth by bonding the damaged sections into an integral part. It also limits water penetration into the structure of the bridge. Ingress of water to steel reinforcement or framework causes corrosion. Hence epoxy resin coatings can give substantial protection against such damage. Some other applications of epoxy resin road and bridge coatings include:

1. Repair of scaling concrete roads with 'thin' coatings
2. Repair of 'slippery' but otherwise satisfactory asphalt or concrete road surfaces
3. Fuel spillage protection
4. Lightweight surfacing for bridge decks

5. Waterproof membrane between concrete and an asphalt topping
6. Upgrading additive for bituminous surfacing compositions
7. Road marking and coloured surfaces.

9.3.2.3 Concrete bonding and repair. Use of epoxy resin formulations as concrete bonding and repairing materials finds application over much wider areas than just bridges and roads (Allen and Edwards, 1987). Virtually, any structure with concrete as its primary construction material will develop cracks with age. Epoxy formulations can be coated onto the surface to prevent the initiation of such cracks and also to repair existing cracks. Structural defects occurring in other materials such as wood, brick and metal can also be repaired by epoxy formulations. Special formulations are available from resin suppliers.

9.3.3 Moulding compounds

Consideration of the application of epoxy resin moulding powders requires comparison with both alternative thermosetting resins and also with casting which is an alternative method of processing liquid epoxy resins. Epoxy resins can be moulded (Lee, 1988) by either compression or transfer moulding and also liquid-injection moulding. Reaction injection moulding also offers processing advantages (Becker, 1979; Manzione, 1989).

The superiority of epoxy resins compared with conventional phenol, urea and melamine formaldehyde resins is that, during cure, volatile products are not evolved and hence their moulding shrinkage is much lower. Also, they can be formulated so that at the moulding temperature they have lower viscosities so that much lower moulding pressures are required. For conventional P/F, U/F or M/F resins the moulding pressure is 1000 to 5000 psi (7–20 MPa) whilst for epoxy resin moulding powders the pressure can be very low, always less than 1000 psi. Thus presses and auxilary equipment can be simpler, and the moulding tools will also have to withstand lower stresses. However, because of the low viscosity of the epoxy resin at moulding temperatures there may be a problem due to excessive flash and 'tight' fitting moulds are essential. Some of the advantages of epoxy moulding powders compared with other thermosetting resins may be listed as follows.

1. Better dimensional tolerances and stability
2. Better electrical and mechanical properties
3. Higher chemical and heat resistance
4. Lower water absorption (depends on hardener and filler).

The disadvantages as well as higher cost include the need for closer mould temperature control, mould release and ejection problems. Mould release can be difficult because of the adhesive properties of the epoxy resin, which of course is often an advantage since the epoxy resins bond well to metal inserts. At the moulding temperature the cured product is flexible and ejector pins can cause damage to the moulding. This can be avoided by suitable design so that excessive local stresses are avoided during ejection of the moulded part.

Compared with casting the moulding of epoxy resins offers the advantages of simpler handling, higher filler loadings and also fibrous reinforcement. Its disadvantages include higher capital costs and storage problems with the uncured moulding powder. Their shelf-life may be limited due to incipient cure if the storage temperature is not low, for example it may be only several weeks at 22°C but much longer in sealed packages refrigerated at 7°C.

Epoxy moulding powders are formulated with a mixture of solid or liquid BPA resins which may be blended with epoxy novolacs. The latter improves the flow properties and has higher functionality and hence decreases gel and cure time. The curing agents are often diaminodiphenol methane (DDM) or acid anhydrides (PA or THPA), but others may be used for specific purposes, for instance chlorine-containing curing agents have been suggested for improving flame resistance. The moulding powders contain suitable particulate fillers, such as silica flour, and can be reinforced with fibres, 1.5–25 mm ($\frac{1}{16}$ to 1 inch) long. Zinc stearate is often included as a release agent, to assist ejection of the cured product from the mould. Thermal conductivity can be increased by use of metal powder as fillers, but should be avoided when the electrical properties are important. The density of the filled resins ranges from about 1.6 to 2.2. Lower densities, 0.75–1.00, can be obtained by the incorporation of microballons, composed of either clay, U/F or P/F resins or glass (Lee and Neville, 1967a). A major application of moulding powders is the encapsulation of electrical equipment and electronic devices. It is essential that during moulding the resin flows so that the component is not damaged or displaced. Typical moulded parts include deposited carbon resistors; chokes; ceramic, polyester and glass capacitors; metallized thin-film resistors; pulse transformers; toroid, solenoid and relay coils; printed circuits, solid-state circuits and many more products (Lee, 1986, 1988). Epoxy resin moulding powders also have applications in chemical plant, such as valves and filter plates where their chemical resistance offers major advantages. Structural components with good thermal and chemical resistance coupled with good mechanical properties and light weight are also moulded from epoxy resins.

9.3.4 *Embedding*

Embedding is used to protect and/or decorate electrical or electronic devices. Allied and related terms are casting, potting, impregnation and encapsulation. Moulding has been discussed in the section on moulding powders. The electrical properties required for electrical applications have been dealt with in chapter 6. Protective functions of the embedding resin include exclusion of moisture, oxygen, salt spray, solvent, other chemicals and microorganisms. The encapsulation of a device also protects it from mechanical shock and vibrations and provides suitable locations for mechanical handling. Alternative materials to epoxy resins include silicones, polyurethanes, polyesters, polysulphides and allyl resins (Lee, 1988). The epoxy resins have low viscosities and shrinkage since volatile compounds are not formed during cure. By suitable formulation it is possible to obtain a range of properties, from the rigid to flexible, elongations at break ranging from 3 to 40%. BPA resins and also mixtures with epoxy novolacs are used for potting. Cycloaliphatic epoxy resins have good dielectric properties and excellent weatherability, which is due to the absence of chlorine and aromatic rings. The cycloaliphatic resins require acid anhydride hardeners and longer high temperature cures. The properties of the resin can be modified by the inclusion of fillers and flexibilizers (chapter 4). Typical particulate fillers include silica, talc, clay and calcium carbonate, mica and chopped glass. Metal powders can be used to increase the thermal conductivity of the product. Lower density products can be produced by the inclusion of microspheres of saran, P/F resin or glass.

9.3.5 Miscellaneous

It is almost an impossible task to list all the existing applications of epoxy resins since there are currently many new developments. However, it is worth mentioning several areas in which epoxy resins are used, although on a small scale, but nevertheless for important specialized applications, such as textile finishes (Tanaka and Shiozaki, 1988).

9.3.5.1 Oil field applications. Epoxy resins are used in drilling operations to consolidate poor well formations, especially where the geological structures are primarily sandy. Resin compositions are adsorbed onto the well walls and subsequently cross-linked. Such treatment is particularly effective in minimizing sand production, which could eventually result in the well being discarded.

9.3.5.2 As reactant to form epoxy-based polymers. Epoxy resins are very reactive polymers. Their reactions with other chemicals as well as polymers can result in numerous new ranges of materials so far as these chemicals and polymers have functionalities smaller than three. For example, reacting epoxy resin with acrylic acids yields products, the epoxy acrylate resin, which are curable by UV irradiation.

9.3.5.3 Epoxy resin foams. Blown foams can be formed with epoxy resin formulations where a blowing agent is present. During the initial stages of cure such an agent either evolves a gas or evaporates to leave voids in the cured network. For example, hydrazide derivatives can be mixed into the epoxy formulation; with the heat generated by cure or provided by heating elements, they decompose to form nitrogen which is the blowing agent.

9.3.5.4 Epoxy resin stabilizers. Epoxy compounds also find application as stabilizers and plasticizers for other polymeric materials. For instance poly (vinyl chloride), PVC, readily dehydrochlorinates especially at elevated temperatures. The dehydrochlorination reactions are complex and appear to be autocatalytic (Hawkins, 1989). Thus, for PVC to be processed it is essential to add stabilizer which 'sequesters' the hydrochloric acid that is formed. The epoxy group reacts readily with the HCl as it is formed and also there is a synergistic effect when metal salts are present as well (Stephenson, 1989). Not only are BPA type epoxy resins effective stabilizers for PVC but also epoxidized oils, esters and fatty acids will both function as stabilizers and plasticizers.

References

Allen, R.T.L. and Edwards, S.C. (eds) (1987) *The Repair of Concrete Structures*, Blackie, Glasgow, UK.

Ash, M. and Ash, I. (1980–1983) *Encyclopedia of Surfactants*, Vol. 1–3. Chemical Publishing Co., New York.

Barton, A.F.M. (1983a) *CRC Handbook of Solubility Parameters and Other Cohesion Parameters*, Ch. 1, 2. CRC Press, Boca Raton, USA, pp. 1–21. (1983b) Many references to epoxy resins, see Index.

Becker, W.E. (1979) *Reaction Injection Moulding* Van Nostrand Reinhold Co., New York.

Bonneau, M. and Burkhart, A. (1991) *MEPPE Focus* (Jan.), pp. 1–6.

Brewer, G.E.F. (1985) *Ency. Polym. Sci. Eng. (Suppl.)* Wiley-Interscience, New York, pp. 675–687.

Burkhart, A. (1988) *Hybrid Circuit Technology,* pp. 222–228.

Chen, X.M., Ellis, B. and Seddon, A.B. (1992) Paper prepared for publication but submission delayed until after patent application.

Ciba–Geigy (1988) *Araldite Resins and Hardeners for Surface Coating Systems*, Ciba–Geigy Plastics, Cambridge, UK.

Coeling, K.J. (1985) *Ency. Polym. Sci. Eng.*, Vol. 3, Wiley-Interscience, New York, pp. 567–575.

Ellis, B., Chen, X.M. and Seddon, A.B. (1991) British Patent Application.

Emerson, J.A., Sparapany, J.J., Martin, A.R., Bonneau, M.R. and Burkhart, A. (1990) *IEEE, 40th Components and Technology Conference* Vol. 1, pp. 600–605.

Grulke, E.A. (1989) *Polymer Handbook*, VII 3rd edn ed. Brandrup, J. and Immergat, E.H. Wiley, New York, pp. 519–559.

Hawkins, W.C. (1989) *Enc. Polym. Sci Eng.* 2nd edn Vol. 15, Wiley, New York, pp. 560–563.

Lee, H. and Neville, K. (1967) *Handbook of Epoxy Resins*, McGraw-Hill, New York, pp. 1–16. (1967b) Ch. 23, pp. 1–5, Ch. 24, 7–40.

Lee, S.M. (1986) Ency. Polym. Sci. and Eng. Wiley-Interscience, New York, Vol. 5., pp. 792–828, see p. 818.

Lee, S.M. (1988) *Epoxy Resins, Chemistry and Technology* 2nd edn, ed. May, C.A., pp. 799–810.

Manzione, L.T. (1989) *Enc. Polym. Sci. Eng.* 2nd edn. Wiley-Interscience, New York, pp. 72–100.

Martin, S.R.W. (1970) *Paint Technology Manuals*, Gen. ed. Taylor, C.J.A. and Marks, S., Part 3, collator Keenan, H.W., Chapman and Hall, London, p. 16.
Maslow, P. (1979) In *Plastic Mortars, Sealants and Caulking Compounds*. ed, Seymour, R.B. ACS Symposium Series 113. American Chemical Society, Washington, D.C. pp. 39–60.
Misev, T.A. (1991) *Powder Coatings, Chemistry and Technology*. Wiley, Chichester, UK. (1991a) pp. 304–324.
Nielsen, L.E. (1977) *Polymer Rheology*, Ch. 9. Marcel Dekker, New York, pp. 133–150.
Paul, S.J. (1986) *Surface Coatings, Science and Technology*, Wiley, Chichester, UK.
Potter, W.G. (1970) *Epoxide Resins*, Ch. 8, Iliffe Books, London, pp. 144–167.
Richart, D.S. (1982) *Kirk-Othmer Enc. Chem. Tech.*, 3rd edn, Vol. 19. Wiley-Interscience, New York, pp. 1–27.
Richart, D.S. (1985) *Ency. Polym. Sci. Eng.* Vol. 3, Wiley-Interscience, New York, pp. 575–601.
Scharenberg, R.T. (1985) *Ency. Polym. Sci. and Eng.*, Vol. 3, Wiley-Interscience, New York, p. 552.
Shell (1991) *Technical Bulletins*, Shell Chem. Co., Houston, Texas 77027, USA.
Shinoda, K. and Friberg, S. (1986) *Emulsions and Solubilization,* Wiley, New York.
Somerville, G.R. and Smith, I.T. (1973) *Epoxy Resins, Chemistry and Technology*, 1st edn, Ch. 7, eds May, C.A. and Tanaka, Y., Marcel Dekker, New York, pp. 451–484.
Stephenson, R.C. (1989) *Enc. Polym. Sci. Eng.* 2nd edn (Suppl.) Wiley-Interscience, New York, pp. 867–870.
Tanaka, Y. (1988) In *Epoxy Resins, Chemistry and Technology*, 2nd edn, Ch. 2, Table 49, pp. 212–231.
Tanaka, Y. and Shorzaki, H. (1988) In *Epoxy Resins, Chemistry and Technology*, 2nd edn, Ch. 12. ed. May, C.A. Marcel Dekker, New York, pp. 956–1048.
Tess, R.W. (1988a) *Epoxy Resins, Chemistry and Technology*, 2nd edn Ch. 8. ed. May, C.A. Marcel Dekker, New York, Ch. 8, pp. 719–782. (1988b) see pp. 726–735.
Utracki, (1988) In *Rheological Measurement* eds Collyer, A.A. and Clegg, D.W. Elsevier, Amsterdam.

General index

For reference to specific curing agents and hardeners see the separate index

Index of curing agents and hardeners

www.ingramcontent.com/pod-product-compliance
Lightning Source LLC
LaVergne TN
LVHW011912200926
848484LV00002B/596
* 9 7 8 0 7 5 1 4 0 0 9 5 3 *